Atlas of Confocal Laser Scanning In-vivo Microscopy in Ophthalmology – Principles and Applications in Diagnostic and Therapeutic Ophthalmology

R.F. Guthoff · C. Baudouin · J. Stave

R.F. Guthoff
C. Baudouin
J. Stave

Atlas of Confocal Laser Scanning In-vivo Microscopy in Ophthalmology – Principles and Applications in Diagnostic and Therapeutic Ophthalmology

With 692 Figures, 183 in Color, and 2 Tables

Springer

Prof. Dr. med. Rudolf F. Guthoff
Universitäts-Augenklinik Rostock
Doberaner Straße 140
18057 Rostock, Germany

Prof. Dr. rer. nat. Joachim Stave
Universitäts-Augenklinik Rostock
Doberaner Straße 140
18057 Rostock, Germany

Christophe Baudouin, MD, PhD
Professor of Ophtalmology
Centre Hospitalier National
d'Ophtalmologie des Quinze-Vingts
28, rue de Charenton
75571 Paris Cedex 12, France

ISBN-10 3-662-50061-2
Springer Berlin Heidelberg New York

ISBN-13 978-3-662-50061-3
Springer Berlin Heidelberg New York

DOI 10.1007/978-3-540-32707-3

Springer is a part of Springer Science + Business Media

springer.com

Softcover re-print of the Hardcover 1st edition 2006

Editor: Marion Philipp, Heidelberg, Germany
Desk editor: Martina Himberger, Heidelberg, Germany
Production: LE-TeX Jelonek, Schmidt & Vöckler GbR, Leipzig, Germany
Cover design: Frido Steinen-Broo, EStudio Calamar, Spain
Typesetting and reproduction of the figures: AM-productions GmbH, Wiesloch, Germany

Printed on acid-free paper
24/3100/YL 5 4 3 2 1 0

Acknowledgments

The authors are grateful for the cooperation and detailed contributions of:

Benjamin Bermaoui
Cand. Med.
Schröderstraße 9
18055 Rostock
e-mail: b.bermaoui@gmx.net

Martin Burmeister, MD
Dental Surgeon
Patriotischer Weg 72
18057 Rostock
e-mail: martin.burmeister@gmx.net

Bénédicte Dupas
Orthoptist
Centre Hospitalier National d'Ophtalmologie des Quinze-Vingts
28, Rue de Charenton
75571 Paris, Cedex 12
e-mail: dupas@quinze-vingts.fr

Alexander Eckard, MD
Ophthalmologist
Universitäts-Augenklinik Rostock
Doberaner Straße 140
18057 Rostock
e-mail: alexeckard@gmx.de

Tino Just, MD
ENT Surgeon
Klinik und Poliklinik für Hals-, Nasen-, Ohrenheilkunde der Universität Rostock
Doberaner Straße 137-139
18057 Rostock
e-mail: tino.just@med.uni-rostock.de

Steffi Knappe, MD
Ophthalmologist
Universitäts-Augenklinik Rostock
Doberaner Straße 140
18057 Rostock
e-mail: steffiKN74@gmx.de

Robert Kraak, MD
Ophthalmologist
Universitäts-Augenklinik Rostock
Doberaner Straße 140
18057 Rostock
e-mail: kakra@web.de

Antoine Labbé, MD
Ophthalmologist
Centre Hospitalier National d'Ophtalmologie des Quinze-Vingts
28, Rue de Charenton
75571 Paris, Cedex 12
e-mail: antoinelabbe@hotmail.com

Bertrand Sonigo, MD
Ophthalmologist
Centre Hospitalier National d'Ophtalmologie des Quinze-Vingts
28, Rue de Charenton
75571 Paris, Cedex 12
e-mail: bsonigo@free.fr

Oliver Stachs, PhD
Physicist
Universitäts-Augenklinik Rostock
Doberaner Straße 140
18057 Rostock
e-mail: oliver.stachs@med.uni-rostock.de

Jacek Szaflik, MD
Ophthalmologist
LASER Centrum Mikrochirurgii Oka
ul. Boya-Zelenskiego 4A
00-621 Warsaw
e-mail: jacek@okolaser.com.pl

Thomas Vagt, MD
Ophthalmologist
Universitäts-Augenklinik Rostock
Doberaner Straße 140
18057 Rostock
e-mail: thomasvagt@web.de

Hans-Peter Vick, MD
Professor of Ophthalmology
Universitäts-Augenklinik Rostock
Doberaner Straße 140
18057 Rostock

Andrey Zhivov, MD
Ophthalmologist
Universitäts-Augenklinik Rostock
Doberaner Straße 140
18057 Rostock
e-mail: andyzhyvov@yahoo.com

Furthermore, this book could not have been completed without the stimulating contributions of Mrs. Beate Stroteich who wrote and rewrote all the text passages, Mr. David Beatty who created a scientific English version out of a mixture of linguistic elements and the continuous discussion of many colleagues inside and outside the University Eye Department who asked questions and sent patients for further evaluation and management.

We also would like to thank Mrs. Marion Philipp and Mrs. Martina Himberger from Springer who accompanied us during the editorial and publishing process with great patience.

Foreword

Early and reliable diagnosis in ophthalmology usually depends on the recognition of minute changes of normal structures. The functional consequences due to such pathologic alterations are the only aspect of concern to the patient. However, ophthalmologists look for reproducible morphologic criteria essential in defining health and disease.

Clinico-pathologic correlations including "biomicroscopy" and "biocytology" have been a very rewarding approach to improve definitions of the phenotype. The spectrum of methods range from angiography and macroscopic observation in vivo to images obtained from fixed and stained tissue in the laboratory using standard light and electron microscopy, as well as localizing molecular biology approaches. In addition to these established methods, refinements of high resolution in vivo imaging techniques, such as ultrasound biomicroscopy (USB) and optical coherent tomography (OCT), as well as computer tomography, nuclear resonance tomography and positron tomography, enhance the sensitivity of detecting changes in structure.

Rudolf F. Guthoff and his group in Rostock have opened a new door in our profession, making us aware of the potential of confocal laser scanning in vivo microscopy. They have expanded the limitations of clinical observation – in vivo ophthalmic histopathology – which doubtlessly will permit more accurate and earlier diagnosis in the future.

Until recently in vivo confocal microscopy, despite convincing images published, remained a tool for research laboratories with an interest in clinical applications of prototype instruments. The Heidelberg Retina Tomograph HRT II, a fully developed and widely accepted tool for optic disc analysis, is the electronic and software foundation for the Rostock Cornea Module. This high magnification confocal anterior segment microscope brings new insight to the scientifically interested clinician.

In vivo confocal microscopy started to compete with and complement established methods of imaging the living structure to deliver information based on cellular and subcellular tissue elements.

We do not yet fully understand how living texture is displayed without the various staining techniques that conventional histopathology is based on. Every ophthalmic clinician should be curious to learn how this new method will allow the integration of personal knowledge and experience into the present diagnostic armament. The authors are the leading experts in this methodology. The outline of the book is convincing, the text concise and readable, and the figures are excellent, as you would expect from a Springer publication.

This book should be available in every ophthalmic library and on the desk of all ophthalmologists dealing with anterior segment diseases of the eye. Each reader will be rewarded by learning how to look beyond established diagnostic horizons.

Prof. (emer.) Dr. Dr. h. c. mult.
G.O.H. Naumann, FRCOphth (Hon.)
Universitäts-Augenklinik mit Poliklinik
Schwabachanlage 6 (Kopfklinikum)
91054 Erlangen, Germany
e-mail:
FritzNaumann@augen.med.uni-erlangen.de

Preface

This atlas and textbook summarizes our experience in the rather new field of clinical in vivo micromorphology. Stimulated by the work of Maurice, Jester, Bourne, Beuermann, Masters, Mathers, Thaer and others, our long-lasting cooperation is based on the mutual interest to push forward the frontiers of biomicroscopy using and combining recently available technical principles such as laser scanning, fast digital signal capture and processing and techniques for image stabilization. As a prerequisite there was the Heidelberg Retina Tomograph system for optic disc analysis with more than 5000 users worldwide and the engineering skill and dedication of physicists and engineers carefully listened to the wishes and dreams of the clinicians. So we are already on the way to "in vivo biopsy," as postulated by Teruo Nishida.

The content of the book reflects the common efforts of clinicians and physicists from the Rostock University Eye Department and the Centre Hospitalier National d'Ophtalmologie des Quinze-Vingts in Paris. Both groups share the enthusiasm to incorporate in vivo microscopy in the clinical decision making process by imaging, measuring and quantifying tissue parameters to achieve a quick and exact diagnosis and to monitor treatment in order to find the optimal path for patient recovery.

In a time when molecular biology, genomics and proteomics are marking the frontier of biomedical research, in vivo micro morphology could well be the platform where these metabolic changes are displayed clinically first.

Baudouin
Guthoff
Stave

Contents

Introduction

With good reason, biomicroscopy in ophthalmology is dominated by slit-lamp examinations. The foundations laid by Gullstrand were outlined in exemplary fashion in Vogt's *Lehrbuch und Atlas der Spaltlampenmikroskopie des lebenden Auges* [*Textbook and Atlas of Slit-Lamp Microscopy of the Living Eye*], first published in 1921 [92]. The second (1930) edition of that standard textbook placed particular emphasis on information yielded by focal illumination of scatters, described as *Hornhautkörperchen* (corpusculi corneae) [22]. However, the maximum magnification achievable with this technique – approximately ×40 – limited further subdifferentiation and did not reveal clinicomorphologic correlations at the cellular level.

Specular microscopy was first described by Vogt [92] but did not gain popularity until photographic and video techniques allowed documentation and quantification of corneal endothelial cells [5, 8, 9, 41, 52]. Although the specular microscope is useful for in vivo examination of the cornea, its applications are restricted to the endothelial cell layer.

In 1968, the same year that Maurice described the first high-powered specular microscope [52], the first scanning confocal microscope was proposed [66]. This device was characterized by high *z*-axis resolution and provided high-resolution microscopic images of cells within living tissues of patients without the need for fixation or staining.

Alongside this tandem scanning confocal microscope, a slit-scanning confocal design was also produced [47, 54, 71, 78, 86]. Currently, the tandem scanning design is manufactured by Advanced Scanning (New Orleans, LA, USA), and the most recently developed version of the slit-scanning design is produced by Nidek Technologies (Vigonza, Italy). The authors have been accumulating experience with the slit-scanning design since 1994 [71, 78] and have attempted to modify the system to achieve reproducible and quantifiable results [4].

Principles of Confocal In Vivo Microscopy

The development of in vivo confocal laser scanning microscopy in the late 1980s permitted precise three-dimensional (3D) visualization of microstructures of the ocular fundus in particular, with its optic nerve head and the peripapillary retina [75, 93]. Modern digital image processing technology enables quantitative data to be collected noninvasively, rapidly, and with a low level of illumination. The precision of laser scanning ophthalmoscopy in this context is based on the principle of the confocality of the examined object with the light source and the detector plane. A laser light source is focused through a pinhole diaphragm to one point on the object. The reflected laser light is separated by a beam splitter from the incident laser beam path and is deflected through a second confocal diaphragm to reach a photosensitive detector. Because of the confocal design, light originating from outside the focal plane is highly suppressed, and only the object layer located at the focal plane contributes to the image. To build up a two-dimensional (2D) image perpendicular to the optical axis of the device, the laser beam has to scan the sample point by point. This is achieved by introducing two oscillating mirrors into the beam path. Figure 2.1 provides a schematic illustration of this principle. By optical movement of the focal plane, an image can be acquired from a deeper layer of the examined object, thus enabling a data cube to be built up in a successive series.

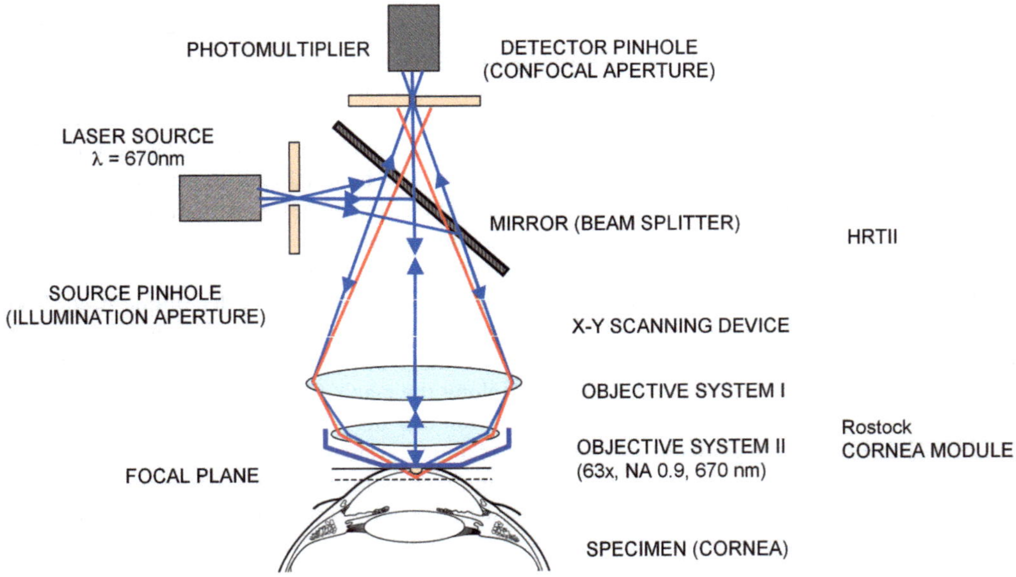

Fig. 2.1 Principle of confocal laser-scanning microscopy

In contrast, in slit-lamp biomicroscopy examination of the cornea, an optical section that is essentially perpendicular to the corneal surface is seen in up to ×50 magnification or, with an additional lens for endothelial viewing (specular microscopy), in up to ×200 magnification. Nowadays, documentation is generally performed using digital photography. All other cellular structures, such as the epithelium, cannot be imaged with this technique because of the high proportion of scattered light.

Optical tomography perpendicular to the incident light path has become possible only with the adaptation of confocal microscopy, as described above, for examining the living eye [7, 11, 12, 37, 55, 86, 87]. This yields images of the endothelium, for example, that are comparable to those obtained with specular microscopy. In this case, the most pronounced source of scattered light is the cytosol of the endothelial cells, with the result that the cell borders appear dark. Only light reflected from the focal plane contributes to the image. In this way, cell structures in the stroma, nerves, and corneal epithelium [95] can also be imaged in fine optical sections.

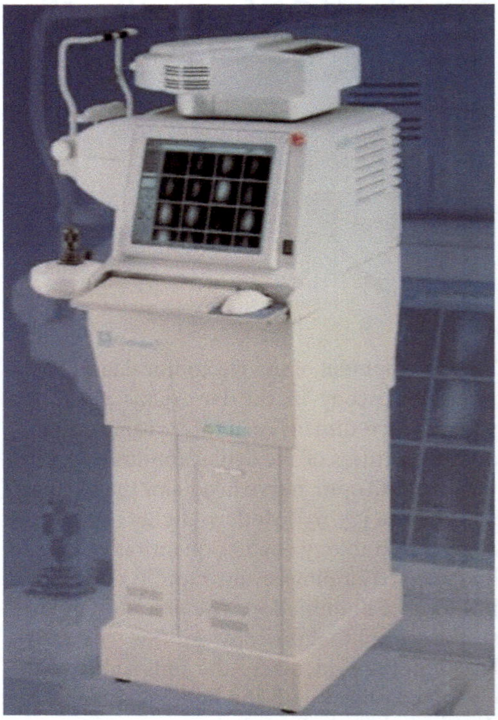

Fig. 2.2 Confocal slit-scanning microscope (ConfoScan 3; Nidek Technologies, Vigonza, Italy)

2.1 Slit-Scanning Techniques

Techniques based on the principles of the rotating Nipkow disk or tandem slit-scanning were used initially for confocal microscopy of the anterior segment of the eye. Figure 2.2 shows a slit-scanning microscope of this type that incorporates the use of a halogen lamp. For corneal assessment these microscopes can be used to image confocal sections with an optical layer thickness of approximately 5–10 µm. Synchronization of slit-scanning [47, 86, 94] with the video rate of a residual light camera yields sharp and motion-independent image sequences at 25 frames per second [46, 48] (Fig. 2.3).

Three-dimensional image distortion due to eye movements with this noncontact microscopy technique can be minimized only by faster image acquisition. However, more rapid movement of the confocal plane along the optical axis (*z*-scan) is accompanied by a loss of resolution. Loose optical coupling to the cornea using a gel (Fig. 2.4) limits the precision of depth information relating to the optical section in the cornea and, hence, the 3D reconstruction of cell structures [57, 65, 78]. Systematic errors such as inhomogeneous image illumination and image distortion also arise as a result of the electromechanical slit-scanning technique used. Linear movement along the *z*-axis during individual image acquisition also produces distortion along the *z*-axis. These fundamental sources of error can be prevented only by a rapid laser scanning system unhampered by mass inertia and incorporating a serial dot-raster technique with stepwise advance of the confocal plane during the *z*-scan.

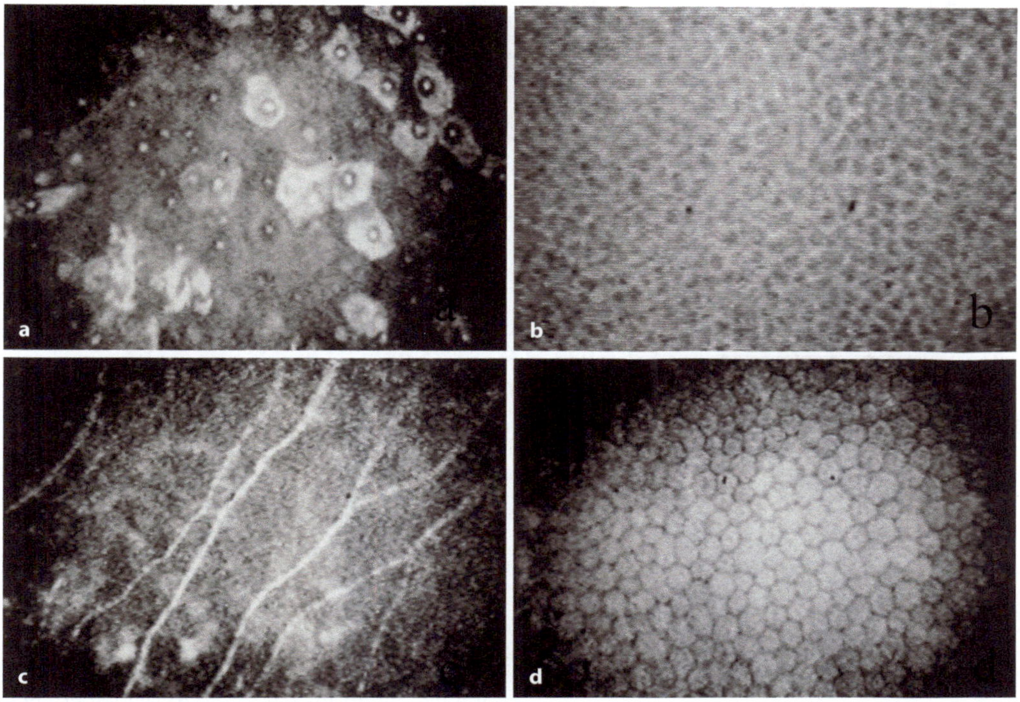

Fig. 2.3 Slit-scanning microscopy. **a** Superficial cells. **b** Basal cells. **c** Nerve plexus with keratocytes and (**d**) endothelium

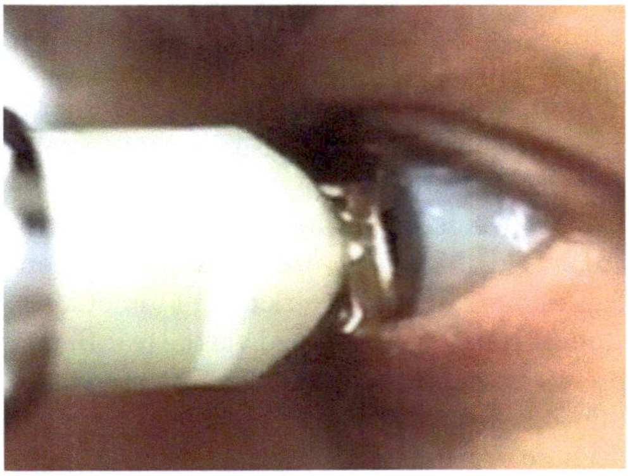

Fig. 2.4 Gel coupling between lens and eye (ConfoScan 3; Nidek Technologies, Vigonza, Italy)

2.2 Laser Scanning Imaging Plus Pachymetry

As an alternative to confocal slit-scanning microscopes, a confocal laser scanning microscope for the anterior segment of the eye was developed at the Rostock Eye Clinic (Germany) on the basis of an already commercially available laser scanning system. Not least because of its compact construction, the Heidelberg Retina Tomograph II (HRT II, Heidelberg Engineering, Germany) was selected as the basic device for a digital confocal corneal laser scanning microscope.

In laser scanning ophthalmoscopy of the posterior segment, the optically refractive media of the eye form part of the optical imaging system. For anterior segment applications, a high-quality microscope lens is positioned between the eye and the device, providing a laser focus less than 1 µm in diameter. The result is a

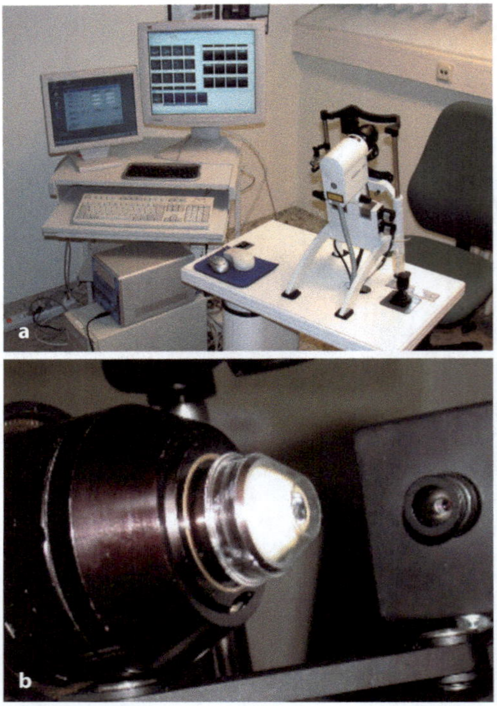

Fig. 2.5 Rostock Cornea Module (RCM; confocal laser scanning microscope). **a** Feasibility study of the fully automated control of the microscope using a separate personal computer, step motors (*x*-, *y*-, *z*-scan), and joystick. **b** RCM (contact) and digital camera

high-resolution, high-speed, digital confocal laser scanning microscope that permits in vivo investigation of the cornea (Fig. 2.5). Movement of the confocal image plane inside the cornea can be achieved manually at the microscope lens or by using the automatic internal *z*-scan function of the HRT II. Laser scanning tomography is consequently possible in the anterior segment of the eye.

This technique permits rapid and reliable visualization and evaluation of all the microstructures in the cornea, including the epithelium, nerves, and keratocytes, as well as the endothelium and bulbar conjunctiva. For the first time, the dendritic (or Langerhans) cells can now also be visualized in vivo with an image quality that permits quantification [83, 97]. In principle, any body surface that can be reached by the lens system is a suitable candidate for examination, with the result that potential applications also exist outside ophthalmology (for instance, for the skin, tongue surface, and oral mucosae).

The original functions of the basic HRT II device for evaluating the optic nerve head in glaucoma are fully retained when the system is modified into the confocal laser microscope. Software adapted to the special requirements of scanning microscopy of the cornea has also been developed. This software permits the acquisition of individual section images, image sequences of section images, and volume images with internal *z*-scan over a distance of approximately 80 µm. The digital properties of the device offer good patient and image data administration and also deliver rapid access to – and hence comparison with – data from previous examinations.

The HRT II has been modified with a lens system attachment known as the Rostock Cornea Module (RCM; J. Stave, utility model number 296 19 361.5, licensed to Heidelberg Engineering, Germany). The module is combined with a manual *z*-axis drive to move the focal plane inside the cornea. This enables a cell layer at any depth to be imaged and, for example, selected as the starting plane for the automated internal *z*-scan. During the examination, pressure-free and centered contact with the cornea can be monitored visually using a color camera.

The distance from the cornea to the microscope is kept stable by use of a single-use contact element in sterile packaging (TomoCap®). Optical coupling is achieved via the tear film or protective gel to the eye. The TomoCap® is a thin cap with a planar contact surface made from polymethyl methacrylate (PMMA) and is coupled optically to the lens with the aid of a gel. Special caps with small tips have been developed for experimental animal work in rats and rabbits (Fig. 2.6).

Confocal laser scanning microscopy attempts to register the reflected laser beam and the intrinsic beam of reflection microscopy and fluorescence microscopy at a very low level, even in thick tissue specimens, using objectives with a high numerical aperture. As a rule, these

Fig. 2.6 TomoCap® contact cap (Rostock Cornea Module). a TomoCap®. b, c Special caps for animal experiments. d Cap with a depression in the cap front. e Prismatic cap for slant imaging of the cornea

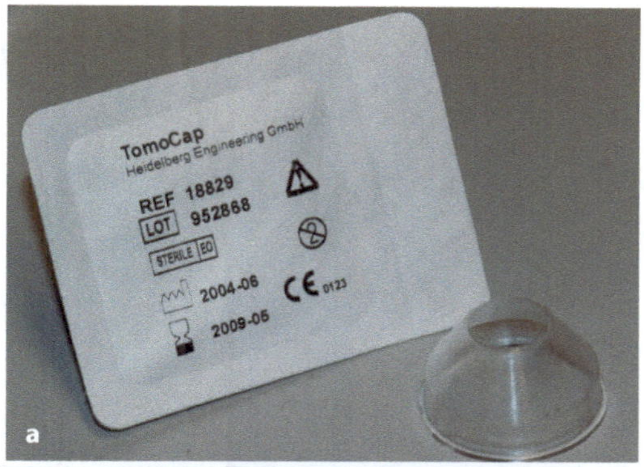

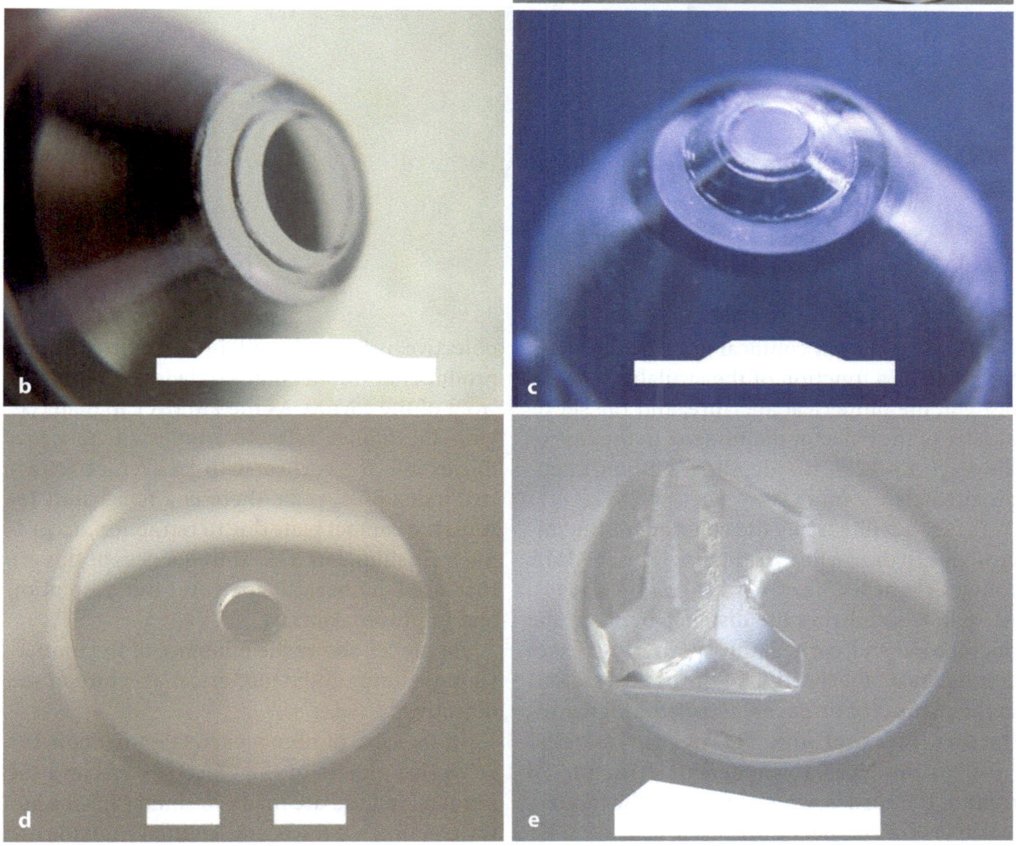

are immersion objectives. The resolution of a microscope is determined using the following equation:

R=0.61λ/NA
(R = resolution, λ = wavelength,
NA = numerical aperture)

Accordingly, the spatial resolution of the objective used (Zeiss ×63/0.95 W, 670 nm, ∞/0, Jena, Germany) and of the carbomer contact gel (Vidisic; Dr.Mann Pharma, Berlin, Germany) with a refractive index of n=1.350 (water n=1.330) yields a value of

R=0.61×0.670/0.95=0.430 μm

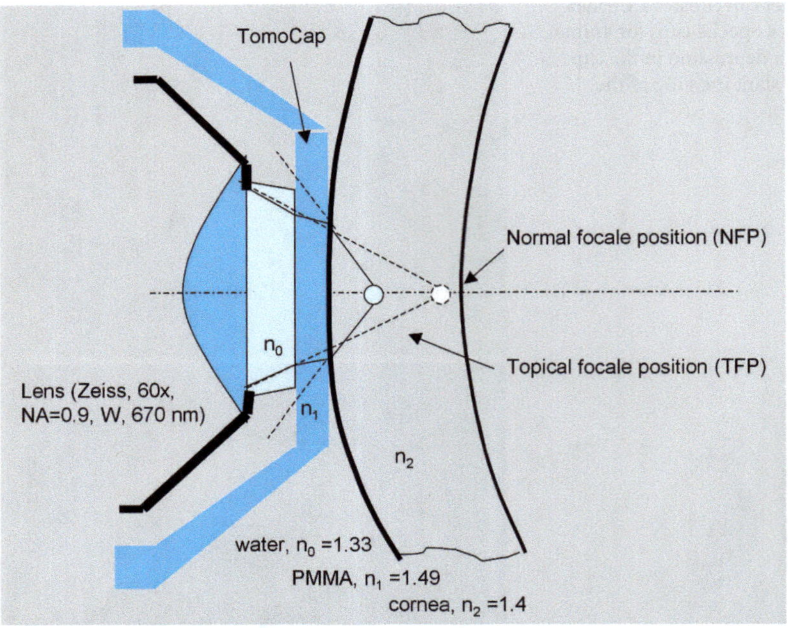

Fig. 2.7 Contact microscopy using the TomoCap® of the Rostock Cornea Module

Depth resolution in confocal microscopy is determined as a function of the available amount of light, the numerical aperture, and the sensible size of the detector diaphragm. In the case of the HRT II-RCM, the depth resolution is approximately 1–2 µm.

Figure 2.7 illustrates the coupling of the microscope to the cornea via the single-use PMMA contact element (Stave, GB 296 12 466.4/1996). As well as stabilizing the distance between the microscope and cornea, thus permitting exact depth data on the location of the optical section in the cornea (optic pachymetry), the higher refraction index of the PMMA cap material (n=1.49) produces an immersion structure in front of the cornea, which increases the numerical aperture and hence the resolution.

The numerical aperture can be calculated from the following equation:

$$NA = n \sin\alpha$$
(angle α = half the angle of aperture)

Accordingly, the limiting factor is the refractive index of the liquid contact medium used (Vidisic; n=1.350): Insertion of the PMMA contact disk of the TomoCap®, which is coupled to the objective using a liquid immersion medium, produces an immersion sandwich structure in front of the cornea. As the PMMA cap material, with the higher refractive index of n_1=1.490, moves into the cornea, with an average refractive index of n_2=1.44, there is a focus shift toward the contact plate due to refraction, according to Snellius' law. The consequences of this are increased numerical aperture and resolution and improved contrast.

Figure 2.8 shows the influence of the immersion medium on the numerical aperture of the objective.

In the case of homogeneous immersion, the refractive indices of the objective lens and the immersion fluid are made so similar for the key wavelength that the beams emitted from the object point pass unbroken through the immersion film and are thus taken up by the object lens. The result is a high light intensity that decisively determines the resolution.

In the special case involving coupling with the PMMA cap with a high refractive index, the numerical aperture is increased by the factor n_2, and thus the limit of resolution is reduced to $1/n_2$. Consequently, when the cap is used, the water-immersion objective (microscopy without the PMMA

Fig. 2.8 Influence of the immersion medium on the numerical aperture of the lens

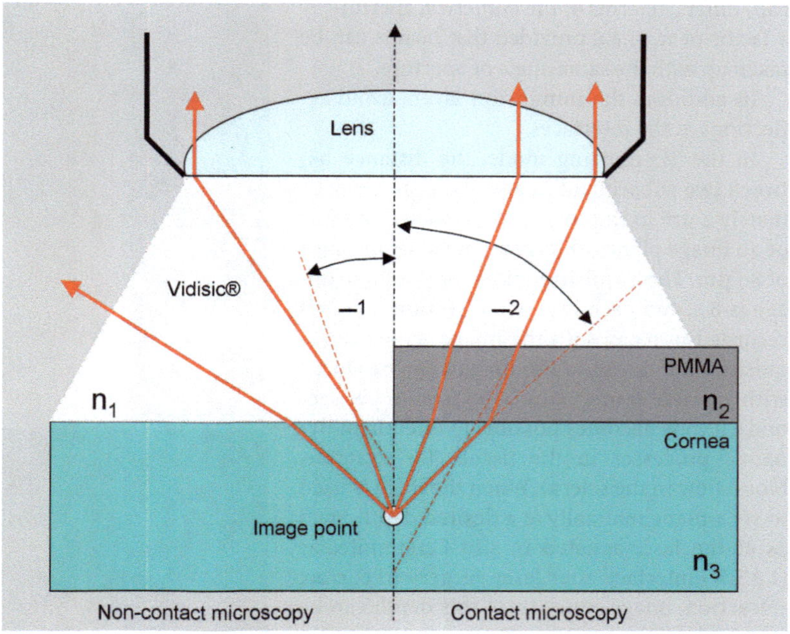

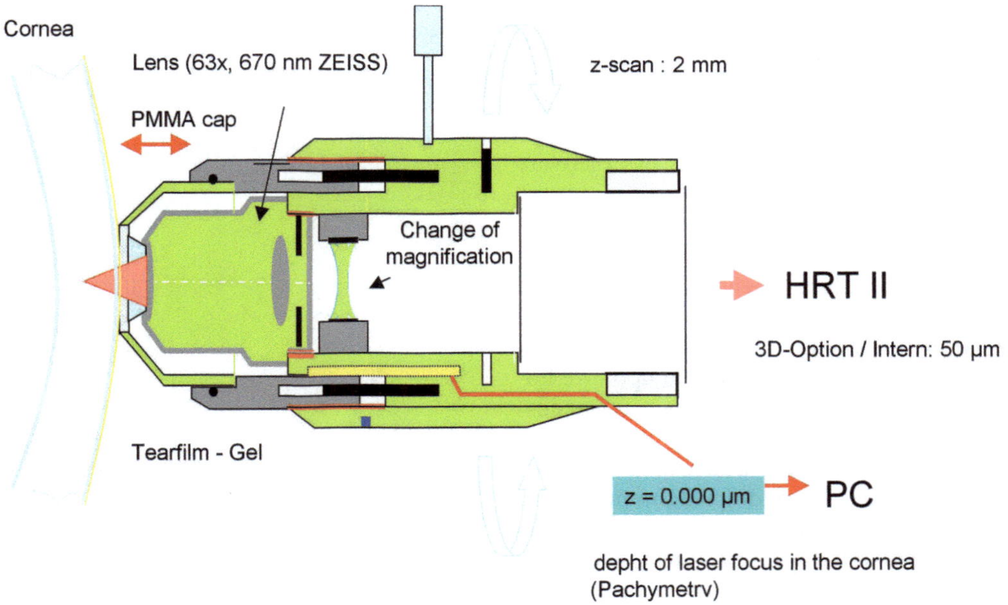

Fig. 2.9 Rostock Cornea Module (RCM)

cap) differs in terms of the numerical aperture by a factor of n_2=1.44, provided that beams can be taken up with the same angle of aperture.

In addition, the immersion layers avoid reflections at the interfaces.

In the 3D imaging mode, the distance between two subsequent image planes is approximately 2 µm in the cornea. A 3D image consists of 40 image planes, thus covering a depth range of 80 µm. The acquisition time for a volume image is 6 s, and each individual section image is recorded in 0.024 s. In the image sequence acquisition mode, up to 100 images can be stored with variable frame rates (1–30 frames per second). It thus becomes possible to document dynamic processes in the tissue (for example, blood flow in the sclera). When the RCM is used to set a plane manually at a desired depth, such as at the laser-assisted in situ keratomileusis (LASIK) interface after laser surgery to correct refraction, image series from this depth can be acquired with almost 100% image yield and precise depth allocation [43, 58, 59, 91].

With the HRT II, *z*-axis movement between images in the internal *z*-scan is performed – for the first time – in a stepwise manner; that is, during acquisition of one section image, the *z*-setting remains constant. This is a major advance and a prerequisite for generating distortion-free images of structure from one plane during the *z*-scan. A crucial prerequisite for undistorted 3D reconstructions has therefore been achieved [35].

A short focal length water-immersion microscope lens with a high numerical aperture was used to achieve high magnification (Achroplan ×63 W/NA 0.95/AA 2.00 mm, 670 nm, Carl Zeiss; alternatively, LUMPLFL ×60 W/NA 0.90/AA 2.00 mm, Olympus). To optimize image quality, the Zeiss microscope lens was customized with a special antireflection coating appropriate for the laser wavelength.

With the aid of an additional lens positioned between the HRT II and the microscope lens, the field of view of the scanning system (fixed at 15° with the HRT II) is reduced to approximately 7.5° to allow for the necessary magnification (Fig. 2.9). Depending on the microscope lens and additional lens used, the size of the field of view in the contact technique can be 250 µm ×

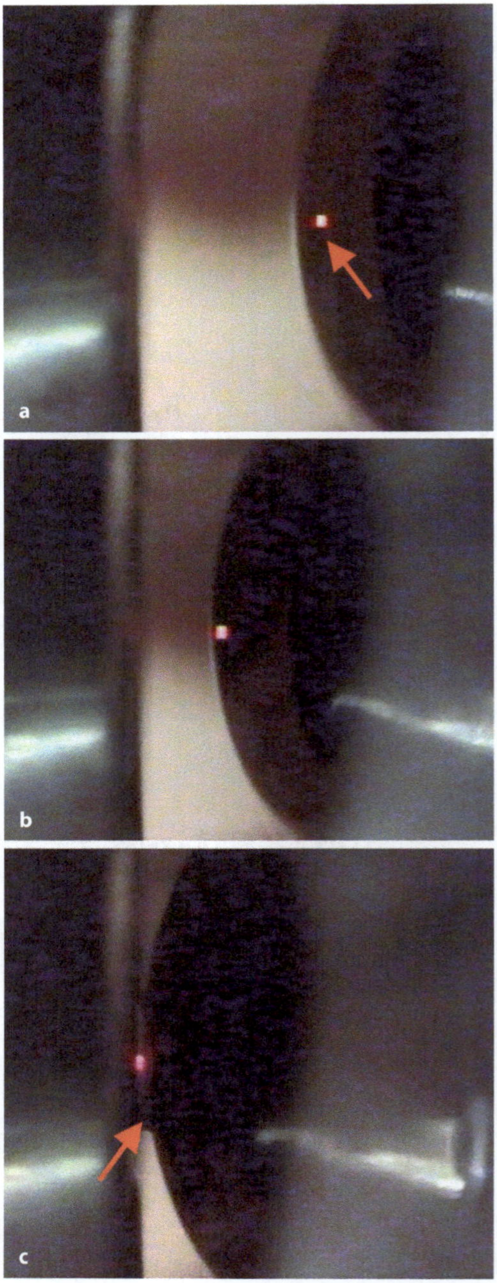

Fig. 2.10 Camera-controlled contact procedure: laser reflexes (*arrow*) on the cornea, producing the immersion gel bridge (*arrow*) between the objective lens/cap and the cornea (**a–c**)

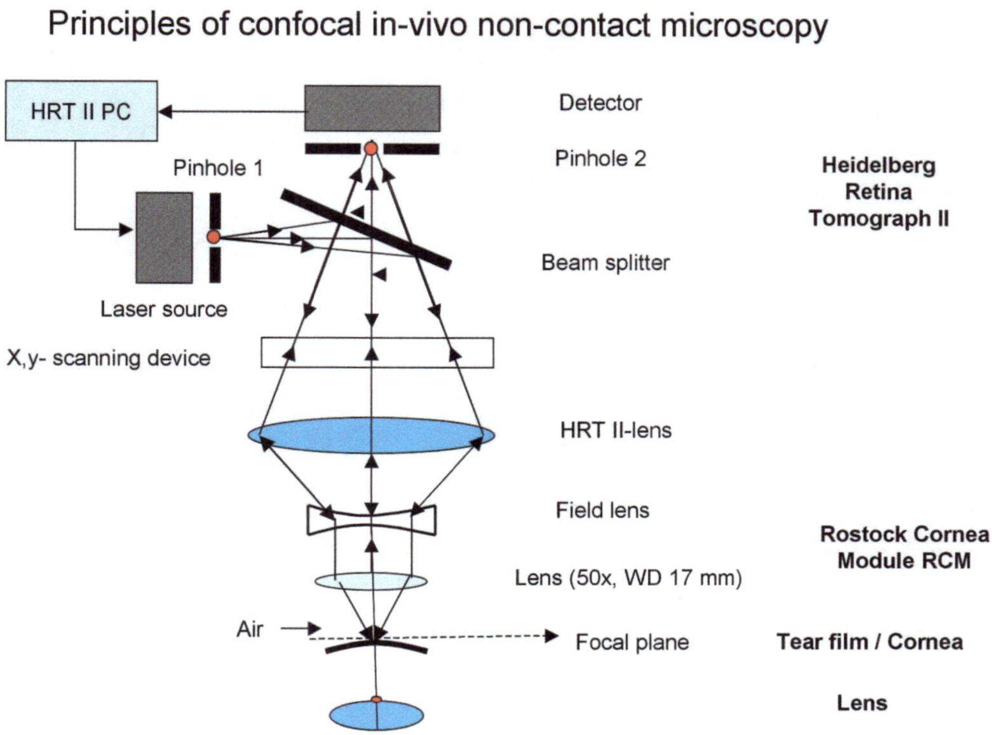

Fig. 2.11 Principles of confocal in vivo noncontact microscopy

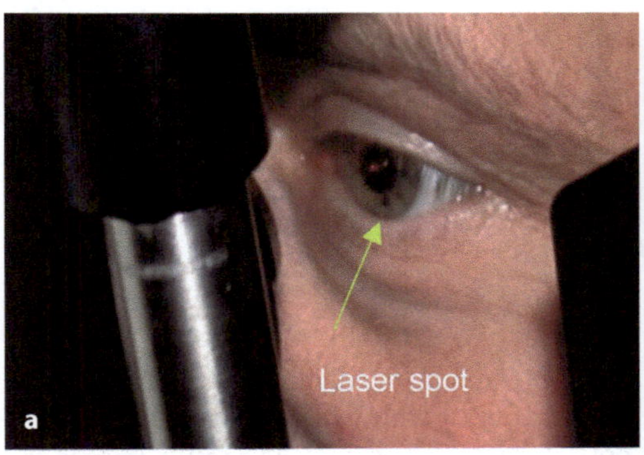

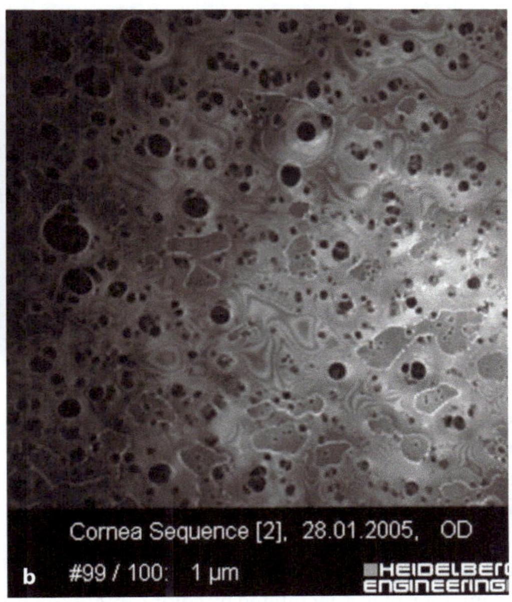

Fig. 2.12 Confocal in vivo noncontact microscopy. **a** Position of the laser light reflex in the center of the cornea during noncontact examination. **b** Example of a tear film image (ointment in the eye)

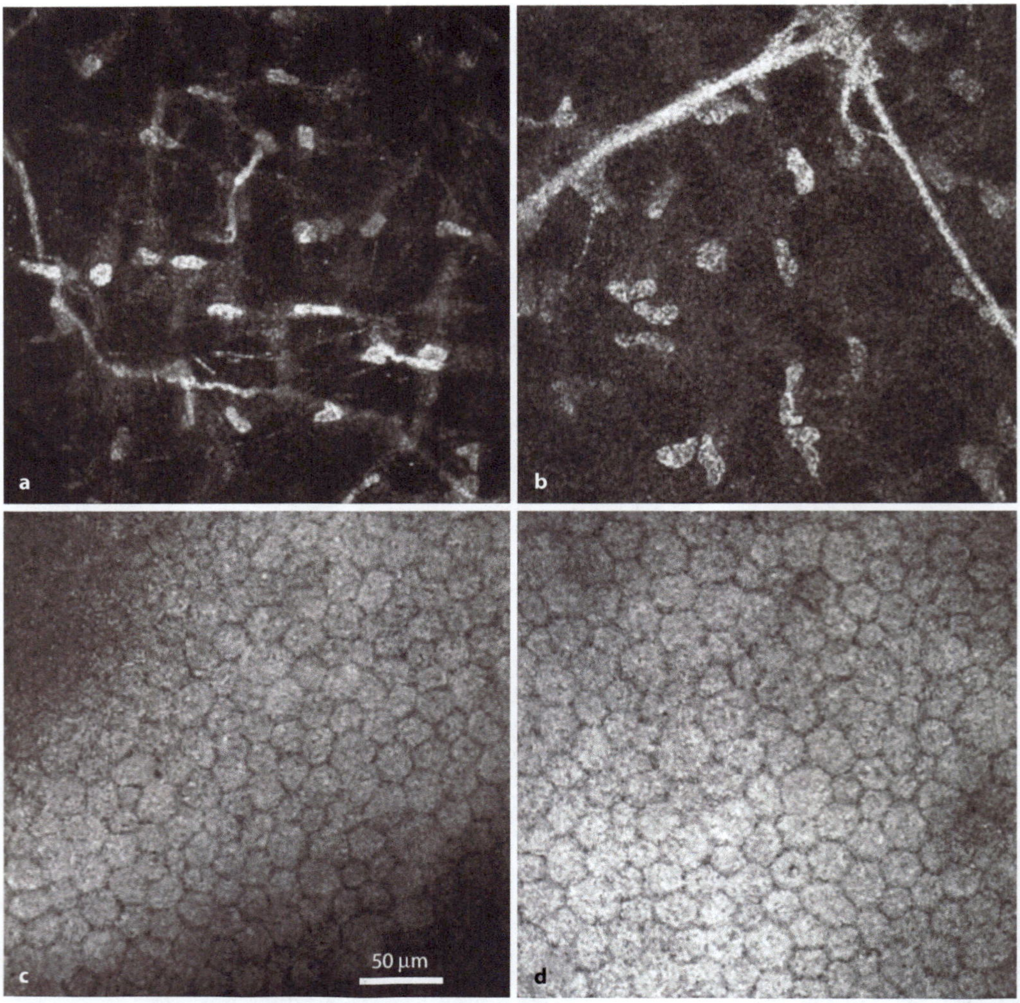

Fig. 2.13 Confocal in vivo noncontact microscopy: **a, b** Posterior stroma (keratocytes/nerve). **c, d** Endothelium

250 µm, 400 µm × 400 µm, or 500 µm × 500 µm. Dry microscope lenses can be used in noncontact work, such as to image the tear film [79].

In particular, the compact construction of the HRT II simplifies its use as a confocal in vivo microscope because the view is virtually unobscured when monitoring the patient and bringing the microscope up to the cornea.

The precise perpendicular positioning of the cornea in front of the microscope within the micrometer range is facilitated by color camera control. By observing the laser reflex on the cornea, even when bringing the microscope to the eye, the user can make a lateral or vertical correction so that contact with the cornea is exactly in the optical axis (Fig. 2.10 a–c). As a result, images are captured only of cell structures that are in a plane parallel to the surface; that is, transverse sectional images are acquired. The contact technique guarantees a fixed distance between the microscope and the cornea. The precise movement of the focal plane through the cornea with simultaneous digital recording of depth position relative to the superficial cells of the epithelium (at the corneal surface) thereby makes exact pachymetry possible (Figs. 2.11–2.13).

2.3 Basics of Image Formation in Confocal In Vivo Microscopy

The laser light that is emitted constitutes an electromagnetic wave with a defined wavelength lambda (λ). This light wave is modified when it passes through the cornea. Part of it migrates unchanged through all layers (transmission). At interfaces with changing refraction indices (n_D), the wave alters direction because of scatter and refraction. Light scattering is the basis of image formation in confocal microscopy in which only back-scattered light is used for the image. The amount of back-scattered light is very small because scatter occurs predominantly in a forward direction [18]. Light-scattering interfaces are found in the cornea, such as at the junction between cytoplasm or extracellular fluid, with n_D=1.35–1.38, and lipid-rich membranes, with n_D=1.47, in the form of cell borders, cell nucleus membranes, and mitochondrial membranes [77] (Fig. 2.14).

The amount of back-scattered light depends on the structure of the interface surfaces. Rough surfaces scatter light in a broadly diffuse pattern, whereas directed waves with narrow scatter cones are formed on smooth structures. In addition, the confocal image is influenced by the number as well as by the size and orientation of scattering organelles or particles. Objects whose diameters are of the same order of magnitude as the wavelength of the laser light display Mie scatter, and much smaller molecules display Rayleigh scatter, which is directed backward to a greater extent than Mie scatter is. Different orientations of elongated cell organelles can place different particle cross-sections in the path of the incident light beam, the result being that the light is back-scattered to varying degrees. A high proportion of cell organelles also increases the amount of back-scattered light [6, 18].

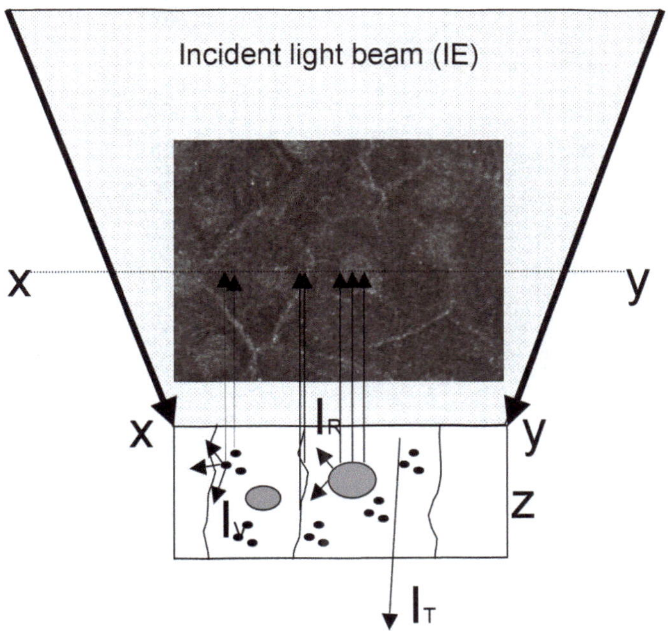

Fig. 2.14 The confocal image of one section (*x*, *y*) is produced by the sum of the back-scattered light intensities (*Ir*) from the focal depth range (*z*). (*Iv* forward scatter, *It* transmission)

2.4 Noncontact Confocal Laser Scanning Microscopy

Noncontact laser scanning microscopy offers the advantage of contact-free high-resolution imaging of the tear film in particular [79] and also of the posterior corneal structures, such as the keratocytes, and of the endothelium. This becomes especially important in the cellular assessment of surface wound healing.

For noncontact microscopy with the HRT II–RCM, the water-immersion objective with the contact cap is removed and replaced with long focal length dry objectives. As well as imaging

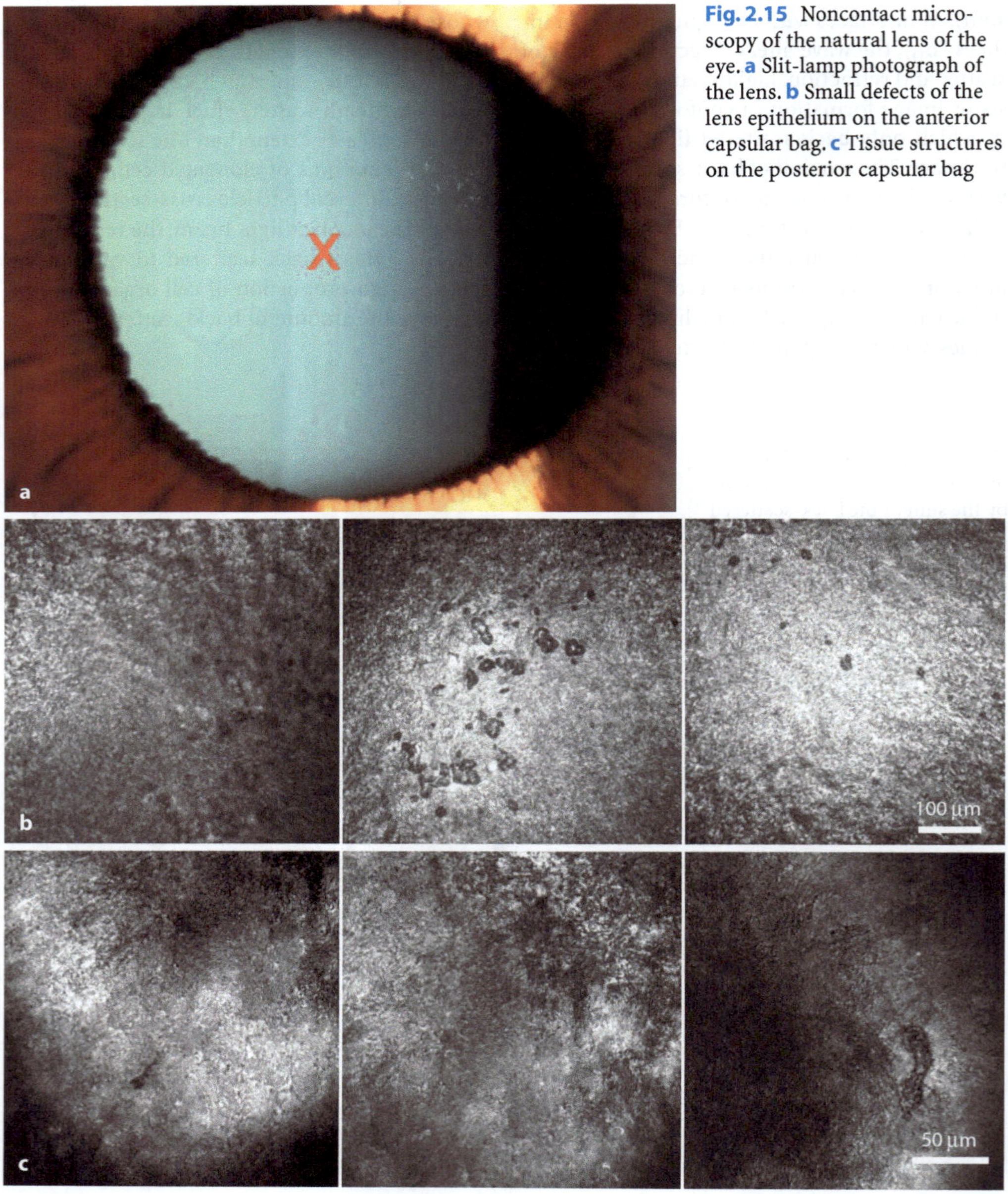

Fig. 2.15 Noncontact microscopy of the natural lens of the eye. **a** Slit-lamp photograph of the lens. **b** Small defects of the lens epithelium on the anterior capsular bag. **c** Tissue structures on the posterior capsular bag

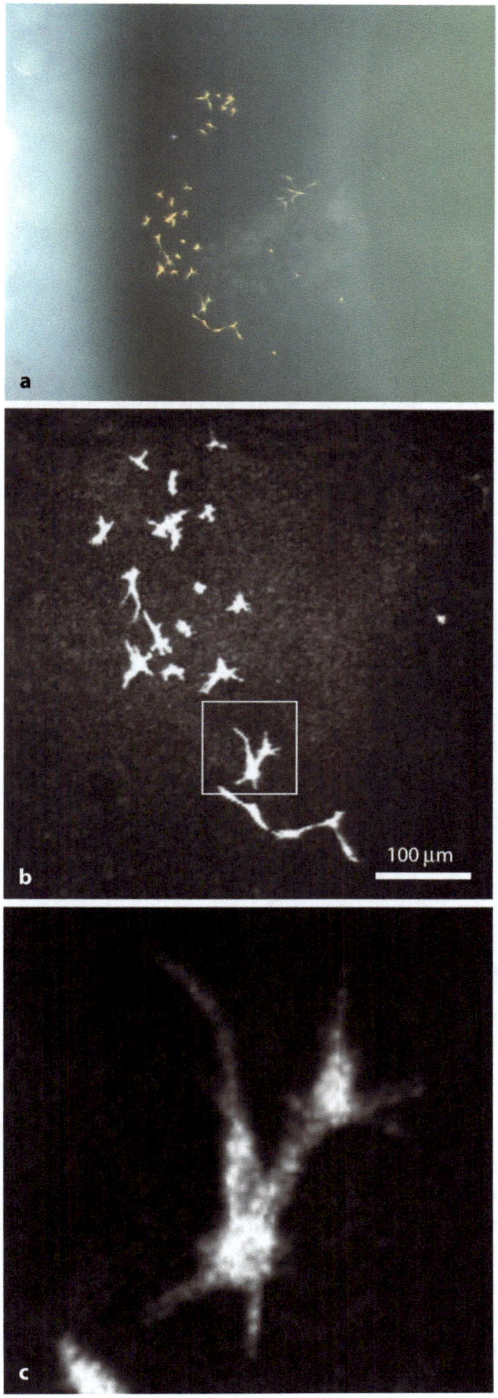

Fig. 2.16 Pigment deposition on the anterior capsular bag. **a** Slit-lamp photograph. **b** Confocal laser scanning microscopy. **c** Higher magnification (manual laser intensity regulation)

the corneal structures mentioned above, it is also possible to perform microscopy of the structures of the natural lens as far as the posterior aspect of the capsule when objectives with super-long focal length and high numeric aperture are used (Fig. 2.15). It is also possible to image secondary cataract formation following intraocular lens implantation.

The dry microscope lens Nikon ×50, 0.45 CF Plan, SLWD (17 mm) in conjunction with the RCM and the HRT II has been found to be most efficient for visualizing corneal structures. To visualize lenses/intraocular lens structures, it is advisable to use the Nikon ×20, 0.35 Plan, ELWD (14 mm) or Nikon ×10, 0.21 L Plan, SLWD (17 mm) because the refraction of the cornea produces additional magnification (Fig. 2.16).

It is necessary to reduce the laser power when imaging the tear film because of its high reflectivity, which causes overmodulation of the HRT II image intensifier. Because the present technical status of the HRT II does not permit this, and a software modification was also unsuccessful in this regard, the use of a detachable neutral glass filter on the dry objective has proved highly effective (Fig. 2.17).

The rapid image sequence with this device also enables dynamic processes to be recorded, such as break-up of the tear film in patients with dry eye or the changes occurring following application of ointments or various medicines, with the ability to determine their retention time in the tear film.

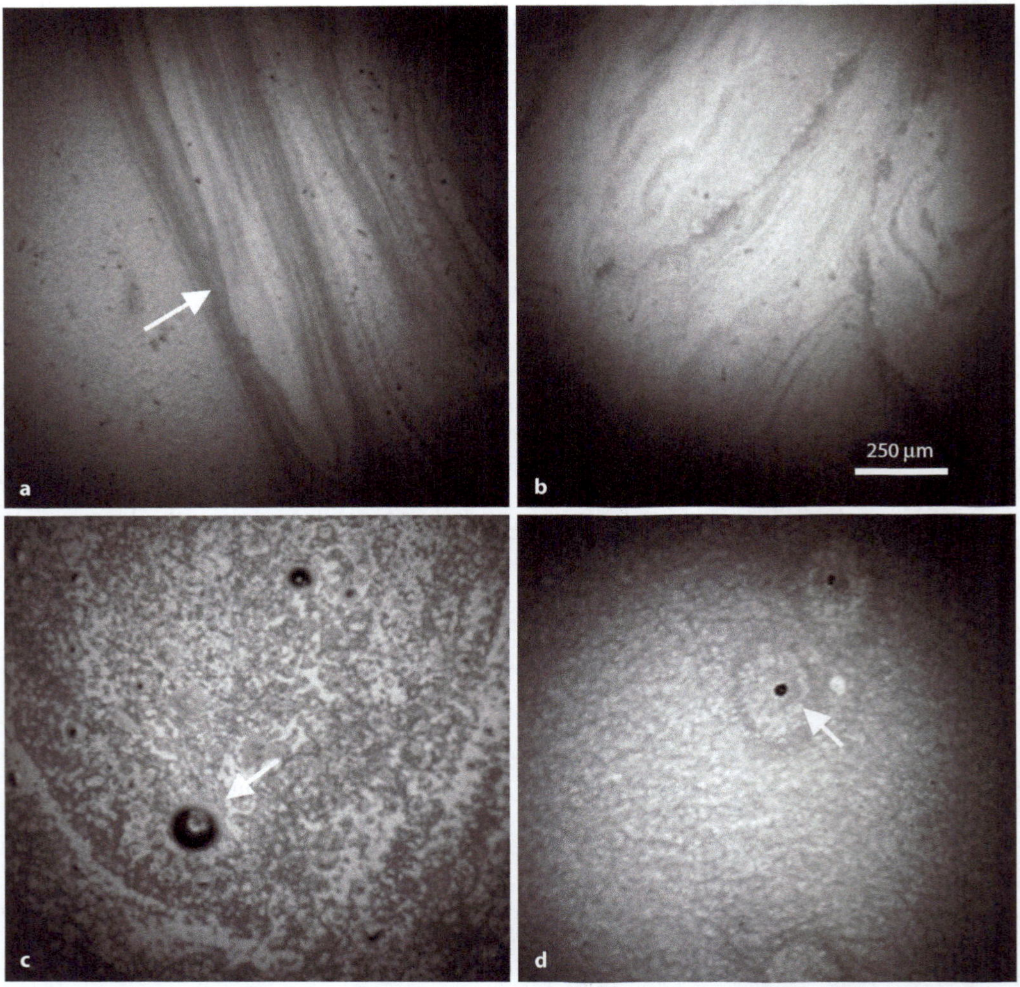

Fig. 2.17 Noncontact laser scanning microscopy of the physiologic tear film. **a, b** Lipid stripes on the tear film after opening the eye. **c** Air bubbles (*arrow*) in the tear film. **d** Meibomian glandular expressions (*arrow*) in the tear film

2.5 Confocal Fluorescence Microscopy

In conventional microscopy, the possibility of using stains to visualize specific anatomical structures provides major information gains. This is particularly true when fluorescence microscopy or immunohistochemistry techniques are used. Because these methods are well-suited for investigating the functional status of tissues ex vivo, they are also of interest for in vivo microscopy. However, problems arise because of the necessity for real-time studies and the selection of suitable nontoxic vital stains. Nevertheless, the first successful steps have already been taken on the road to confocal in vivo fluorescence microscopy of the anterior segments of the eye.

Both the Heidelberg Retina Angiograph (HRA) Classic and the HRA II, which were developed for fluorescence angiography of the ocular fundus, have been modified with the RCM in such a way that this device also permits the

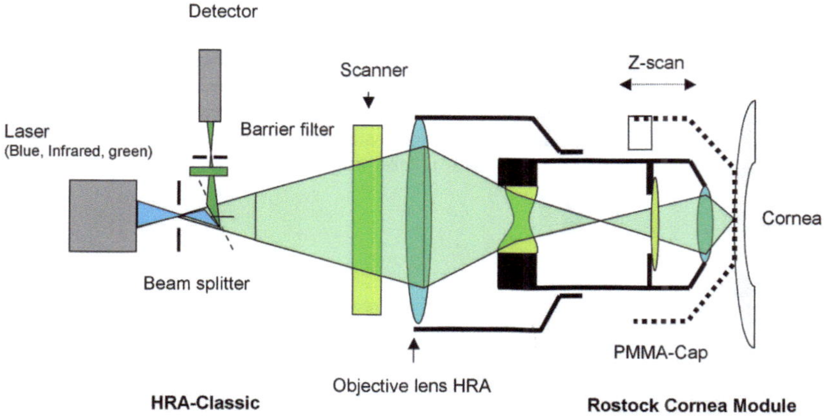

Fig. 2.18 Heidelberg Retina Angiograph (HRA)–Rostock Cornea Module (RCM). Fluorescence mode: *blue* argon laser line. Reflection mode: *green* argon laser line

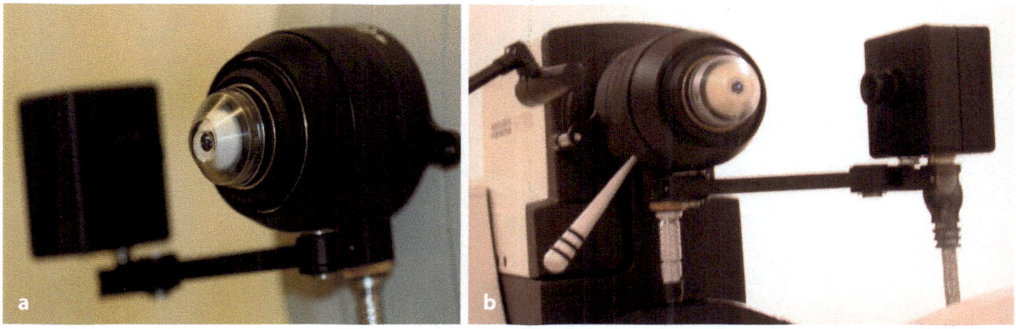

Fig. 2.19 Confocal fluorescence laser scanning microscopes: Rostock Cornea Module (RCM)–Heidelberg Retina Angiograph (HRA) Classic and the HRA II (Heidelberg Engineering, Germany). **a** RCM combined with the HRA Classic (contact). **b** RCM combined with the new fluorescence retina angiograph HRA II (contact)

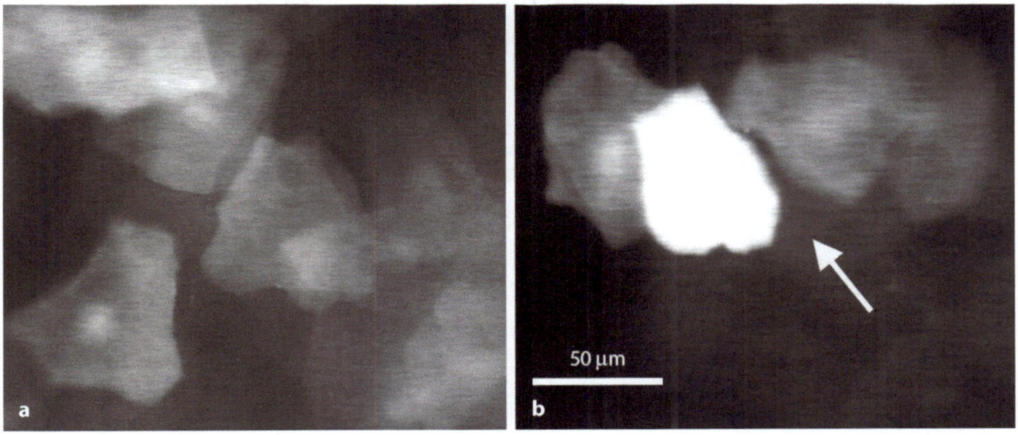

Fig. 2.20 Confocal fluorescence microscopy of the corneal epithelium from a pig's eye (Rostock Cornea Module–Heidelberg Retina Angiograph Classic). **a** Autofluorescence of superficial cells. **b** Fluorescein-stained superficial cell (*arrow*)

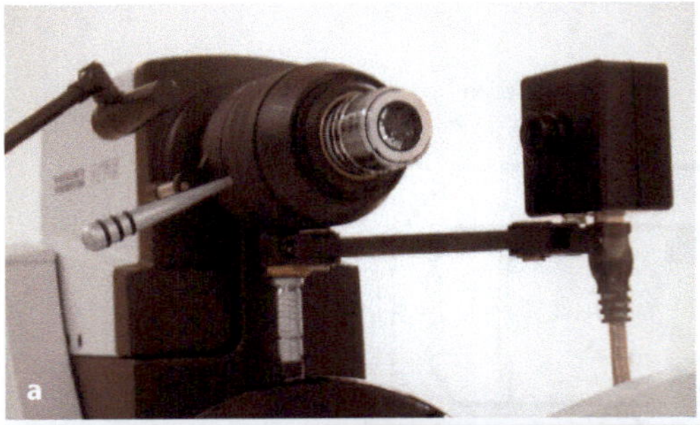

Fig. 2.21 Confocal noncontact fluorescence laser-scanning microscopes [Rostock Cornea Module (RCM)–Heidelberg Retina Angiograph (HRA) II]. **a** RCM (noncontact) adapted for the HRA II. **b** Noncontact micrograph of sodium fluorescein-stained epithelium after laser-assisted in situ keratomileusis (LASIK)

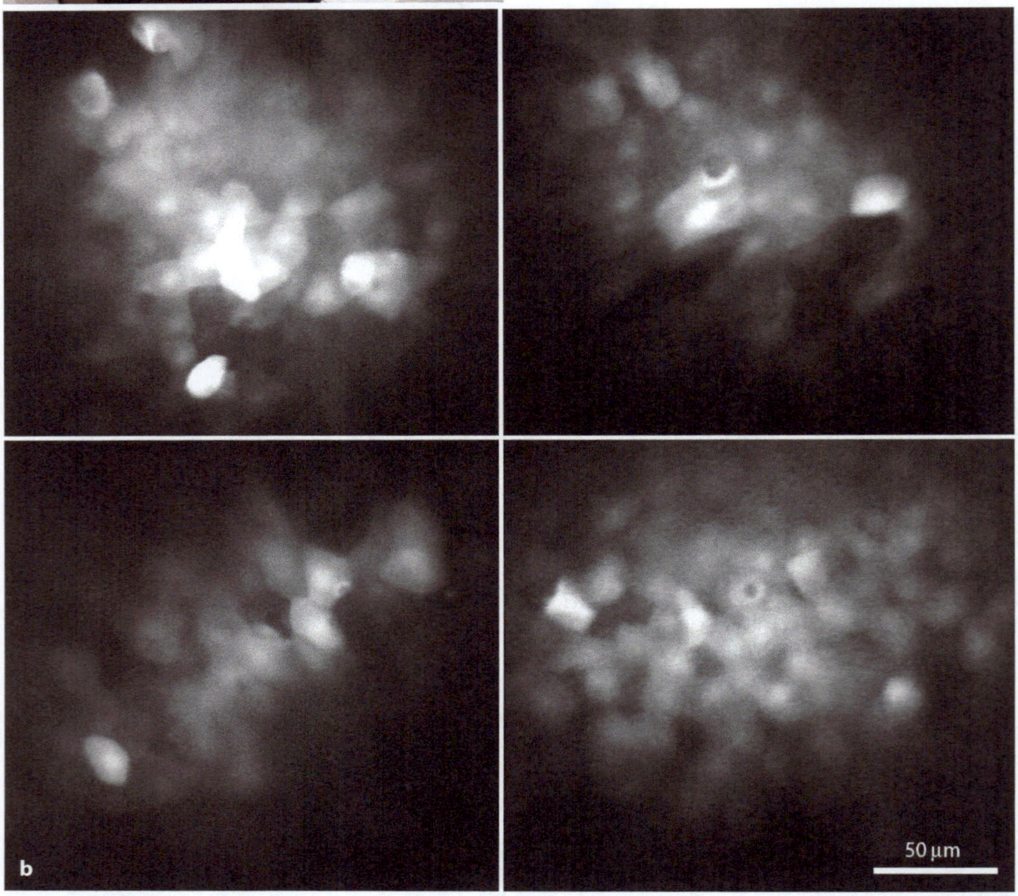

laser focus to be moved to the plane of the anterior segment of the eye. As a result, it becomes possible to visualize cellular structures of the cornea and conjunctiva by recording intrinsic fluorescence or stimulated fluorescence after labeling with sodium fluorescein (NaF).

Figure 2.18 is a schematic illustration of the combined HRA and RCM. Figure 2.19 shows the HRA Classic and its successor, the HRA II, combined with the RCM as a confocal contact microscope.

The functional principle here is the same as with the modification of the HRT II. The wavelength of the integrated argon laser can be switched, for example, with the HRA. As a result, in addition to a reflection image (wavelength

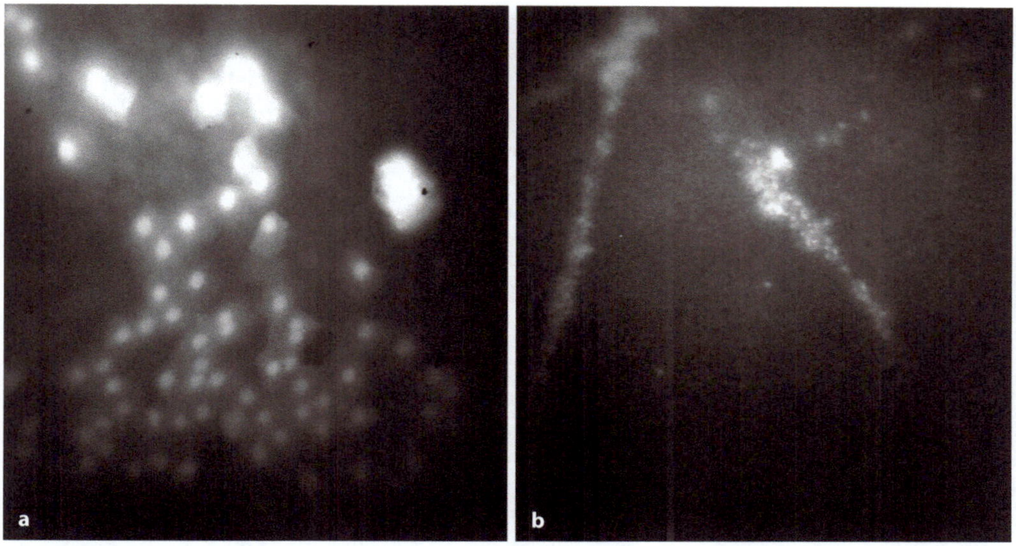

Fig. 2.22 Confocal noncontact fluorescence laser scanning microscopes (Rostock Cornea Module–Heidelberg Retina Angiograph II). **a, b** Epithelium stained with sodium fluorescein (SE Thilo; Alcon, Germany) after mechanical contact with a tonometer system

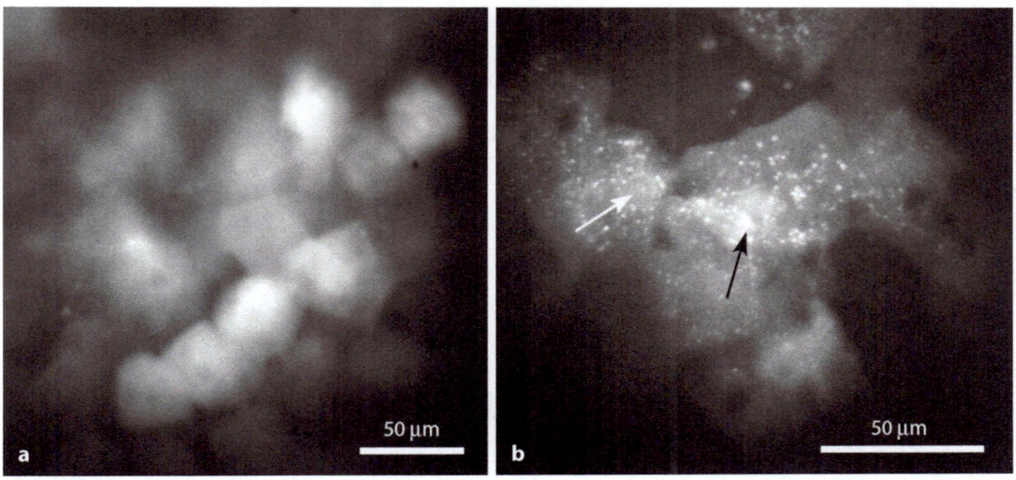

Fig. 2.23 Superficial cells (pig's eye) after pharmacological destruction of the cell membrane and staining with fluorescein (SE Thilo; Alcon, Germany). The fluorescein penetrates the cell membrane (*arrow*) and is concentrated near the nucleus. **a** Untreated. **b** After treatment

514 nm, 150 mW/m^2), it is also possible to perform stimulation with blue laser light (wavelength 488 nm, 1.5 W/m^2). By adding in a blocking filter (500 nm), it is possible to visualize the fluorescence of NaF-labeled intracellular structures (Figs. 2.20–2.24).

Noncontact fluorescence microscopy utilizes dry objectives with an ultralong working distance (Nikon ×20, 0.35, WD; Nikon ×50, 0.45, WD). For studies using the contact method, a water-immersion objective (×63, Zeiss) is used, which (as in investigations with the HRT II–

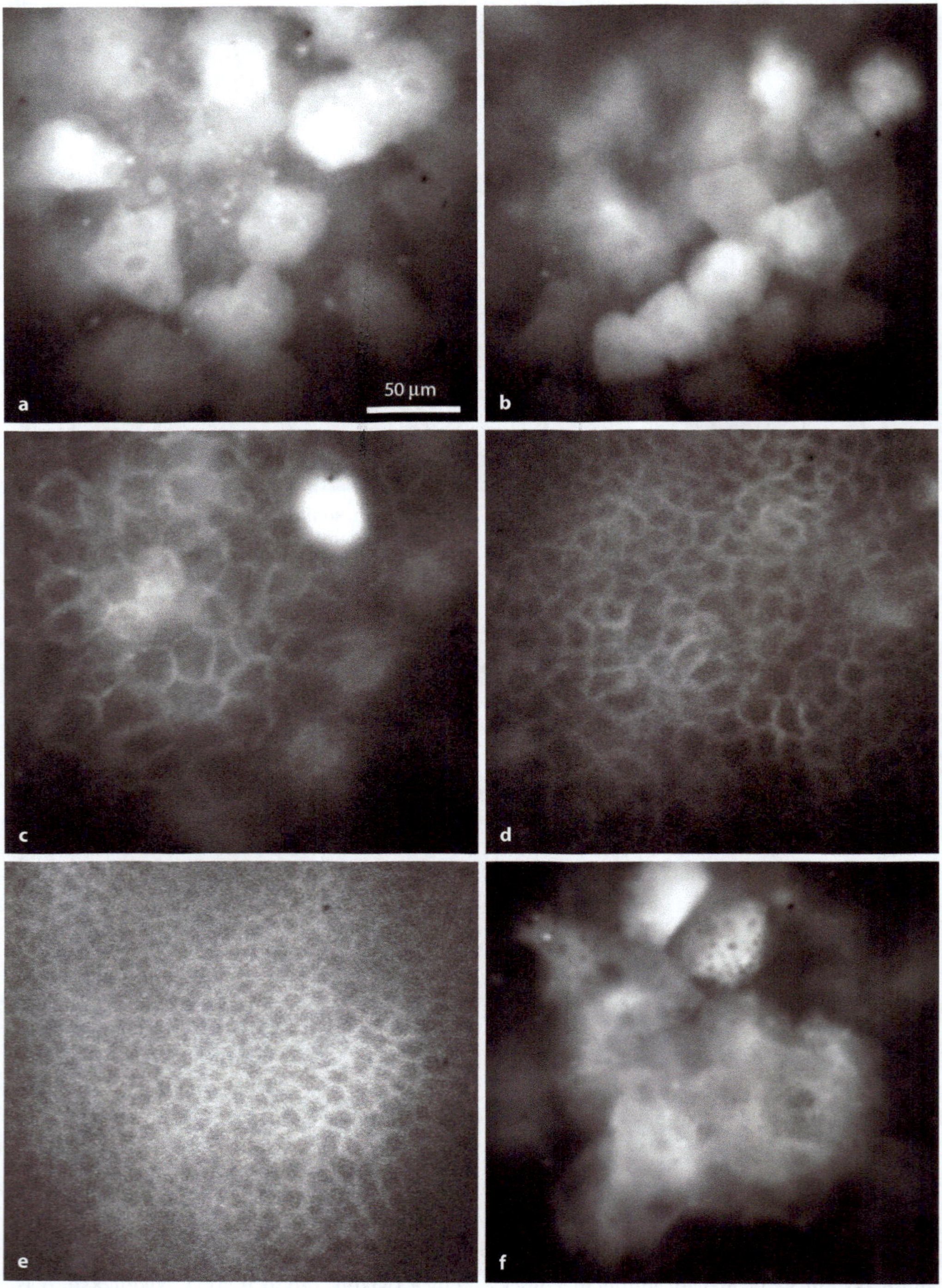

Fig. 2.24 Fluorescein penetration into the cornea: confocal ex vivo contact microscopy with the Rostock Cornea Module–Heidelberg Retina Angiograph II. Porcine cornea following fluorescein staining (Fluo SE Thilo for 1 min; irrigation with 0.9% sodium chloride). **a, b** Superficial cells. **c, d** Intermediate (wing) cells. **e** Basal cells. **f** Superficial cells after 60 min

RCM) is optically coupled to this by the plastic TomoCap® incorporating a plane PMMA disk on the epithelial side and a protective gel (Vidisic). Corneal staining was performed with fluorescein SE Thilo eye drops (Alcon, Germany). Three different magnification stages can be selected (contact methods: 150×150 µm, 250×250 µm, and 400×400 µm; noncontact methods: 450×450 µm, 750×750 µm, and 1,200×1,200 µm). Using this design, it is possible to verify previous findings concerning the staining characteristics and distribution spaces after corneal staining in the in vivo situation.

Figure 2.21 illustrates the HRA II with the RCM as a noncontact fluorescence microscope as well as fluorescence images with varyingly pronounced NaF-stained corneal epithelial cells from a female patient following LASIK surgery. Figure 2.22 shows the epithelium of a patient's eye following multiple contact with a damaged contact cap. Figures 2.23 and 2.24 show ex vivo images of a porcine cornea from a test series following pharmacological lesion induction and staining with fluorescein (Fluorescein SE Thilo, Alcon).

General Anatomical Considerations

The **corneal epithelium** consists of five to six layers of nucleated cells that can be subdivided functionally and morphologically into three zones:

- **Superficial cells** are approximately 50 µm in frontal diameter and approximately 5 µm thick. About one-seventh of these cells are lost by desquamation within 24 h. Before detachment, the cytoplasm and nucleus undergo changes in their optical characteristics.
- **Intermediate cells** are 50 µm in diameter and 10 µm thick. These cells form a contiguous polygonal, wing-shaped pattern (wing cells).
- **Columnar basal cells** have a flat basal surface adjacent to Bowman's membrane, a frontal height of approximately 20 µm, and a frontal diameter of 8–10 µm. Like endothelial cells, they can be quantified accurately because of their defined location in relation to the basement membrane (Fig. 3.1).

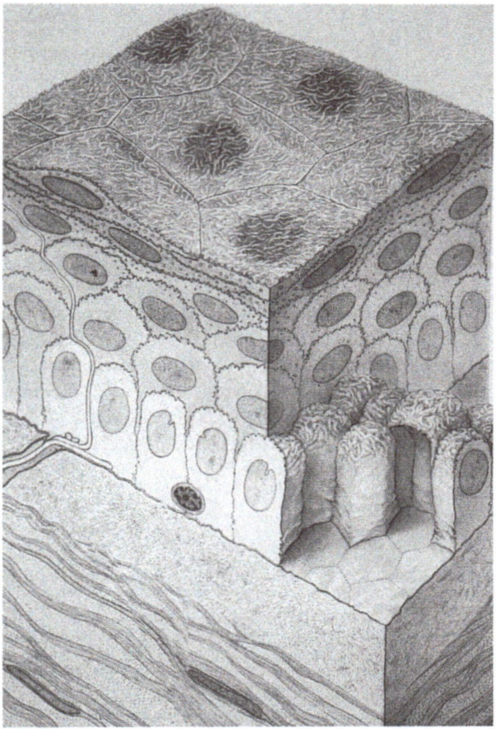

Fig. 3.1 Schematic illustration of the corneal epithelium and upper corneal stroma

Bowman's membrane – which is clearly distinct histologically from the epithelial basement membrane – is 10–16 µm thick and remains amorphous on light microscopy. Its location on in vivo confocal microscopy is well defined by the subepithelial plexus (SEP).

The **stroma** accounts for some 90% of total corneal volume. Ninety-five percent of the stroma consists of amorphous ground substance (glycoproteins, glycosaminoglycans: keratan sulfate and chondroitin sulfate) and collagen fibers. The remaining 5% of stromal volume is accounted for by cellular structures known as keratocytes, which are specialized fibroblasts.

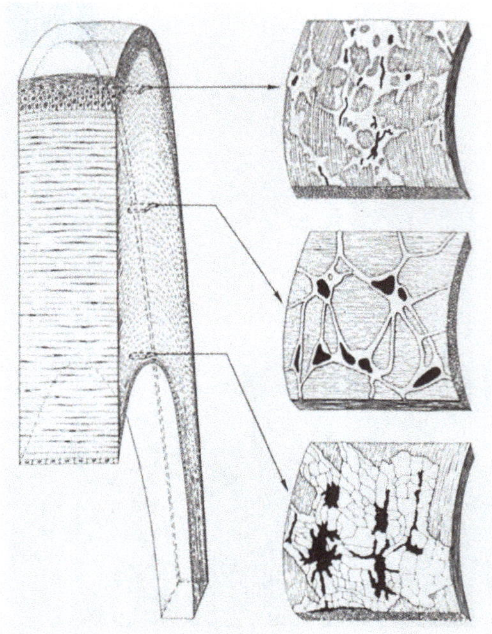

Fig. 3.2 Schematic illustration of the layered structure of the human cornea. The differently shaped keratocyte nuclei can be distinguished on in vivo confocal microscopy (adapted from Krstić RV. *Human Microscopic Anatomy: An Atlas for Students of Medicine and Biology*. Berlin, Heidelberg, New York: Springer-Verlag, 1991 [39])

Besides the nerves, their irregularly shaped nuclei are the only well-defined sources of scattered light in corneal stroma detected on confocal microscopy. Their widely branching cytoplasmic extensions are not visible in the healthy cornea (Fig. 3.2).

The cornea is the most densely innervated tissue in the human body. It is supplied by the terminal branches of the ophthalmic nerve in the form of 30–60 nonmyelinated **ciliary nerves.** In the limbus region these are seen as whitish, filigree-like structures; their complex stromal and epithelial branchings are not visible by slit-lamp microscopy but are relatively clear on confocal microscopy.

Like Bowman's membrane, **Descemet's membrane,** which should be regarded as the basement membrane of the endothelium, remains amorphous on light microscopy. It is 6–10 μm thick. On confocal microscopy it is defined optically by the easily identifiable endothelial cells.

The **endothelium** consists of about 500,000 hexagonal cells approximately 20 μm in diameter and 5 μm thick and with large, flattened central nuclei. The high concentration of cell organelles is indicative of very intensive metabolic activity.

Ex Vivo Applications

Ex vivo confocal imaging using complex microscope systems has become a prominent and integral feature of modern-day scientific technology. Ex vivo conditions enable the process of image acquisition to be optimized and standardized, thus yielding optimal resolution and contrast. The information obtained in this way may be helpful when similar structures have to be identified and interpreted under in vivo conditions.

The following examples illustrating blood constituents, pathogenic microorganisms, and ocular tissue structures demonstrate the resolution of the RCM under optimized experimental conditions.

4.1 Blood Components

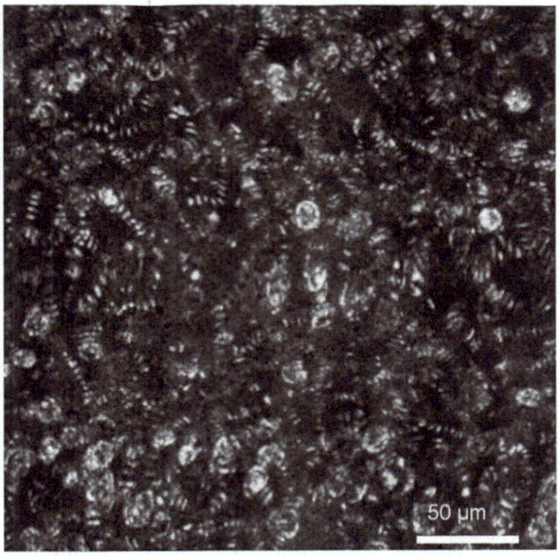

Fig. 4.1 Ex vivo: Blood, concentrated; rouleaux phenomenon

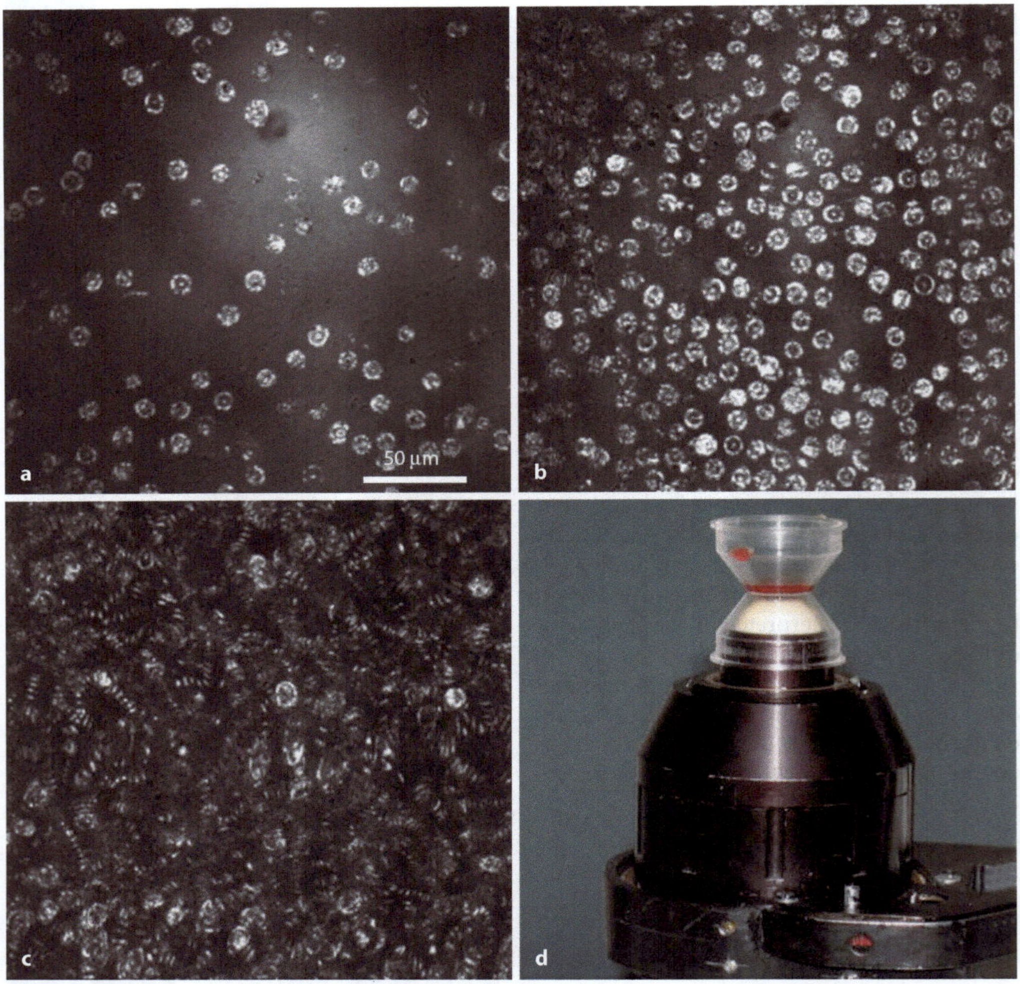

Fig. 4.2 Blood. a–c Erythrocytes, different concentrations. d Vertical imaging with the Rostock Cornea Module–Heidelberg Retina Angiograph II (TomoCap®)

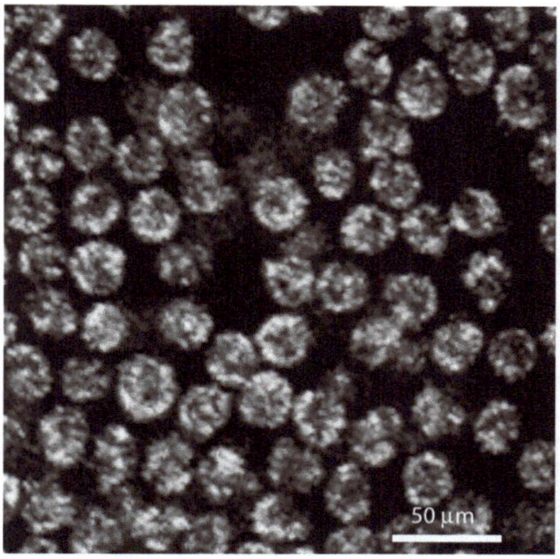

Fig. 4.3 Leukocytes: diluted separation

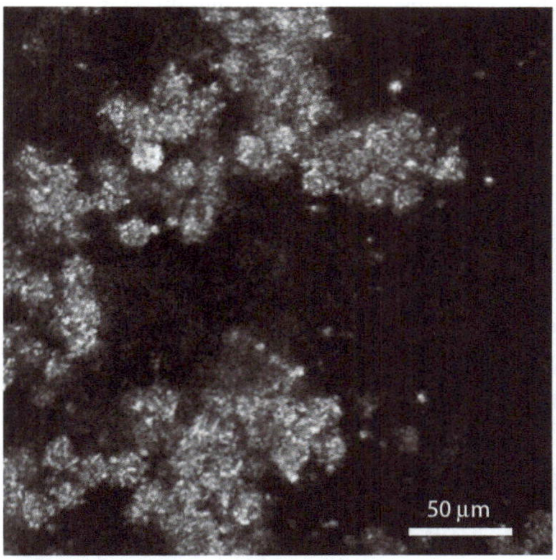

Fig. 4.4 Lymphocytes: diluted separation

4.2 Pathogenic Microorganisms

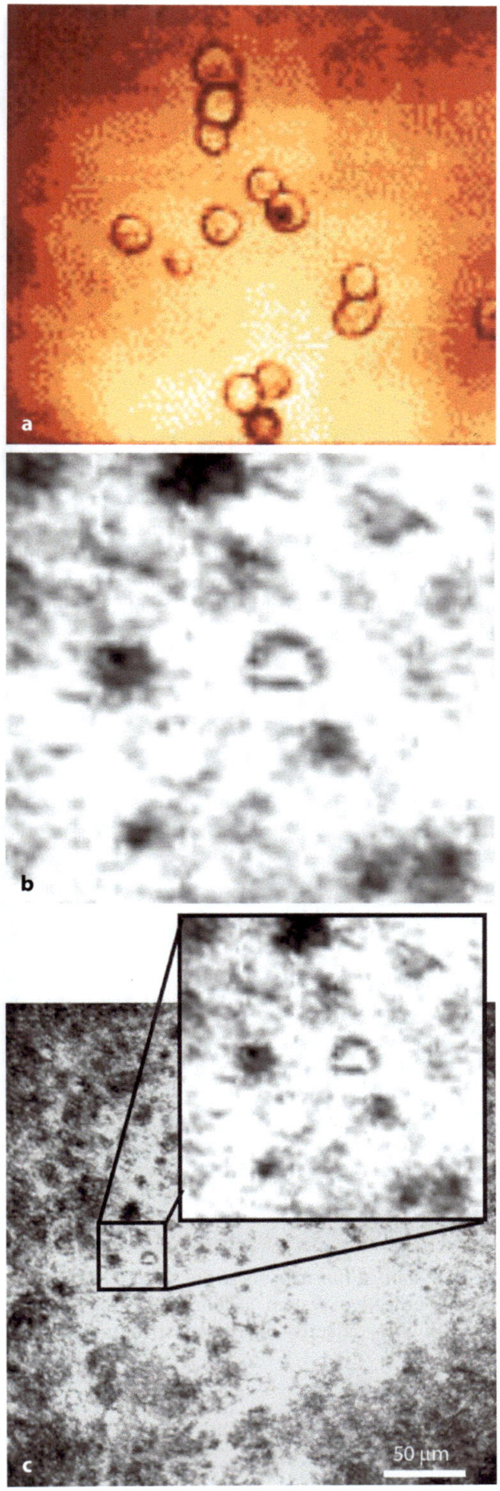

Fig. 4.5 *Acanthamoeba*. Ex vivo (**a**) and in vivo (**b**) confocal images showing microcysts (cystic stage of life cycle), which are round and up to 10 µm with double-walled structure. **c** In vivo confocal image of anterior stroma (depth 93 µm) with several Acanthamoeba microcysts

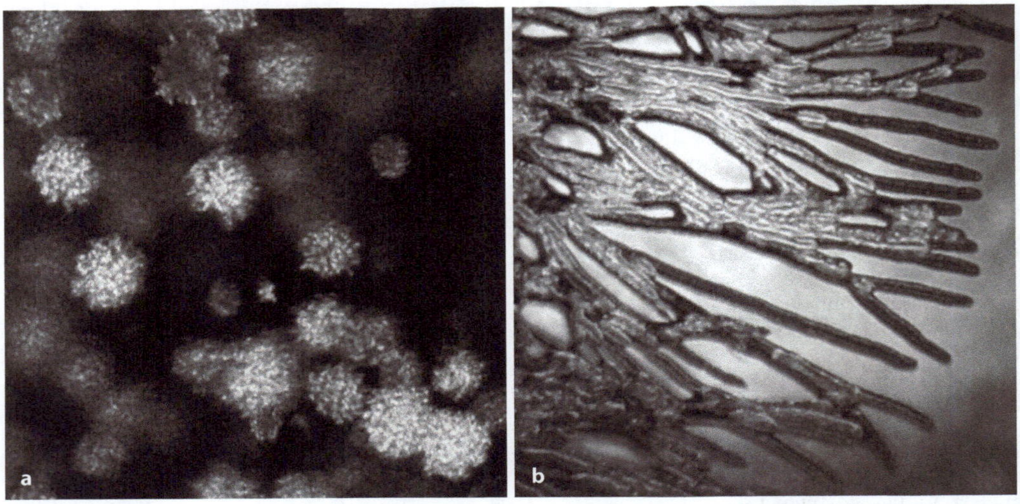

Fig. 4.6 Noncontact confocal microscopy (Nikon ×50 object lens) of *Aspergillus terreus*. **a** Highly reflective conidial heads with conidophores terminating in phialides are visible. **b** Noncontact confocal microscopy (Nikon ×20 object lens) of *Aspergillus terreus*. The picture shows a cross-section of branched fungal mycelium growing into the agar

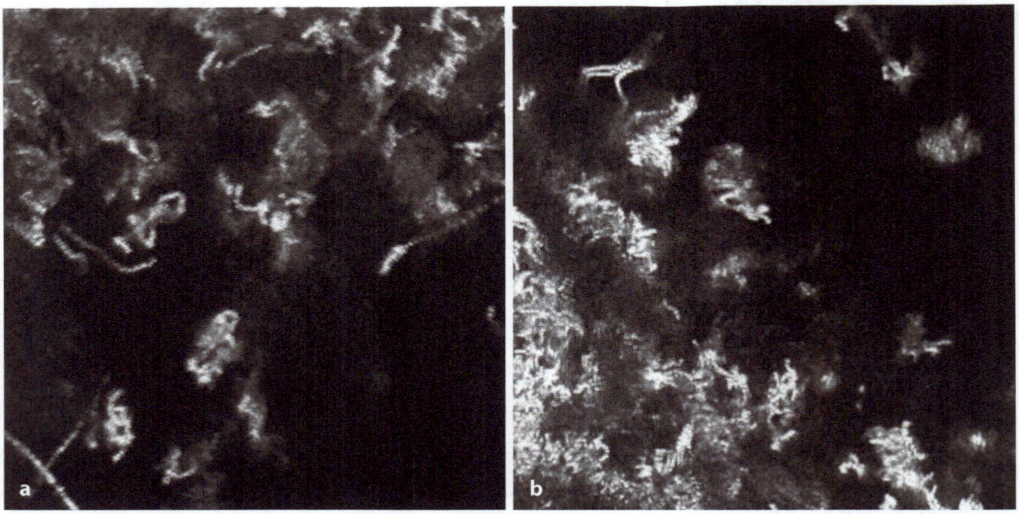

Fig. 4.7 *Penicillium chrysogenum*. **a, b** With a noncontact objective system (Nikon ×50), the culture shows highly reflective brushlike structures at the level of the agar. These consist of metulae-carrying phialides and conidophores

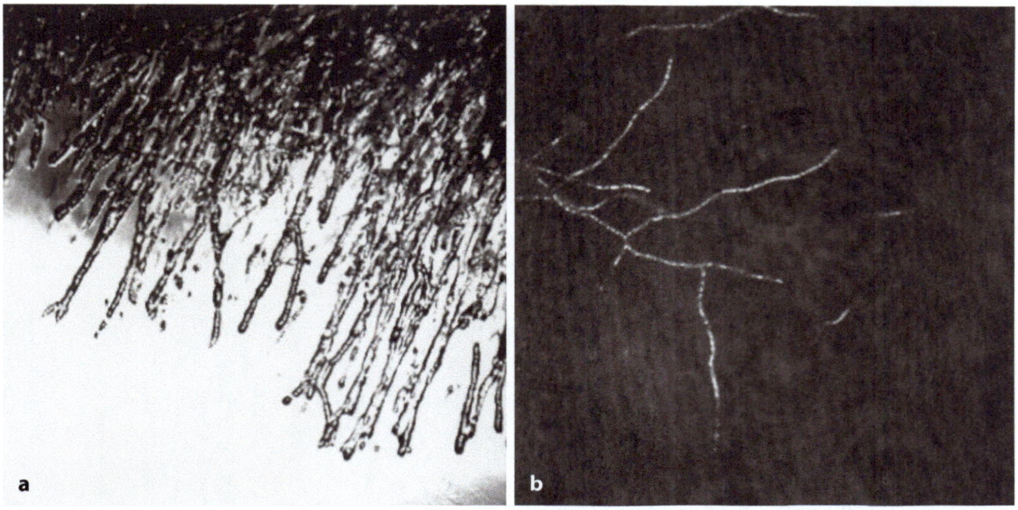

Fig. 4.8 *Penicillium chrysogenum*. **a** The picture obtained with a noncontact objective system (Nikon ×20) shows hyaline hyphae growing into the transparent agar. **b** Contact confocal microscopy of highly reflective hyaline hyphae growing into the agar

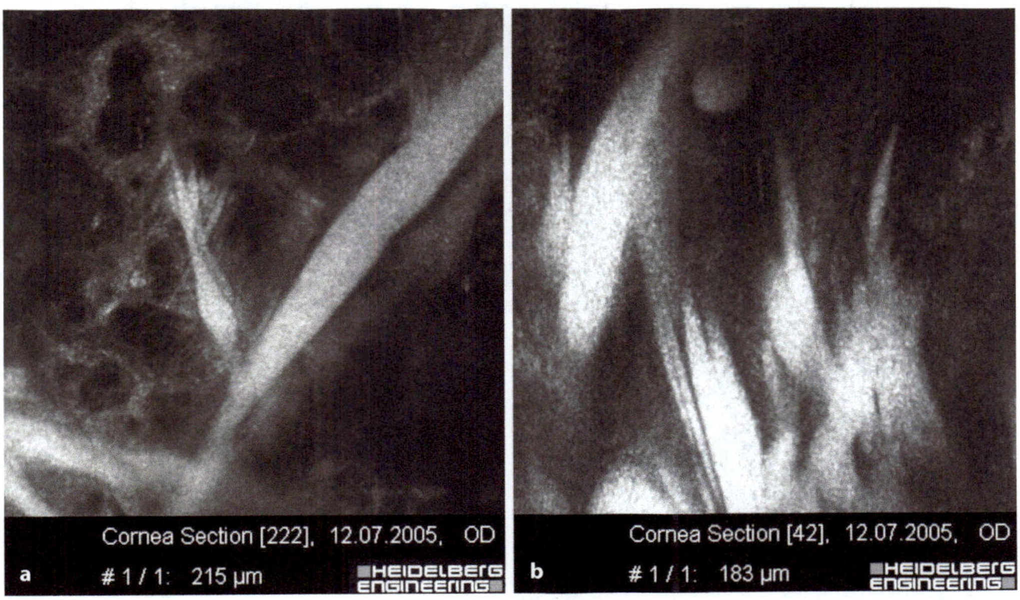

Fig. 4.9 Reflective opacities in a single stromal plane (depth of focus 180–220 µm) forming branched, needle-like patterns

4.3 Ocular Tissue Structures

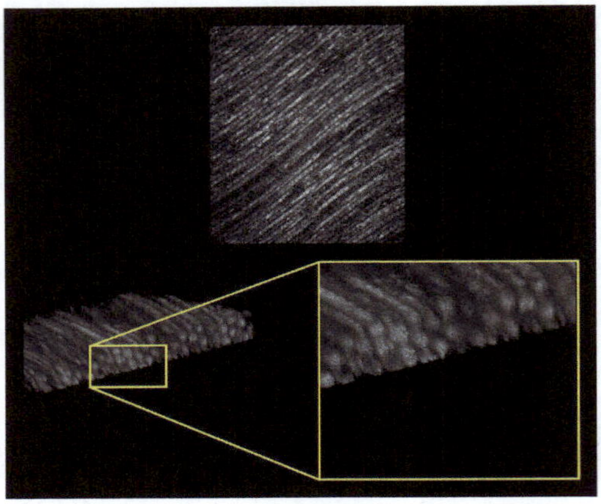

Fig. 4.10 Two-dimensional image and three-dimensional reconstruction of human lens fibers (ex vivo contact mode)

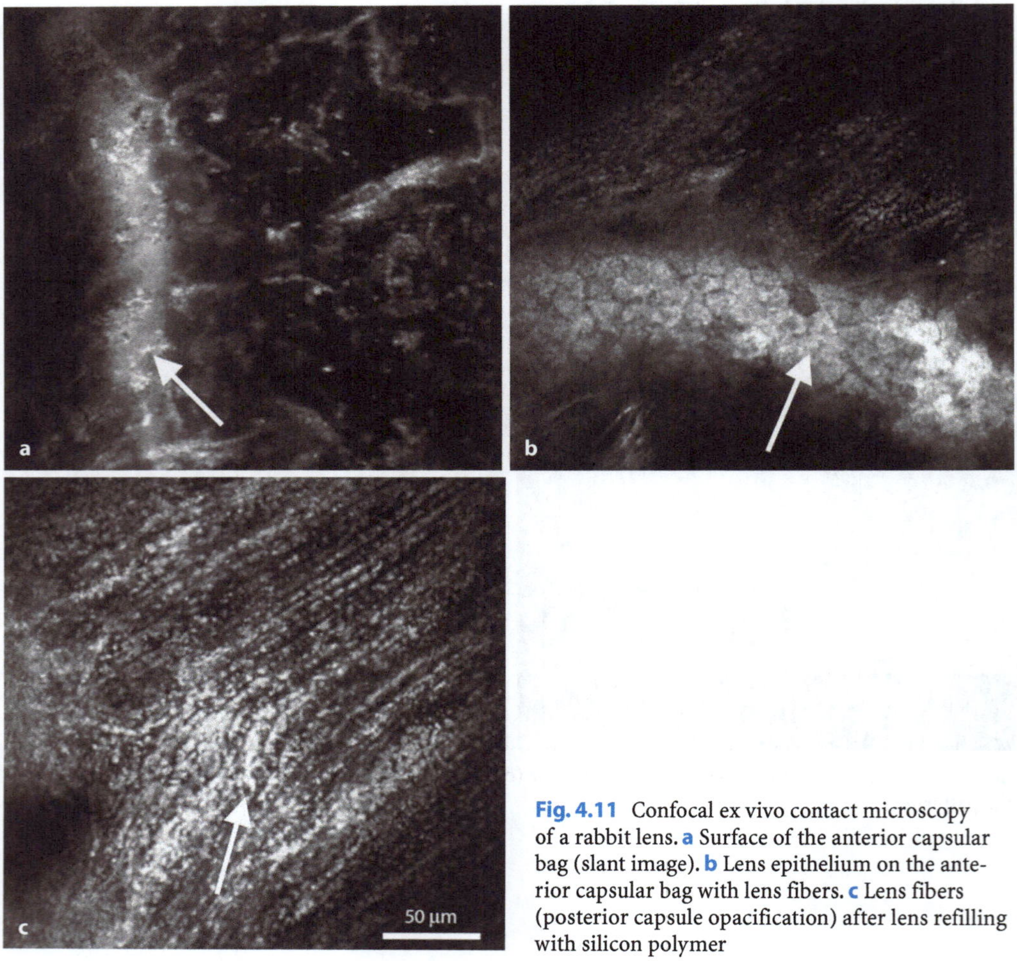

Fig. 4.11 Confocal ex vivo contact microscopy of a rabbit lens. **a** Surface of the anterior capsular bag (slant image). **b** Lens epithelium on the anterior capsular bag with lens fibers. **c** Lens fibers (posterior capsule opacification) after lens refilling with silicon polymer

Confocal Laser Scanning In Vivo Microscopy

The layered structure of the ocular coats can be visualized with high contrast using the RCM attachment with the HRT II and, because of the good quality of depth resolution, can be imaged in optical sections a few micrometers thick. The description of the normal in vivo microanatomy will proceed inward, beginning with the outer structures.

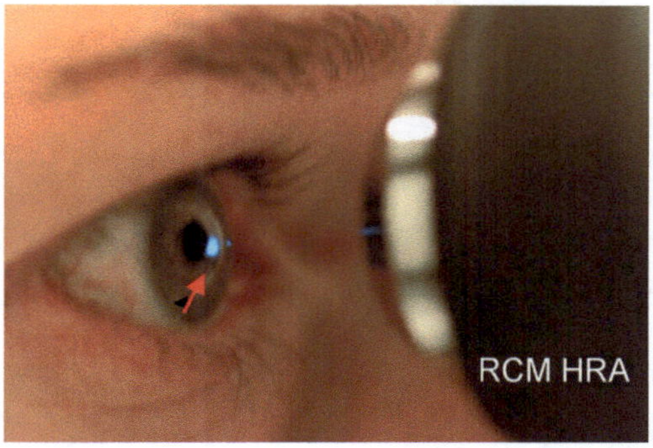

Fig. 5.1 Noncontact microscopy. Scanning laser reflex (*arrow*) on the cornea (argon laser/Heidelberg Retina Angiograph Classic)

5.1 Tear Film

5.1.1 Normal Anatomy

The preocular tear film with its complex fluid structure bathes the cornea and conjunctiva. Tear film structure and function are maintained by a highly differentiated system of secretory, distributive, and excretory interactions [24, 62, 73]. In particular, these function to smooth the corneal surface and maintain its optical clarity. The water content of the cornea is regulated by evaporation and the resultant osmotic gradient. The oxygen in the air is dissolved in the tear fluid and thus supports the aerobic metabolism of the epithelium.

The tear film is 7–10 μm thick and is characterized by a three-layered structure. The external lipid layer, which is produced chiefly by the meibomian glands close to the margin of the eyelids, prevents rapid evaporation of the aqueous layer and renders the surface hydrophobic. The inner mucin layer consists of glycoproteins. Its task is to make the epithelial surface hydrophobic and thus guarantee wettability.

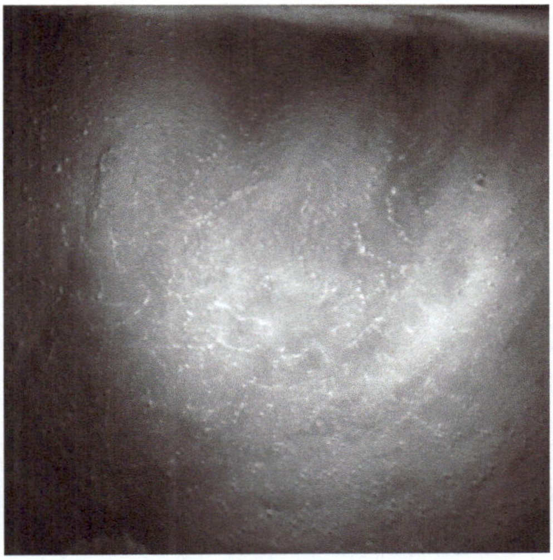

Fig. 5.2 Normal tear film

Replacing the contact system in the confocal laser scanning microscope with a dry objective lens, combined with an 80–90% absorption gray filter to decrease laser intensity (Fig. 5.1), enables the fine structure of the tear film to be imaged (Fig. 5.2). The rapid imaging sequence in the device also permits dynamic processes to be recorded [51, 77, 88].

5.1.2 Pathological Findings

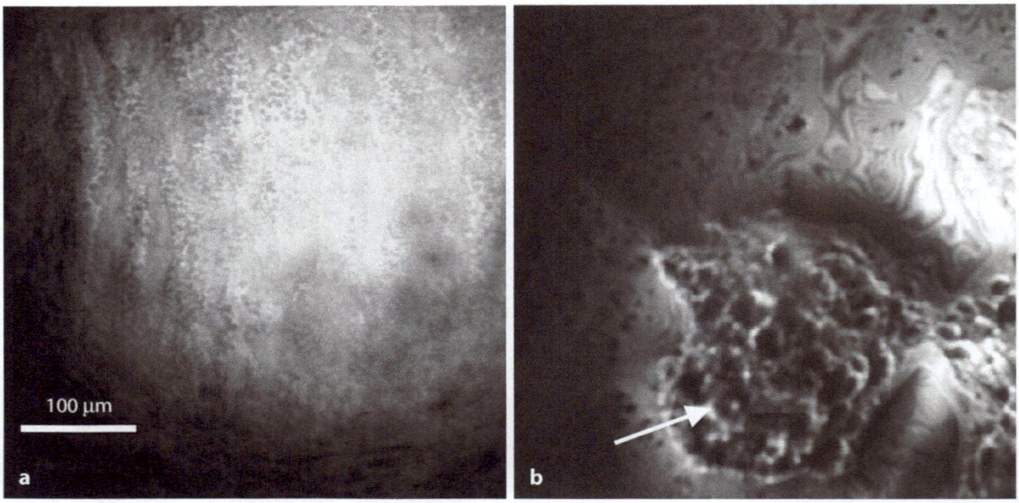

Fig. 5.3 Normal and pathological tear film. **a** Normal tear film, noncontact examination. **b** Tear film with dry spot (*arrow*) in a patient with pathological break-up time; the surface of the superficial cells is visible in the area of the dry spot

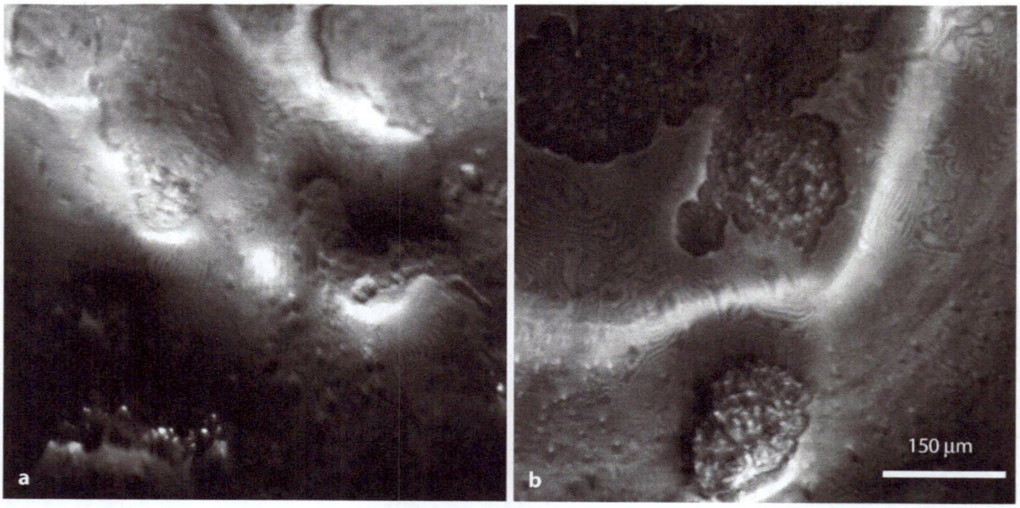

Fig. 5.4 Further dry spots in the tear film (**a, b**; Heidelberg Retina Angiograph, red-free mode)

5.2 Epithelium

5.2.1 Normal Anatomy

5.2.1.1 Superficial Cells (Up to Approximately 50 µm in Diameter)

In the case of the most superficial epithelial cells, bright cell borders and a dark cell nucleus and cytoplasm are readily visualized on confocal laser scanning microscopy. The cells characteristically display a polygonal – often hexagonal – shape. Cells undergoing desquamation are characterized by a highly reflective cytoplasm, in the center of which the brightly appearing (pyknic) cell nucleus with its dark perinuclear space is clearly visible (Fig. 5.5). The average density of superficial cells in the central and peripheral cornea is approximately 850 cells/mm²

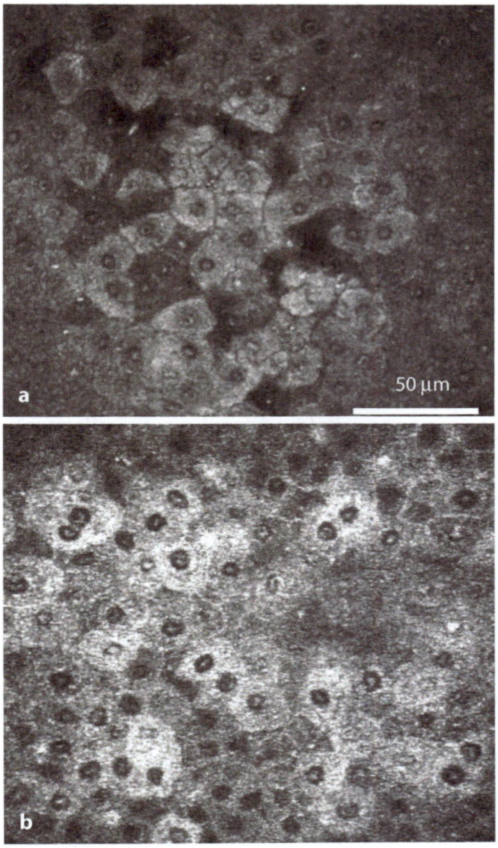

Fig. 5.5 Superficial cells: the cytoplasm and cell nuclei are visualized; cells in the process of desquamation possess a highly reflective cytoplasm, in the center of which the bright (pyknic) cell nucleus with its dark perinuclear space is clearly visible. **a** Uncompressed superficial cells, using a special TomoCap with a central hole. **b** Compressed superficial cells, using the standard TomoCap

5.2.1.2 Intermediate Cells/Wing Cells (Up to Approximately 20 µm in Diameter)

The cells of the intermediate layers are characterized by bright cell borders and a dark cytoplasm. The cell nucleus can be distinguished only with difficulty. In terms of size and appearance, wing cells in healthy subjects exhibit only minimal variation (Fig. 5.6). The average density is approximately 5,000 cells/mm² in the central cornea and approximately 5,500 cells/mm² in the periphery.

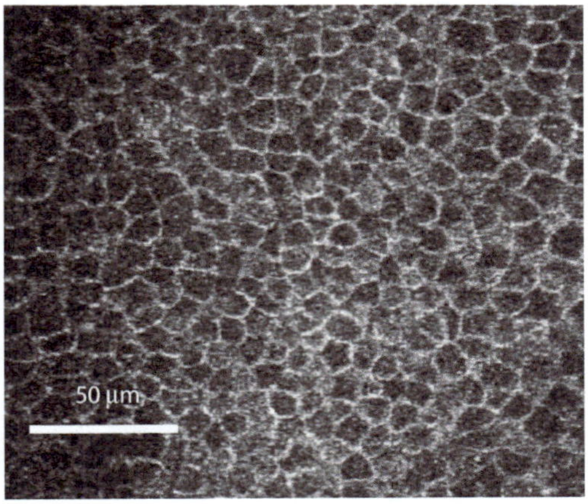

Fig. 5.6 Intermediate cells: the cells of the intermediate layers are characterized by bright cell borders and a dark cytoplasm. The cell nucleus can be identified only with difficulty. The wing cells display only minimal variation in terms of size and appearance

5.2.1.3
Basal Cells
(Up to Approximately 10 µm in Diameter)

The basal cells are located immediately above Bowman's membrane. They present as brightly bordered cells in which the cell nucleus is not visible. Between-cell comparison reveals inhomogeneous reflectivity of the cytoplasm. Like the wing cells above them, the basal cells display only minimal variation in shape and size (Fig. 5.7). The average density is approximately 9,000 cells/mm² in the center of the cornea and 10,000 cells/mm² in the periphery. Therefore, in terms of cell densities in normal subjects, the ratio between superficial cells, intermediate cells, and basal cells is 1:5:10.

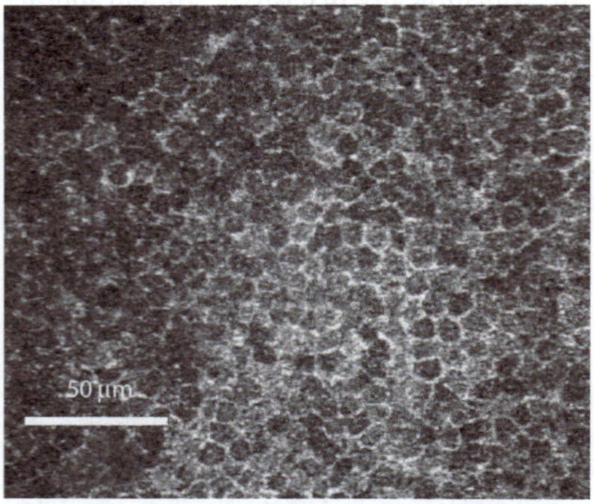

Fig. 5.7 Basal cells. These are regularly arranged cells with bright borders, but the cell nucleus is not visualized. Intercellular comparison reveals inhomogeneous cytoplasmic reflectivity

5.2.1.4
Langerhans Cells

Confocal microscopy permits in vivo evaluation of Langerhans cells (LCs) within the human cornea, with a particular emphasis on cell morphology and cell distribution.

LCs present as bright corpuscular particles with dendritic cell morphology and a diameter of up to 15 µm. LC distribution follows a gradient from low numbers in the center to higher cell densities in the periphery of the cornea. Moreover, confocal in vivo microscopy permits differentiation of LC bodies lacking dendrites, LCs with small dendritic processes forming a local network, and LCs forming a meshwork via long interdigitating dendrites (Fig. 5.8a–c). Whereas almost all the cells located in the periphery of the cornea demonstrate long processes interdigitating with the corneal epithelium, those in the center of the cornea often lack dendrites, most probably underlining their immature phenotype [26]. Immature LCs are equipped to capture antigens, while mature forms are able to sensitize naive T-cells through MHC molecules and secretion of interleukin-12 as well as costimulatory molecules, and thus represent an integral part of the immune system [3].

The average density of LCs in normal subjects is 34±3 cells/mm² (range 0–64 cells/mm²) in the central cornea and 98±8 cells/mm² (range 0–208 cells/mm²) in the periphery [96]. In contact lens wearers, LC density varies from 78±25 cells/mm² (range 0–600 cells/mm²) in the central cornea to 210±24 cells/mm² (range 0–700 cells/mm²) in the periphery. LC densities differ significantly between healthy volunteers and contact lens wearers both in the central (p= 0.03) and in the peripheral cornea (p=0.001), while the gradient of LC density from periphery to center has been found to be almost identical in the two groups (unpublished data).

It has been suggested that LCs participate in immune and inflammatory responses, thereby determining cell-mediated immunity. In light of this theory, the present data on LCs in the human cornea provide a helpful basis for further investigations in ocular pathology.

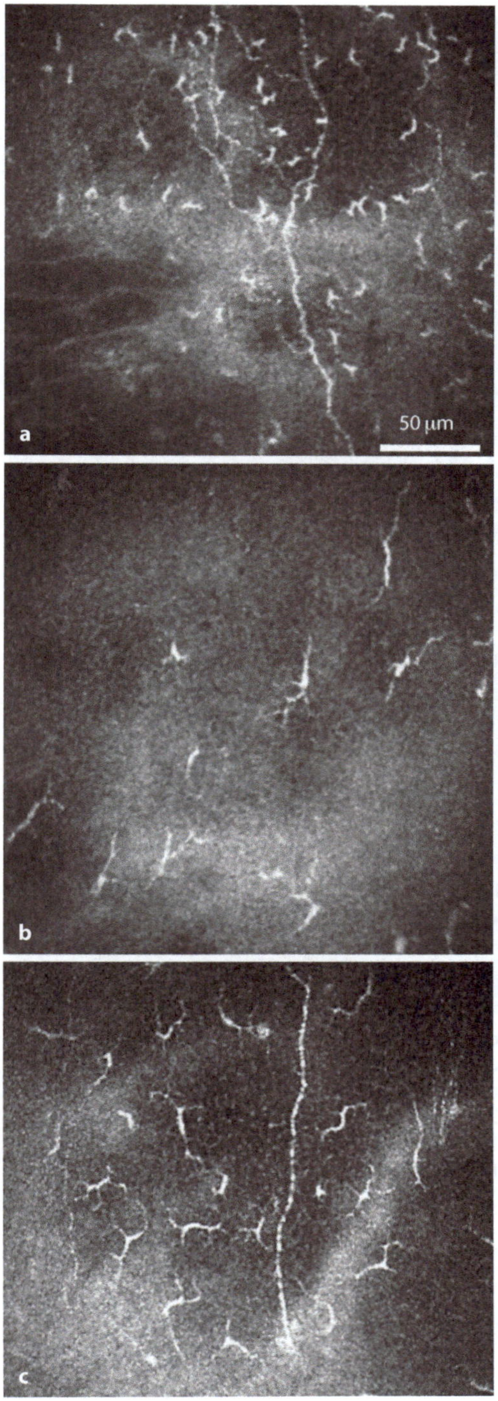

Fig. 5.8 In vivo confocal microscopic images, representing different forms of Langerhans cells. **a** Individual cell bodies without processes. **b** Cells bearing dendrites. **c** Cells arranged in a meshwork via long interdigitating dendrites

5.2.2 Pathological Findings

5.2.2.1 Dry Eye

Disturbances of tear film secretion or tear film structure give rise to a condition known as dry eye. On microscopy such disturbances are evident as altered reflection or dry spots on the epithelium.

As the most important component of the corneal diffusion barrier, the corneal epithelium displays differing permeability for aqueous ionic substances such as NaF. Patients with diabetes, for example, have significantly increased permeability for NaF. NaF also penetrates areas of microerosions and pathologically altered cells. The literature reveals discrepant views concerning the nature of the penetration process. Most authors subscribe to the view that the fluorescein fills the "footprint" spaces vacated by cells that have been lost. Others, however, assume that fluorescein fills the intercellular space. At present, only confocal slit-scanning microscopes and fluorophotometers are used to analyze this phenomenon.

The HRA in combination with the RCM can be used for confocal laser scanning fluorescence microscopy of the microstructure of the corneal epithelium and tear film using contact and non-contact techniques (see Fig. 2.18) with a lateral resolution of 1 µm and up to ×1,000 magnification. The red-free reflection and fluorescence images display the intercellular microstructure with stained cell nuclei and altered cell surfaces and borders. The same area examined on the cornea can be visualized simultaneously in reflection and fluorescence mode. The penetration profile of NaF can be measured with precise depth resolution over a prolonged time using the contact technique. Autofluorescence measurements are also possible. Figure 5.11 shows the tear film and epithelium in fluorescence mode.

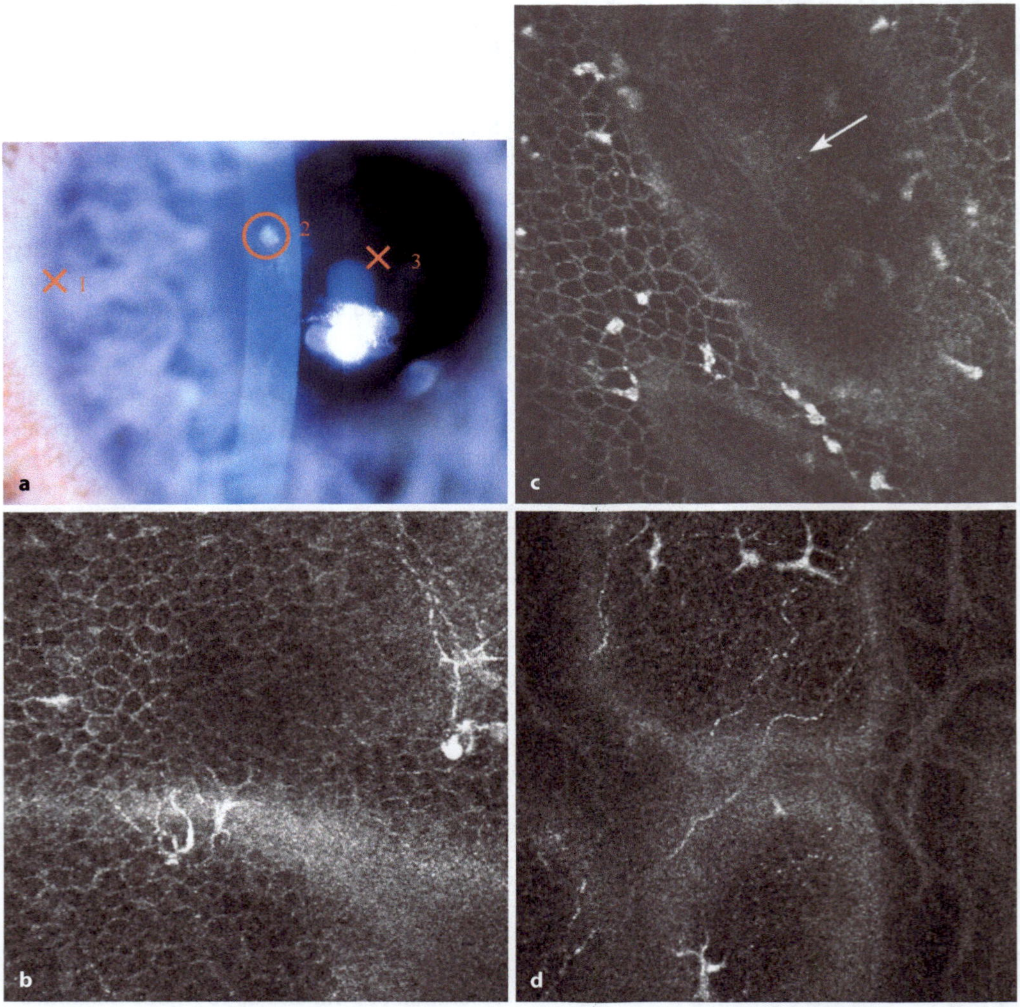

Fig. 5.9 Corneal erosion. **a** Slit-lamp photograph of 25-year-old man on day 2 after foreign body removal. Fluorescein discoloration is minimal, other corneal areas are clear, and there are no signs of intraocular inflammation. **b–d** Confocal images of the corneal periphery. **b** Level of lower intermediate and basal cells and subepithelial nerve plexus with dendritic cells. **c** Intraepithelial defect at the level of Bowman's membrane with leukocyte infiltration. Note the visible anterior stroma (*arrow*), **d** the mature form of Langerhans cells at the levels of the basal cells, and Bowman's membrane in the central cornea

Fig. 5.10 Infiltrate in the central cornea. **a** Slit-lamp photograph of 78-year-old woman with an infiltrate in the central cornea. Minimal signs of fluorescein discoloration and intact peripheral zone. **b, d** Confocal images of the central cornea are characterized by the presence of bright cell bodies, most probably leukocyte infiltration, in all epithelial layers. **c, e** Inflammatory cells are absent at the periphery

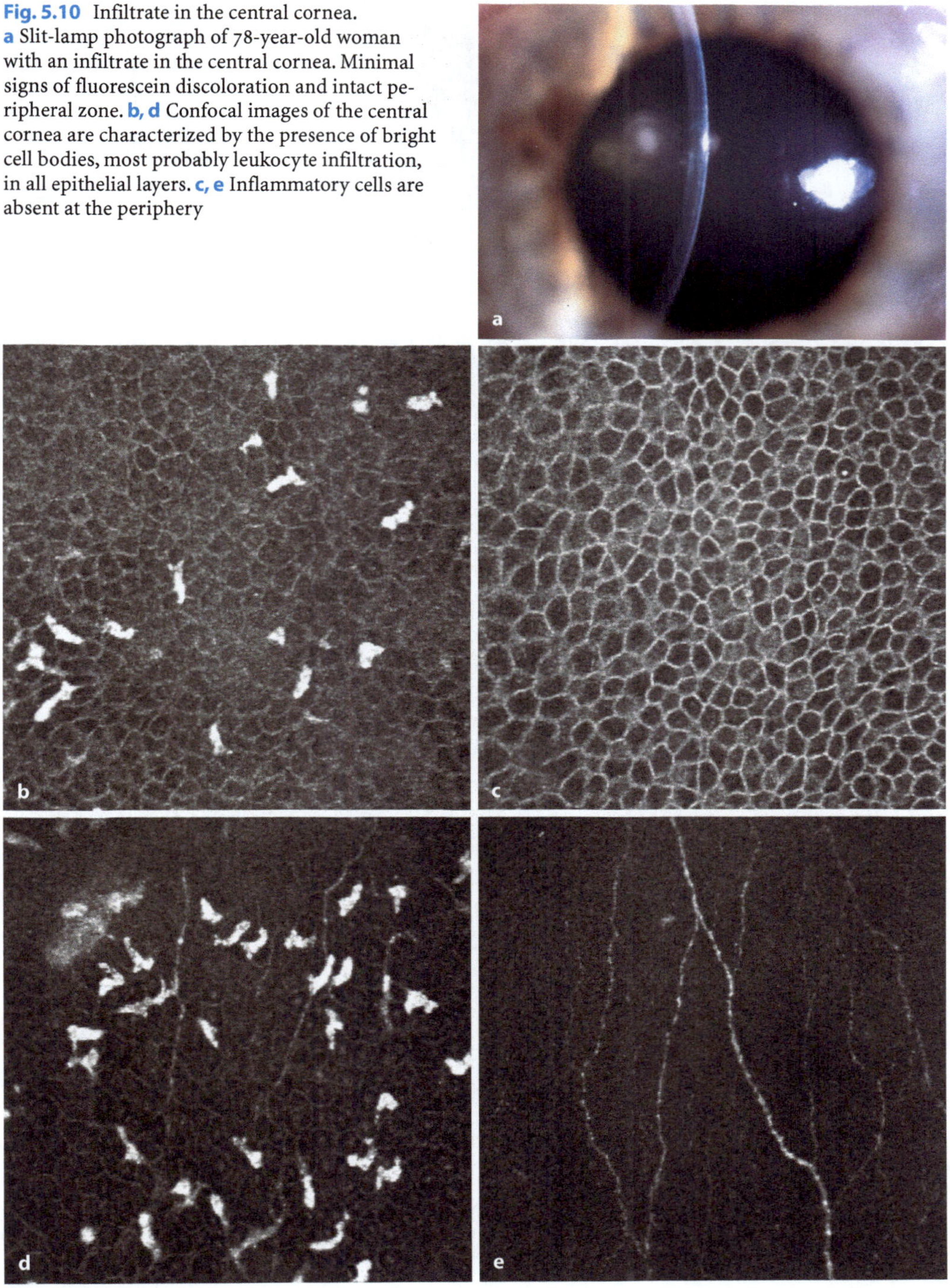

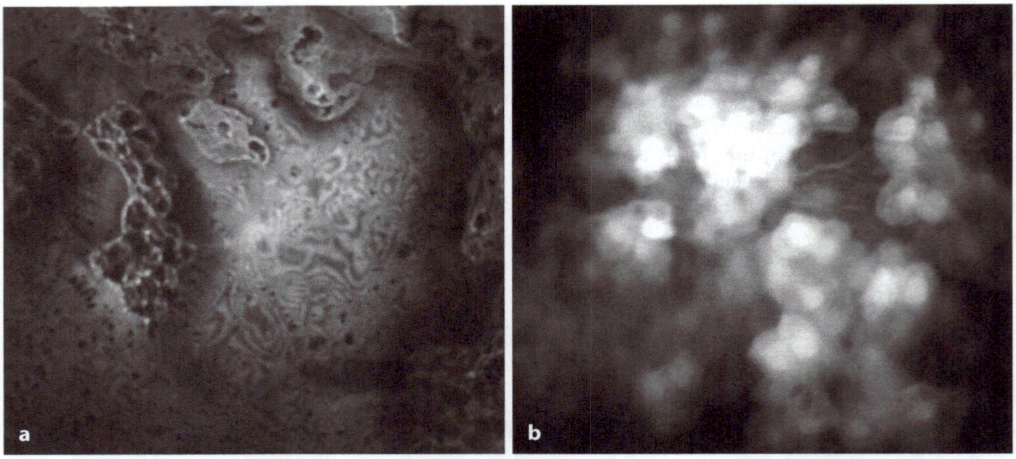

Fig. 5.11 Tear film (Heidelberg Retina Angiograph Classic). **a** Reflection mode: tear film defect causing the surface of the superficial epithelial cells to be visible. **b** Fluorescence mode: individual epithelial cells visible on sodium fluorescein staining

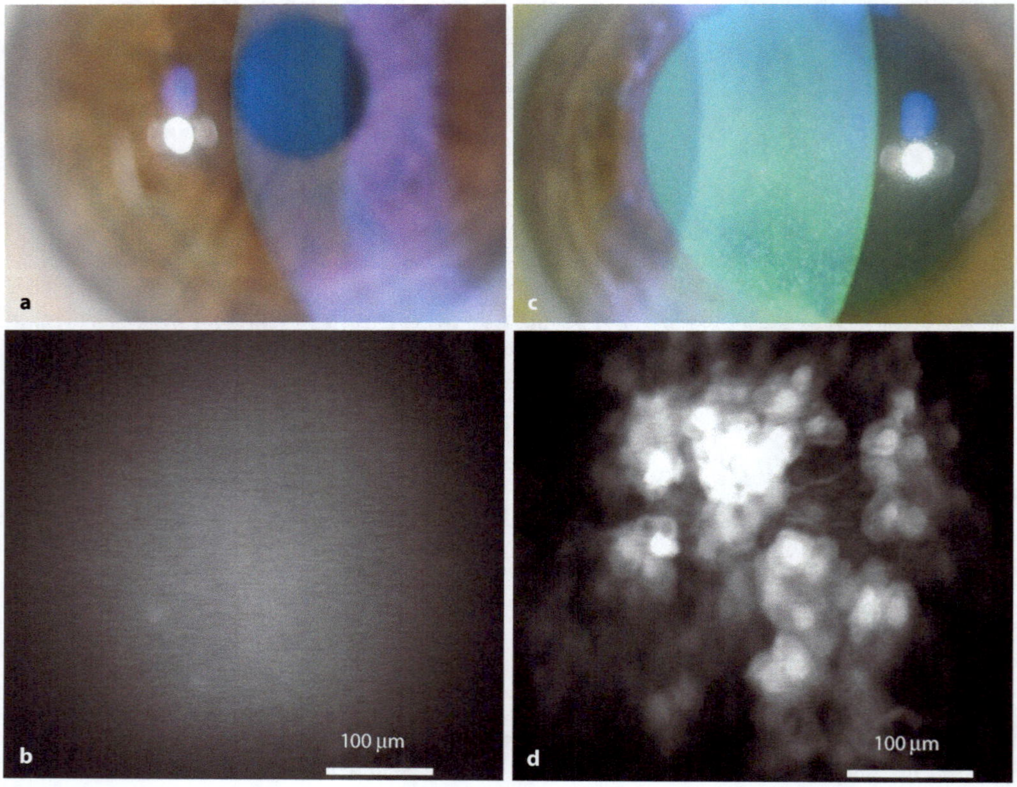

Fig. 5.12 Confocal fluorescence microscopy before and after applanation tonometry. **a** Slit-lamp photograph before applanation tonometry. **b** Fluorescence microscopy: no relevant fluorescein staining. **c** Slit-lamp photograph after applanation tonometry: stippled epithelial lesions. **d** Fluorescence microscopy: fluorescein-stained superficial cells

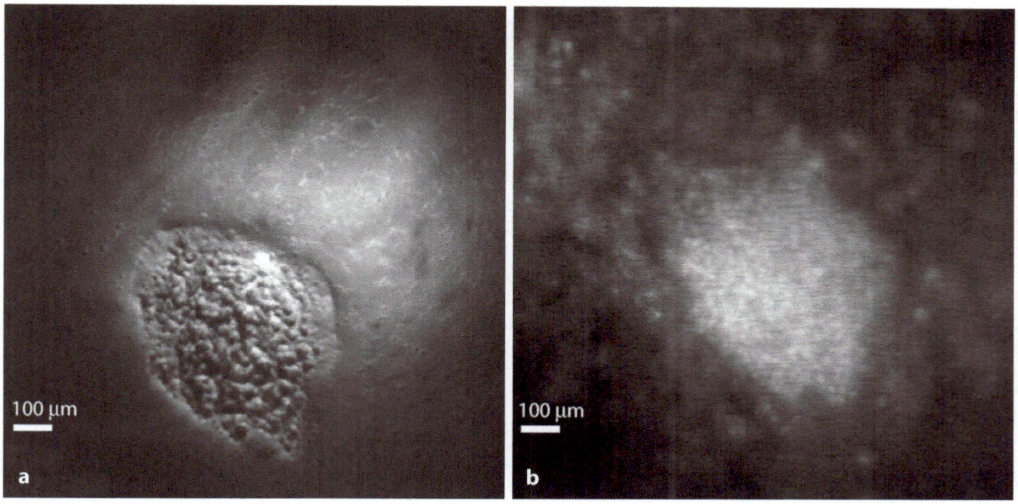

Fig. 5.13 Dry eye. a Dry spot in the tear film; noncontact examination using the Heidelberg Retina Angiograph Classic and the Rostock Cornea Module. b Same area after switching to fluorescence mode. The superficial cells in the region of the dry spot are stained with fluorescein

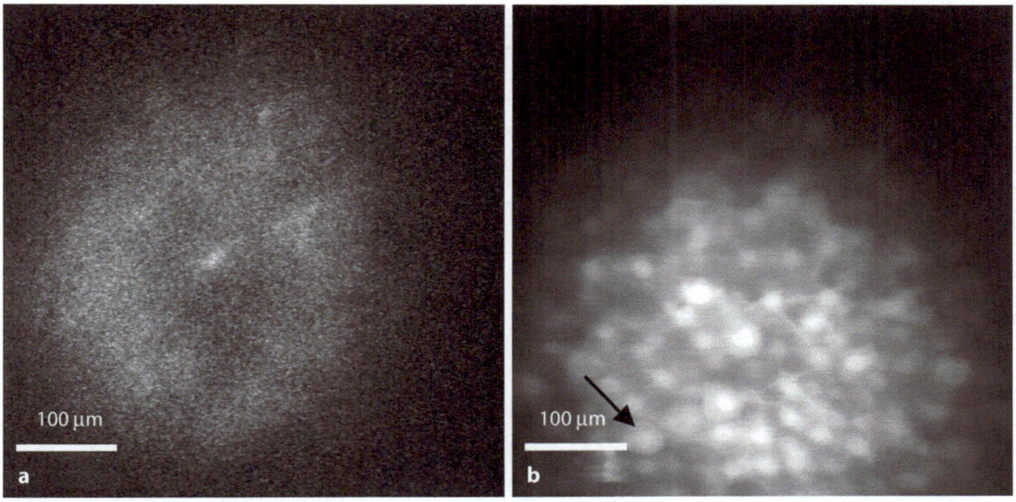

Fig. 5.14 Dry eye (fluorescence mode, contact examination). a Superficial cells of the normal cornea: only a few cells with minimal fluorescence. b Superficial cells of a stippled cornea: many cells are fluorescein-stained, and there is intracellular staining of the cell nuclei (*arrow*)

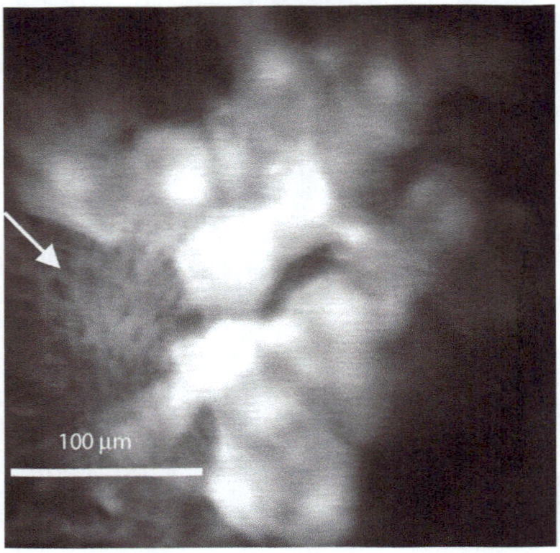

Fig. 5.15 Corneal erosion (fluorescence mode, contact examination). Fluorescein-stained superficial cells (up to 50 µm in size). In the left part of the picture (*arrow*), the borders of the smaller intermediate cells are visible

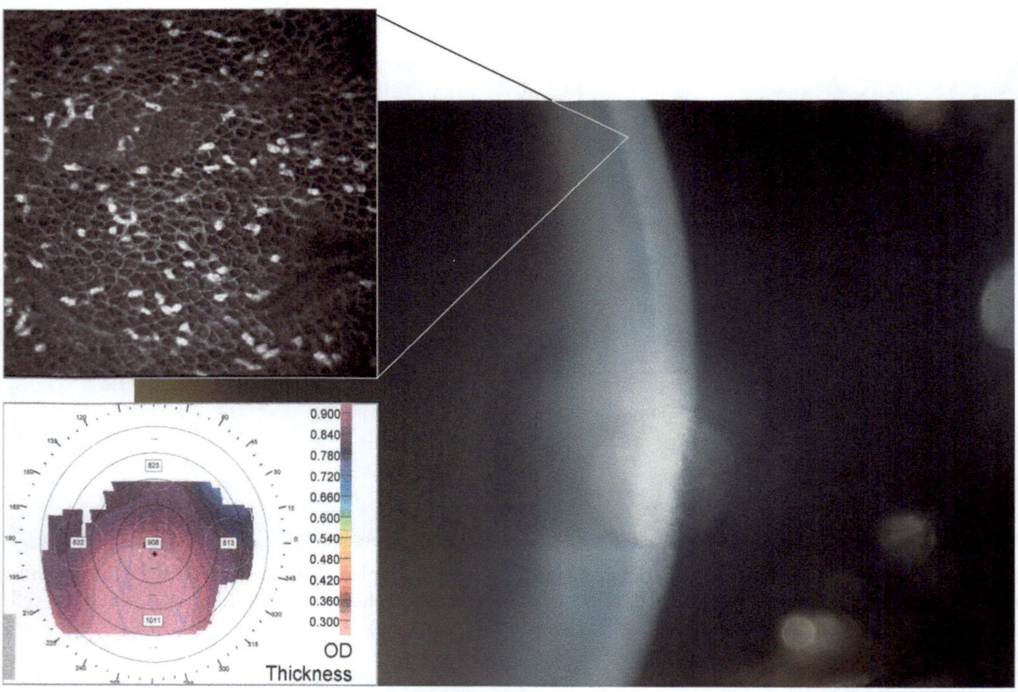

Fig. 5.16 Corneal edema. Confocal images of central cornea. Infiltration of epithelium with antigen-presenting cells and leukocytes; corneal thickness (Orbscan) 908 µm in center

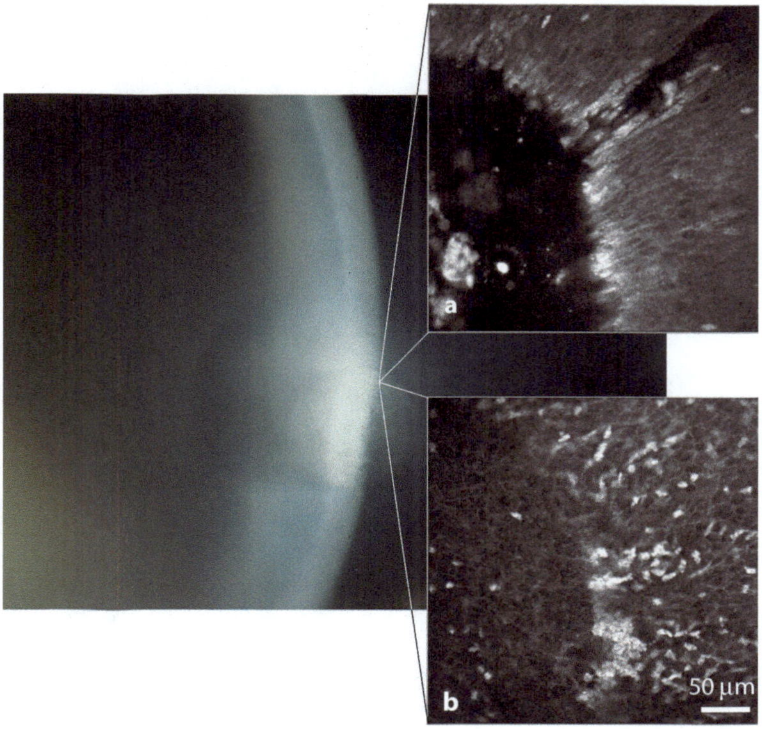

Fig. 5.17 Corneal infiltrate due to contact lens wear. Confocal images of transition area (from normal cornea to ulcer). **a** Edematous changes in epithelium and absence of cells in ulcer zone. **b** Massive leukocyte infiltration in transition area at the level of wing cells

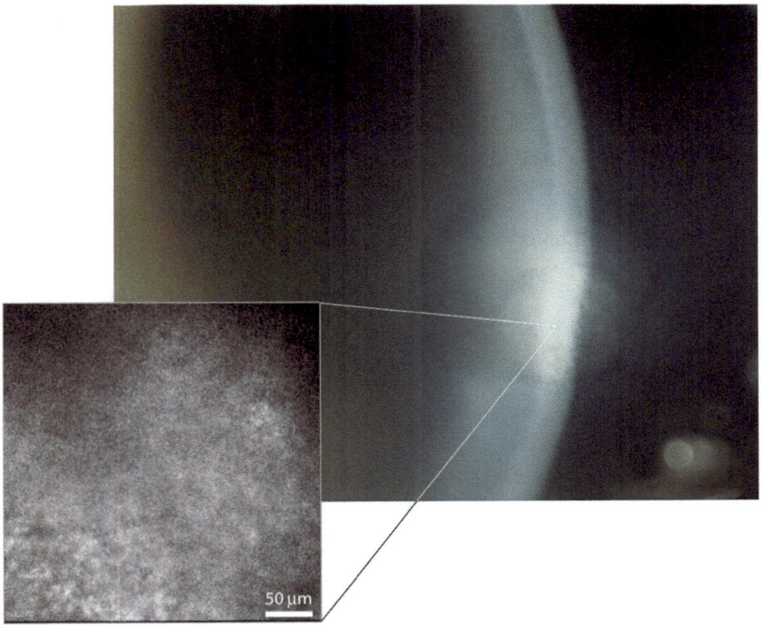

Fig. 5.18 Corneal infiltrate due to contact lens wear. Confocal image of infiltrate zone. Absence of cellular or structural elements in the defect area

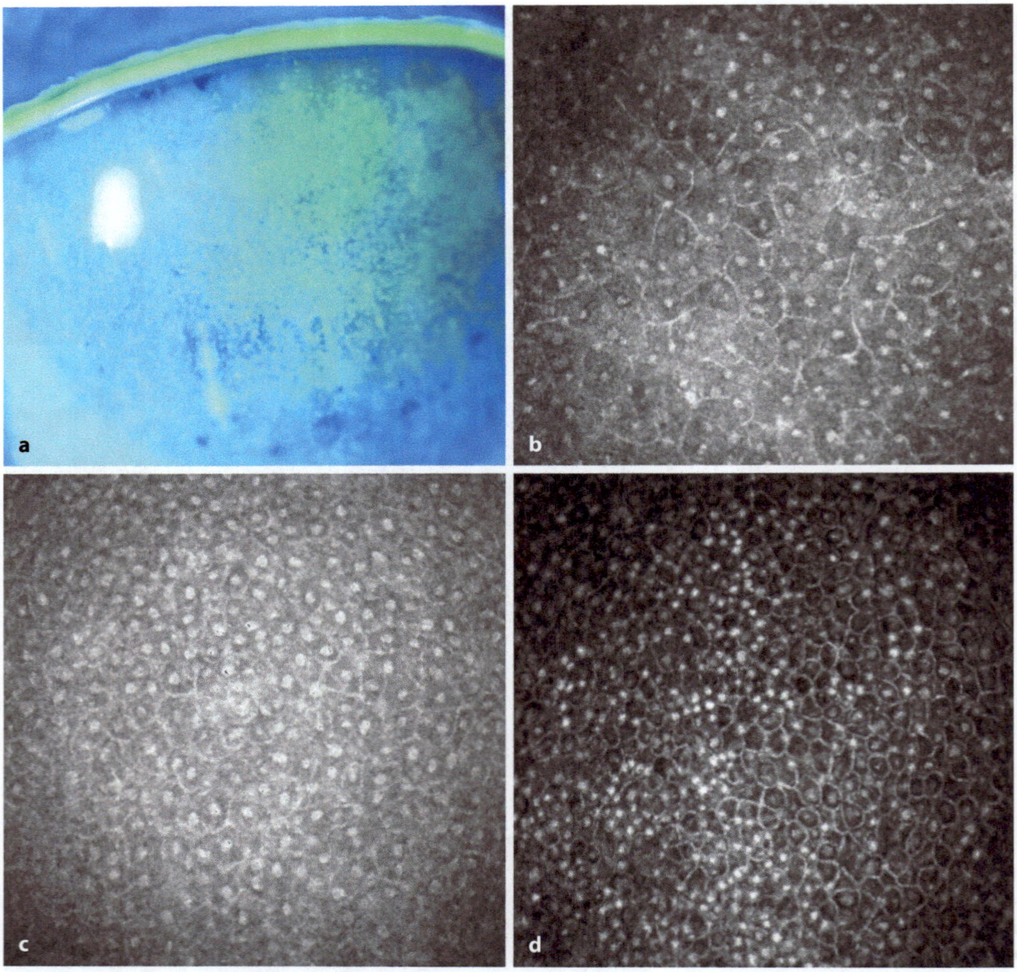

Fig. 5.19 Keratoconjunctivitis sicca. **a** Slit-lamp photograph of a 45-year-old woman with keratoconjunctivitis sicca. Severe punctate keratitis on fluorescein staining. **b–d** Confocal in vivo microscopy images showing corneal epithelial metaplasia with enlarged cells, activated nuclei, and decreased nucleus/cytoplasm ratio

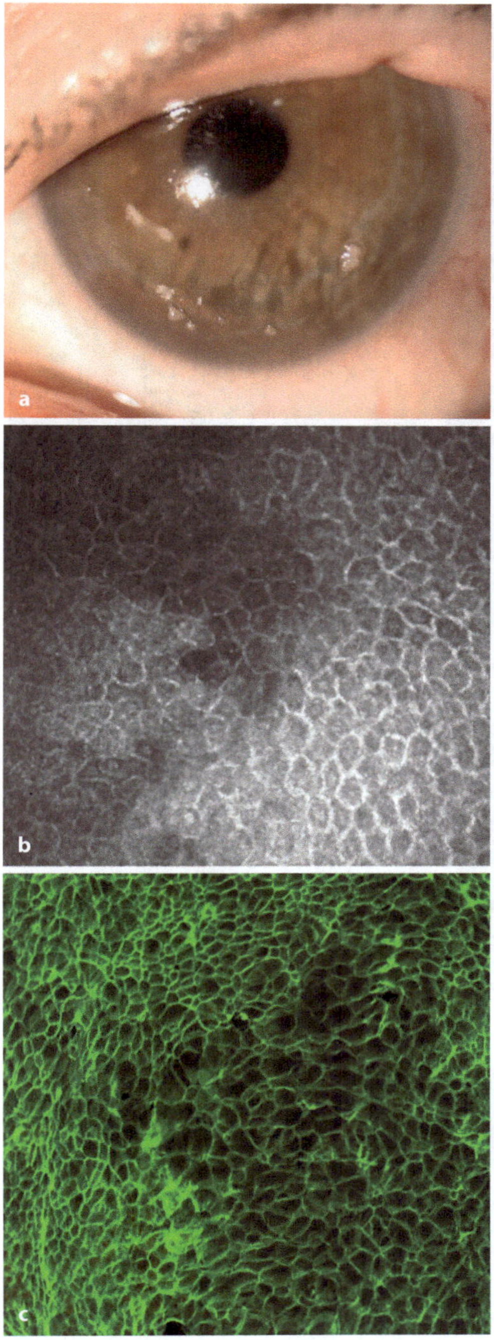

Fig. 5.20 Keratoconjunctivitis sicca in a case of Sjögren's syndrome. **a** Slit-lamp photograph of a 50-year-old woman with severe keratoconjunctivitis sicca. **b** Confocal in-vivo microscopy image of abnormal corneal-conjunctival junction at the limbus. **c** Impression cytology of the limbus showing the same abnormal corneal–conjunctival junction

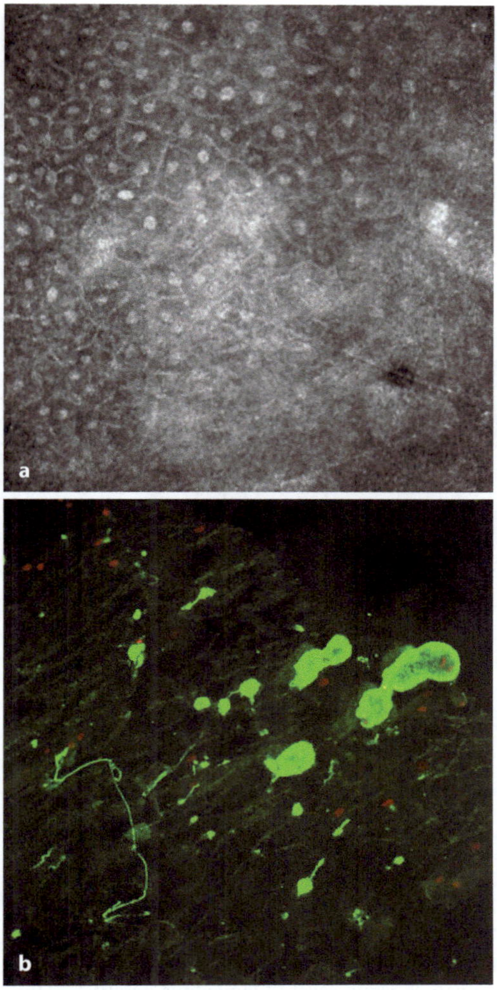

Fig. 5.21 Keratoconjunctivitis sicca in a case of Sjögren's syndrome. **a** Confocal in-vivo microscopy image showing corneal metaplasia. Enlarged corneal cells displaying reflective nuclei and polymorphism. **b** Impression cytology immunostaining: goblet cells (*green*) in the corneal epithelium

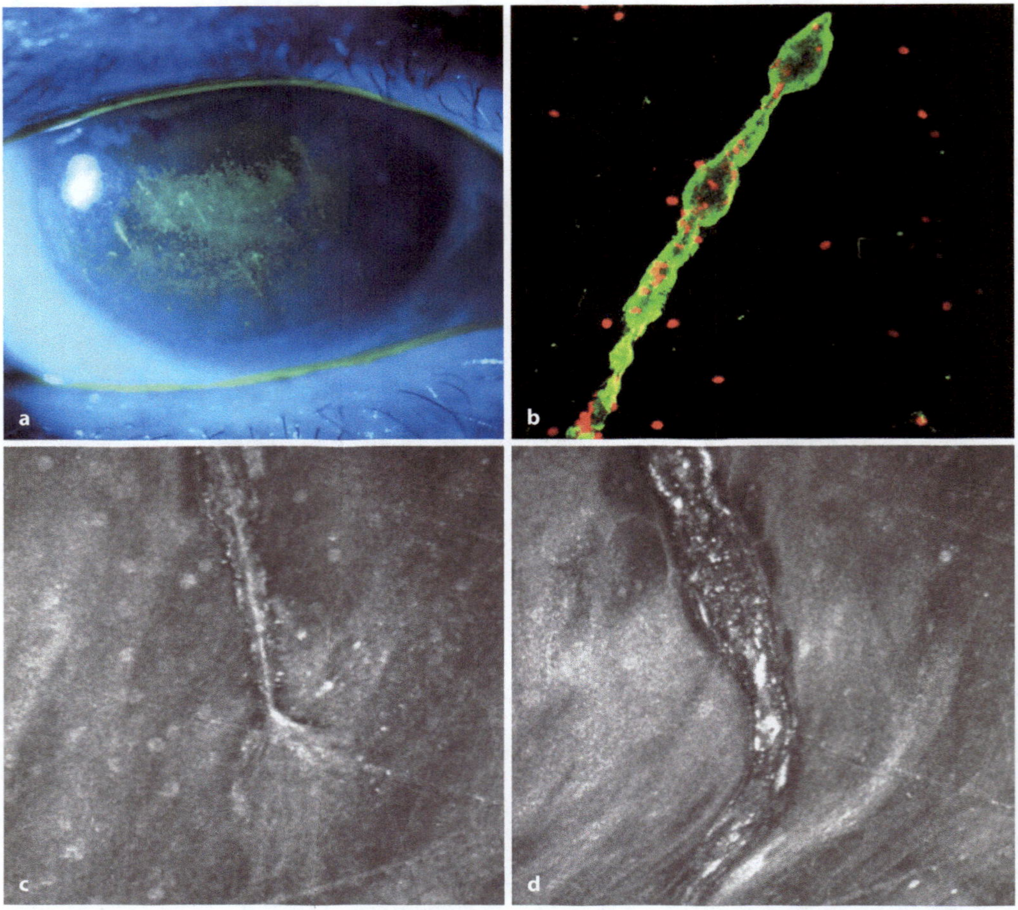

Fig. 5.22 Filamentous keratitis in a case of Sjögren's syndrome. a Slit-lamp photograph of a 51-year-old woman showing severe filamentous keratitis. b Impression cytology and immunostaining of a filament. c, d Confocal in vivo microscopy image of the same filament

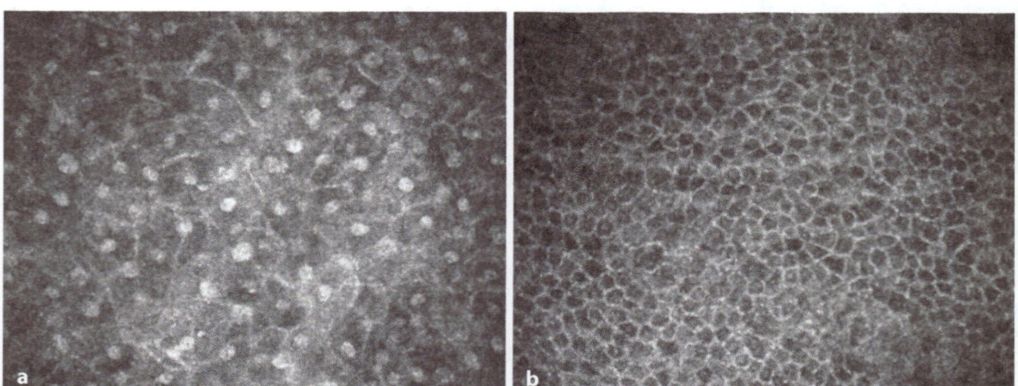

Fig. 5.23 Keratoconjunctivitis sicca. a Corneal metaplasia before treatment: polymorphism, enlarged cells, and reflective nuclei. b After treatment with topical cyclosporin A and clinical improvement: recovery of normal basal corneal cell appearance

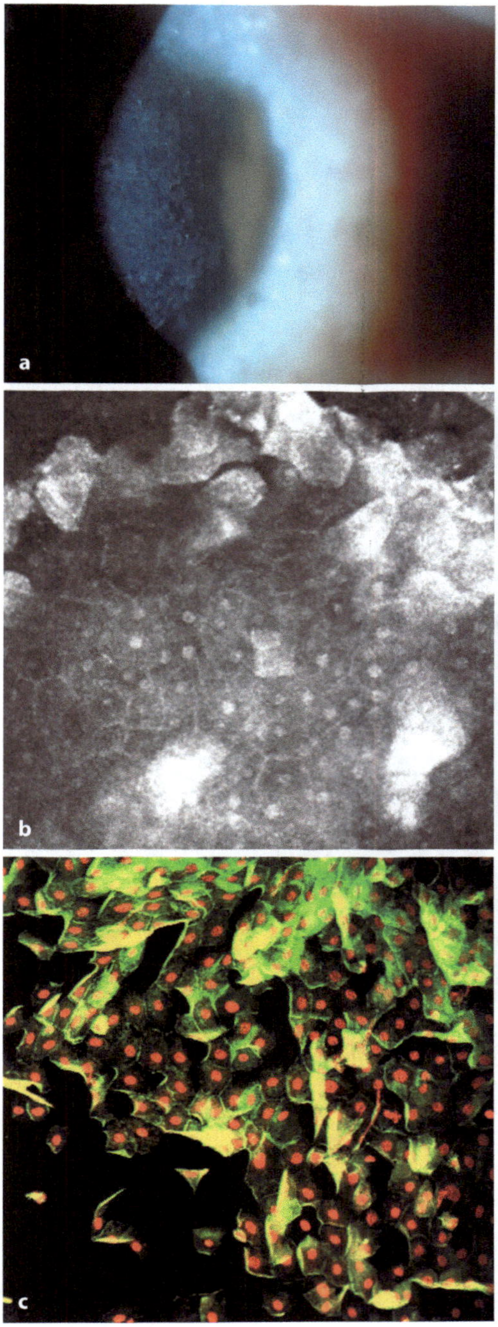

Fig. 5.24 Allergic keratoconjunctivitis: vernal keratoconjunctivitis. **a** Slit-lamp photograph of a 16-year-old boy with severe vernal keratoconjunctivitis causing corneal desiccation and corneal cell desquamation. **b** Confocal in vivo microscopy showing corneal epithelial metaplasia and hyperreflective desquamating superficial corneal cells. **c** Impression cytology image confirming corneal metaplasia

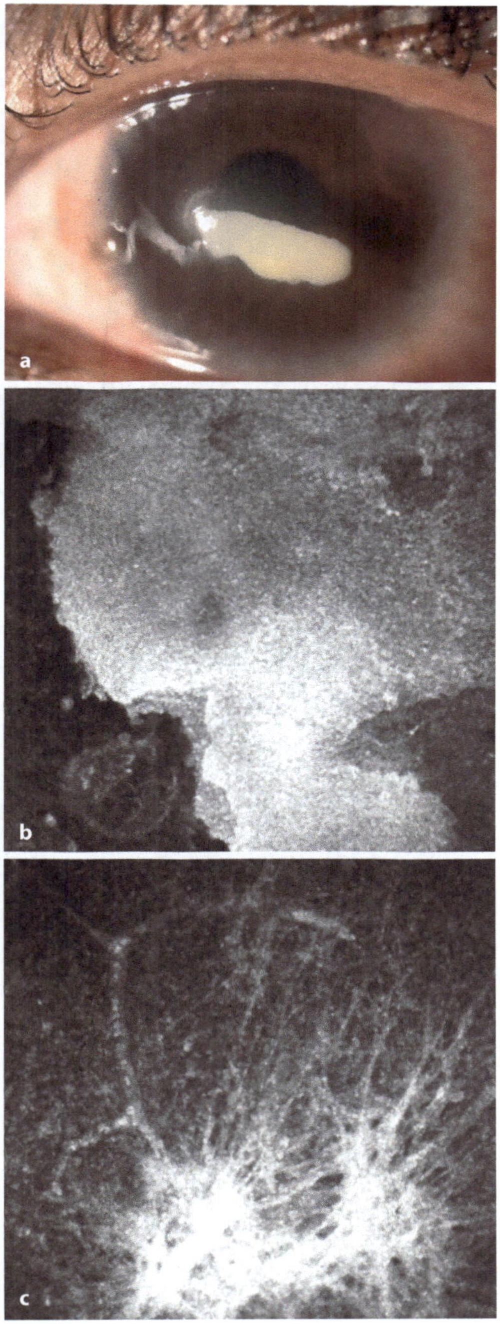

Fig. 5.25 Vernal keratoconjunctivitis. **a** Slit-lamp photograph of the same patient as in Fig. 5.24 1 year later, showing corneal vernal plaque. **b** Confocal in vivo microscopy image of the plaque with hyperreflective area within the corneal epithelium. **c** Confocal in vivo microscopy image of the scar after surgical scraping of the plaque

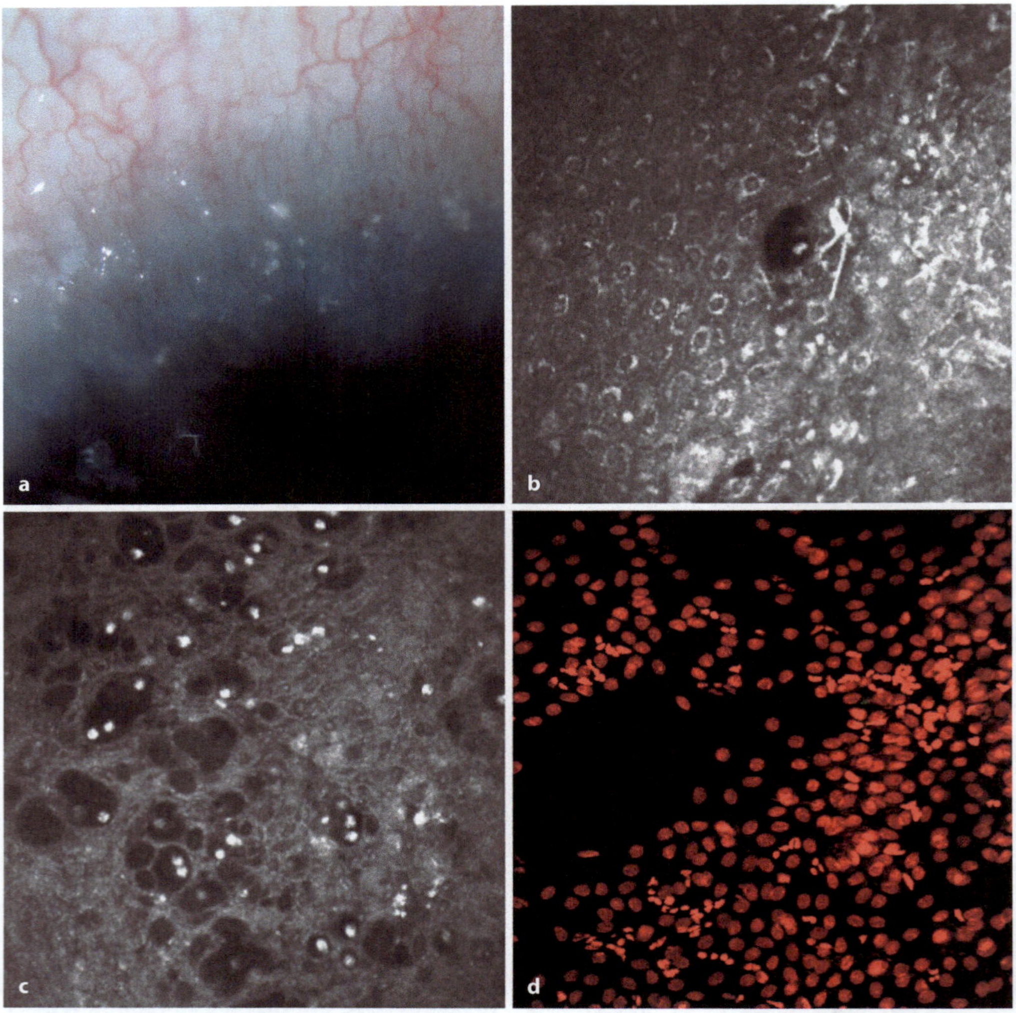

Fig. 5.26 Vernal keratoconjunctivitis. **a** Slit-lamp photograph of a 13-year-old boy with vernal keratoconjunctivitis: Trantas dots. **b** Confocal in vivo microscopy image of the limbus. Epithelial infiltration by dendritic cells. **c** Confocal in vivo microscopy image of Trantas dots: microcysts and inflammatory cells (*hyperreflective cells*). **d** Limbal impression cytology showing numerous inflammatory cells (*small red nuclei*) within the epithelium

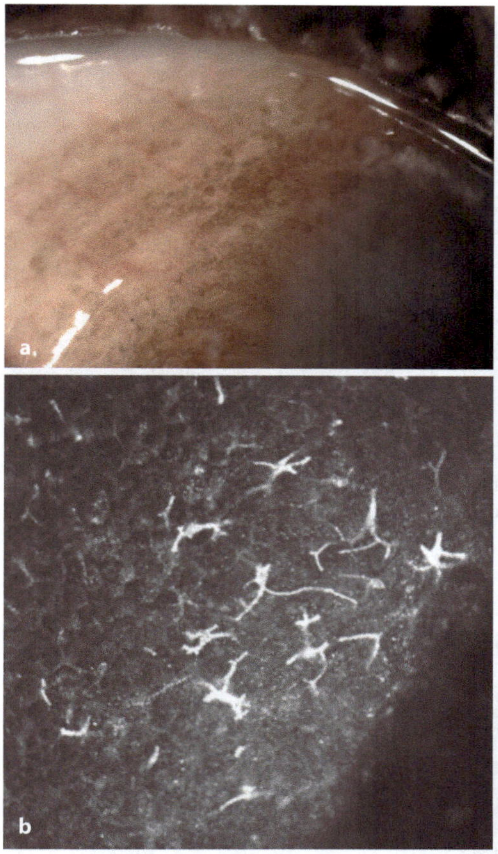

Fig. 5.27 Vernal keratoconjunctivitis. **a** Slit-lamp photograph of an 18-year-old man with vernal keratoconjunctivitis. Trantas dots. **b** Confocal in vivo microscopy image of the limbus. Numerous (*hyperreflective*) dendritic cells within the epithelium

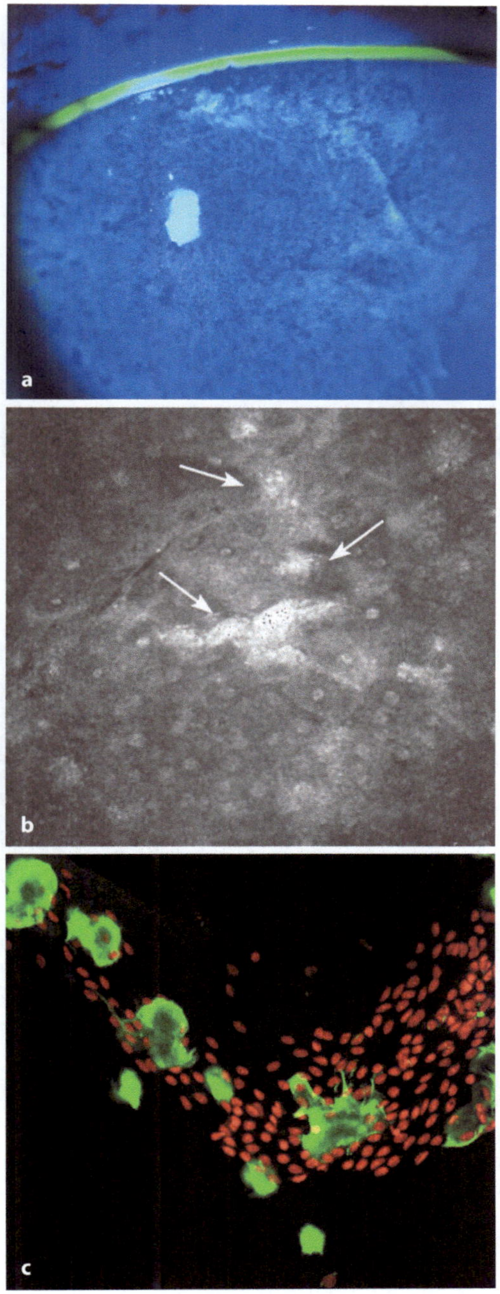

Fig. 5.28 Limbal stem cell deficiency and corneal conjunctivalization. **a** Slit-lamp photograph and fluorescein staining in a patient with corneal conjunctivalization. **b** Confocal in vivo microscopy image of corneal metaplasia with goblet cells within the corneal epithelium. **c** Impression cytology showing goblet cells (*green*) within the corneal epithelium (*red nuclei*)

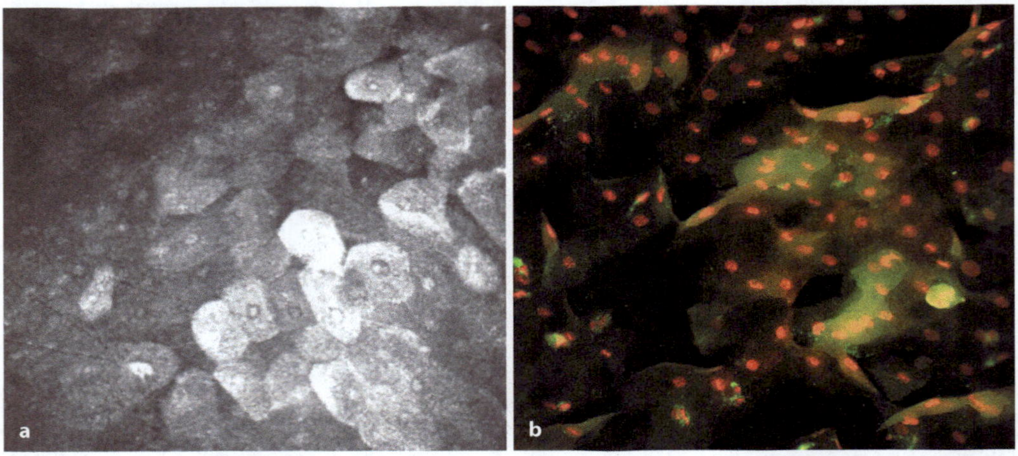

Fig. 5.29 Limbal stem cell deficiency. **a** Corneal metaplasia with hyperreflective enlarged cells similar to conjunctival cells. **b** Impression cytology of the cornea from the same patient

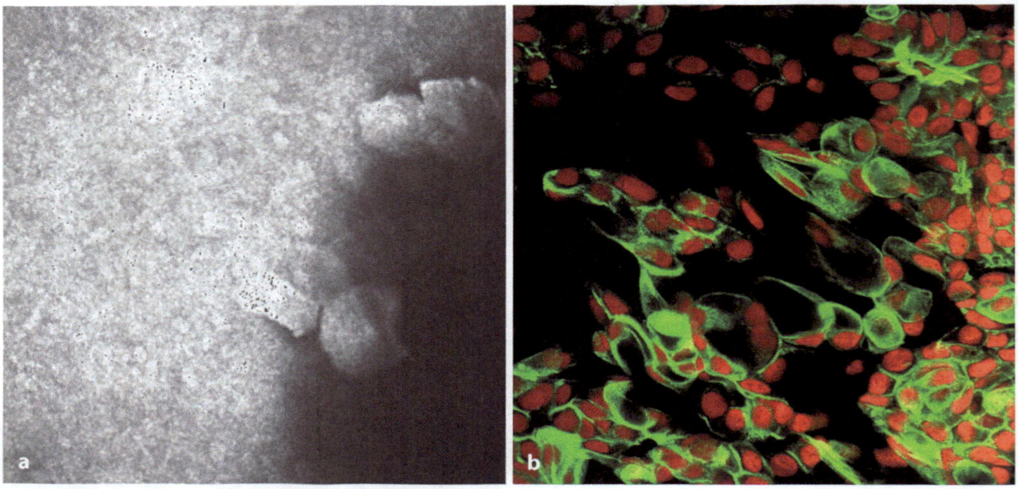

Fig. 5.30 Limbal stem cell deficiency. **a** Conjunctival epithelial and goblet cell migration within the corneal epithelium. **b** Impression cytology from the same patient showing goblet cells (*green*) within the corneal epithelium

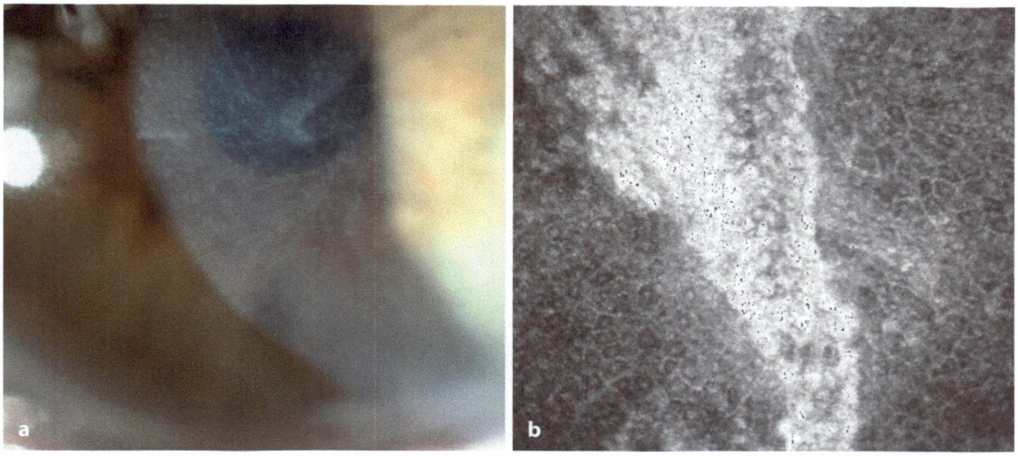

Fig. 5.31 Corneal epitheliopathy. **a** Slit-lamp photograph of a patient with severe ocular rosacea and conjunctival cell migration within the corneal epithelium. **b** Confocal in vivo microscopy image from the same patient. Conjunctival cells (*hyperreflective cells*) migrating into an abnormal corneal epithelium (*hyporeflective cells*)

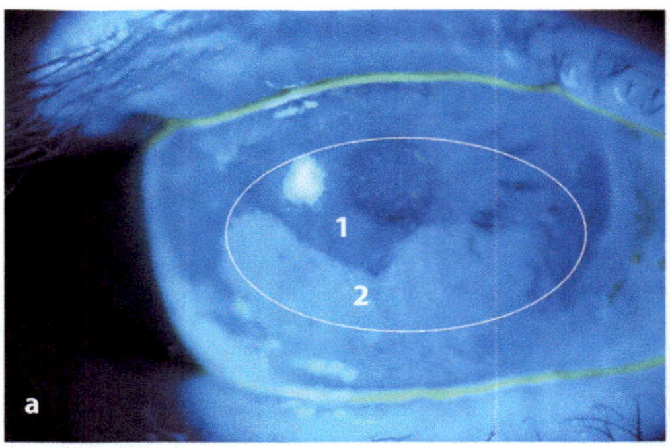

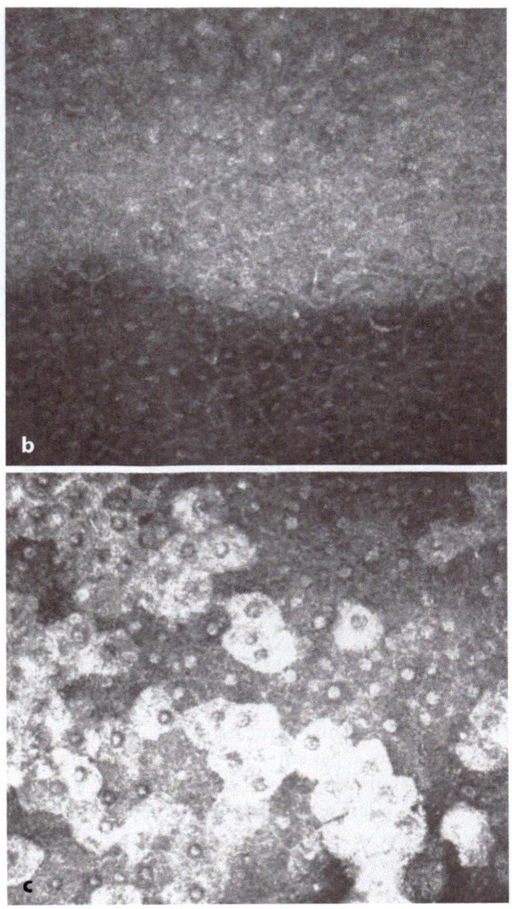

Fig. 5.32 Corneal conjunctivalization in a 55-year-old woman as a complication of severe ocular rosacea. Partial keratinization and superior conjunctivalization (*area 1*) and severe corneal metaplasia (*area 2*). **a** Appearance on slit-lamp examination. **b, c** Confocal in vivo microscopy image

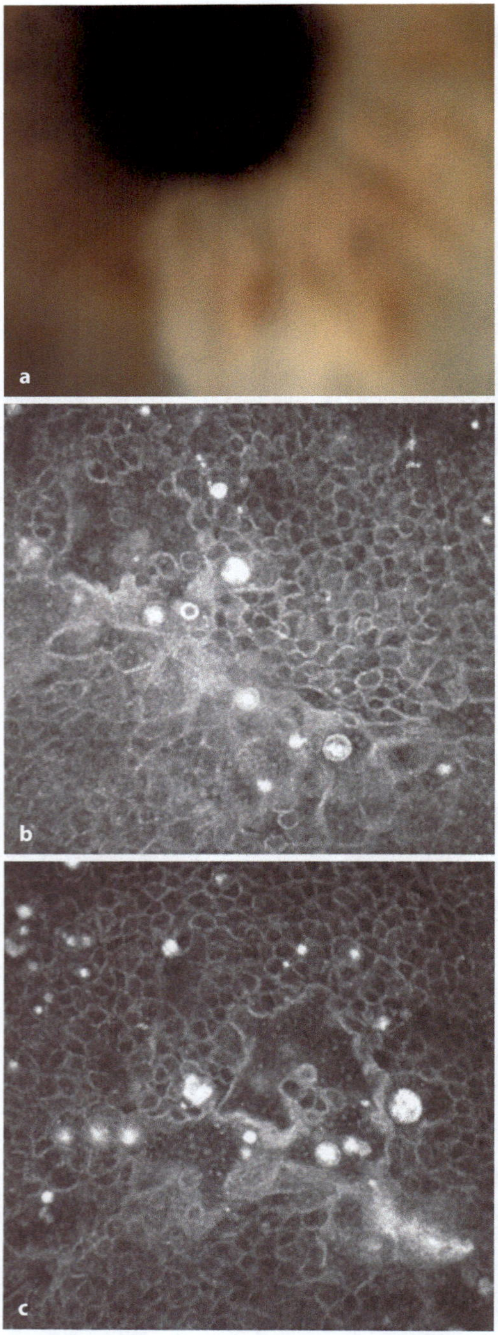

Fig. 5.33 Meesmann's corneal dystrophy. **a** Slit-lamp photograph of a patient with Meesmann's corneal dystrophy. Multiple epithelial cystic lesions. **b, c** Confocal in vivo microscopy images. Microcysts are seen as hyporeflective areas in the basal epithelial layer. Hyperreflective dots are observed inside most of the microcysts

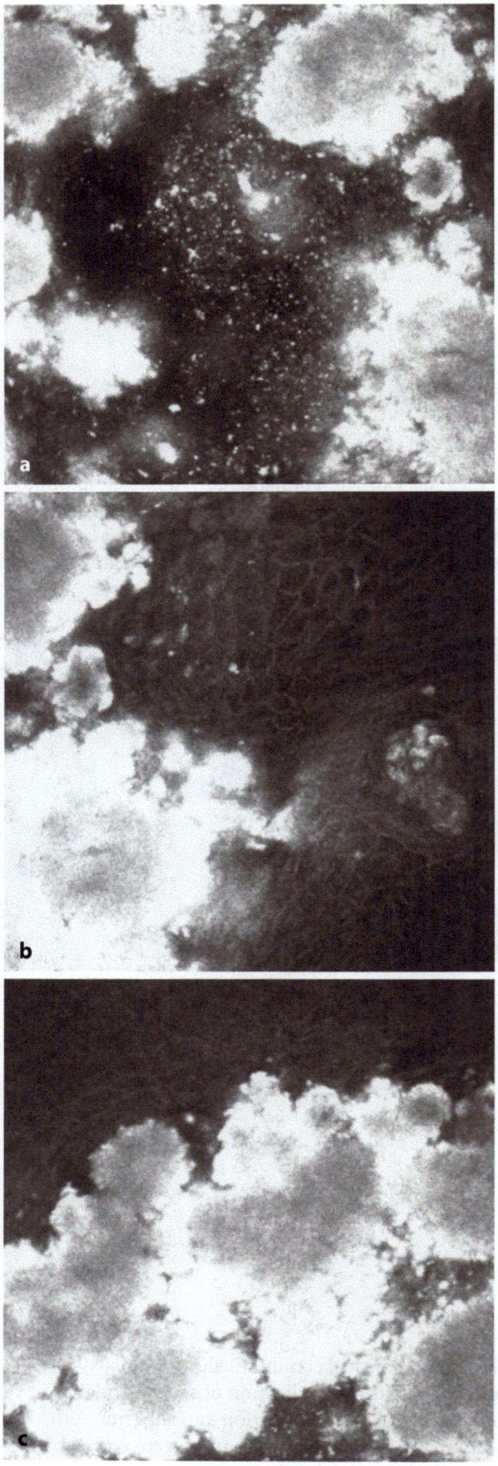

Fig. 5.34 Calcific band keratopathy. Hyperreflective areas in the corneal epithelium

Fig. 5.35 Epithelial basement membrane dystrophy (map-dot-fingerprint dystrophy) with fingerprint-like corneal lesions. **a** Slit-lamp photograph of fingerprint lesions. **b–d** Linear hyperreflective structures corresponding to abnormal basement membrane insinuated into the corneal epithelium. **e** Oblique section clearly showing this aberrant tissue within the corneal epithelium

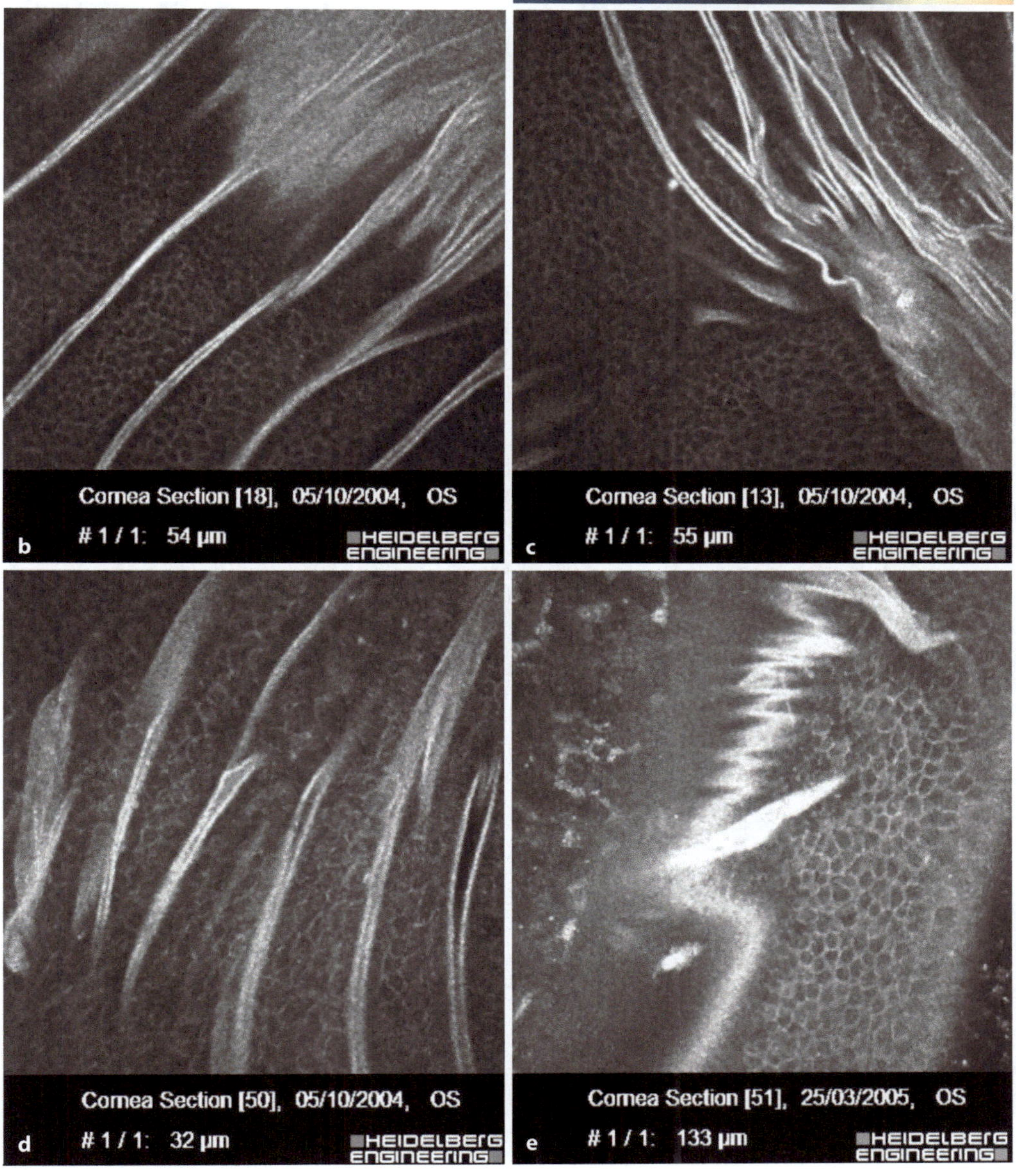

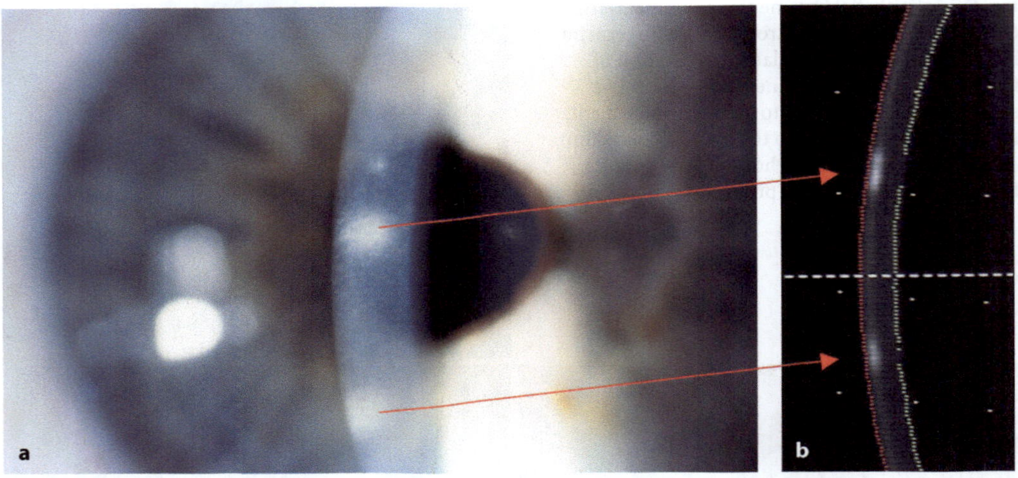

Fig. 5.36 Epidemic keratoconjunctivitis. Slit-lamp microscopy photograph: right eye of a 28-year-old woman on day 14 after the onset of symptoms of epidemic keratoconjunctivitis, showing the subepithelial nummular lesions as fleecy-fused areas of opacity with unclear margins. **a** Slit-lamp microscopy. **b** Pentacam Scheimpflug camera (Oculus Optikgeräte, Wetzlar, Germany)

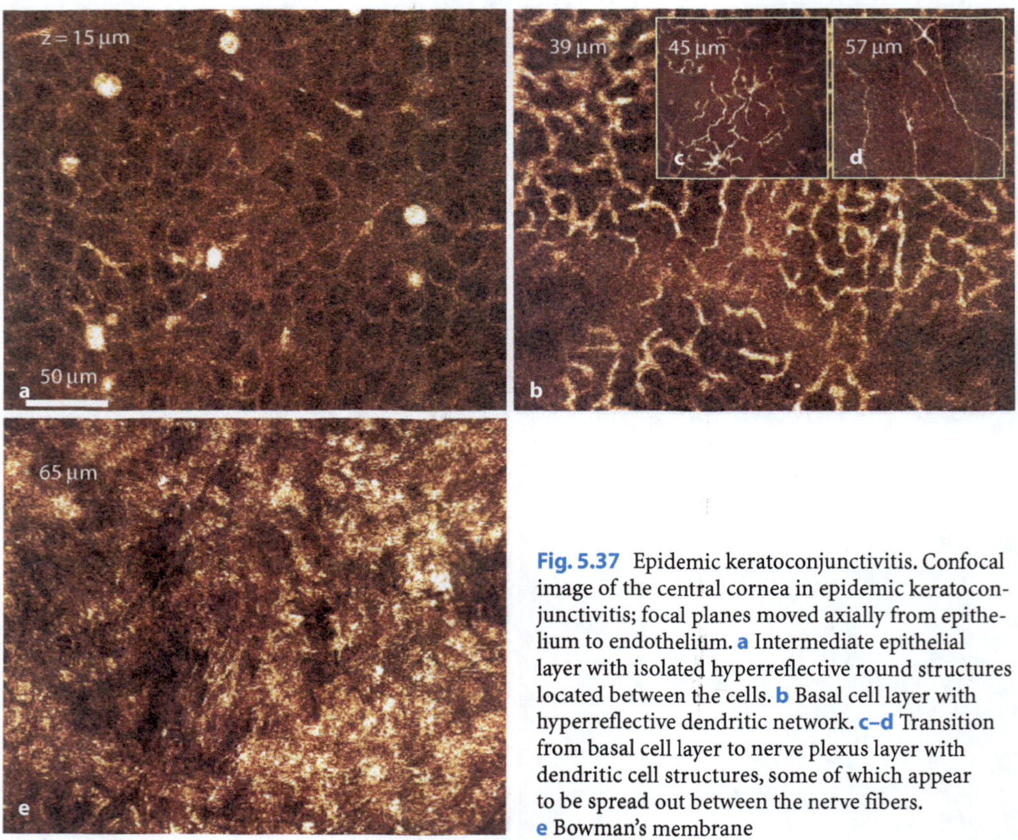

Fig. 5.37 Epidemic keratoconjunctivitis. Confocal image of the central cornea in epidemic keratoconjunctivitis; focal planes moved axially from epithelium to endothelium. **a** Intermediate epithelial layer with isolated hyperreflective round structures located between the cells. **b** Basal cell layer with hyperreflective dendritic network. **c–d** Transition from basal cell layer to nerve plexus layer with dendritic cell structures, some of which appear to be spread out between the nerve fibers. **e** Bowman's membrane

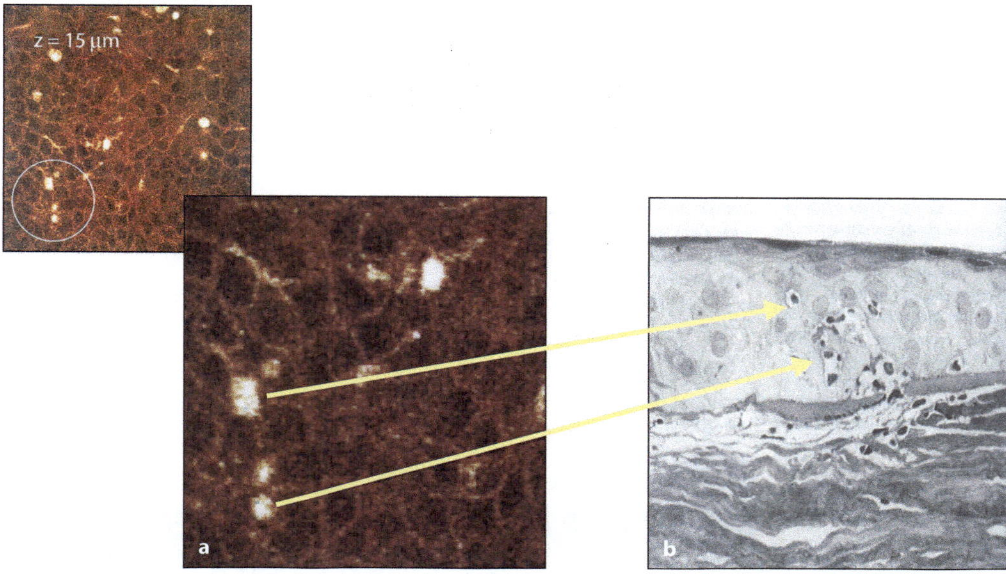

Fig. 5.38 Comparison between (**a**) confocal microscopy images and (**b**) histological sections, (reproduced from Lund O-E, Stefani FH. Corneal histology after epidemic keratoconjunctivitis. *Arch Ophthalmol* 1978;**96**:2085–2088 [45]). The infiltrates within the epithelium could be lymphocytes, histiocytes, or fibroblasts

5.3 Corneal Nerves

5.3.1 Normal Anatomy

The cornea is one of the most sensitive structures in the human body, and even the most minimal contact provokes the lid reflex to protect the eye. This sensitivity is attributable to the large numbers of nerve fibers that pass through the cornea. Furthermore, the corneal nerves exert an influence on the regulation of epithelial integrity and on wound healing [23]. In vivo visualization of these nerve structures is possible by confocal corneal microscopy.

The cornea is innervated primarily by sensory fibers arising from the ophthalmic nerve, a terminal division of the trigeminal nerve. Human corneal nerves are nonmyelinated and vary in thickness between 0.2 µm and 10 µm.

The nerve fiber bundles, which enter the anterior and central stroma in the corneal periphery, run parallel to the corneal surface in a radial pattern before making an abrupt 90° turn in the direction of Bowman's membrane [60]. On confocal corneal microscopy, these nerve fibers mostly present as thick, almost always stretched, highly reflective structures. Frequently, the stromal nerves are found in close proximity to keratocytes. The deep stroma is devoid of nerves that can be visualized on confocal microscopy.

In the anterior stroma, immediately before Bowman's membrane, the nerve fiber bundles display three different patterns. Some of the nerve fibers ramify before reaching Bowman's membrane without penetrating it and form the subepithelial plexus [63] (Fig. 5.39). Other nerves penetrate Bowman's membrane either directly, following a perpendicular or slightly oblique course, or by ramifying into several fine branchlets just before penetration. After they have penetrated Bowman's membrane, they again make a 90° directional change and pass between the basal cell layer of the epithelium and Bowman's membrane toward the corneal

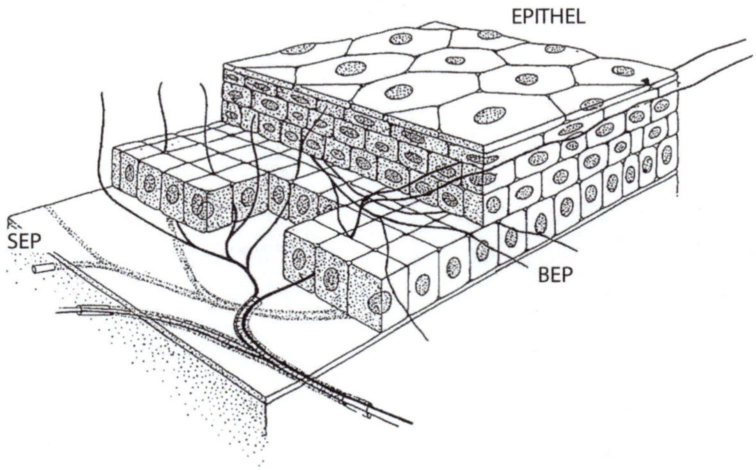

Fig. 5.39 Schematic illustration of the subepithelial plexus (*SEP*) and its branches in the corneal epithelial layers (*BEP* basal epithelial plexus)

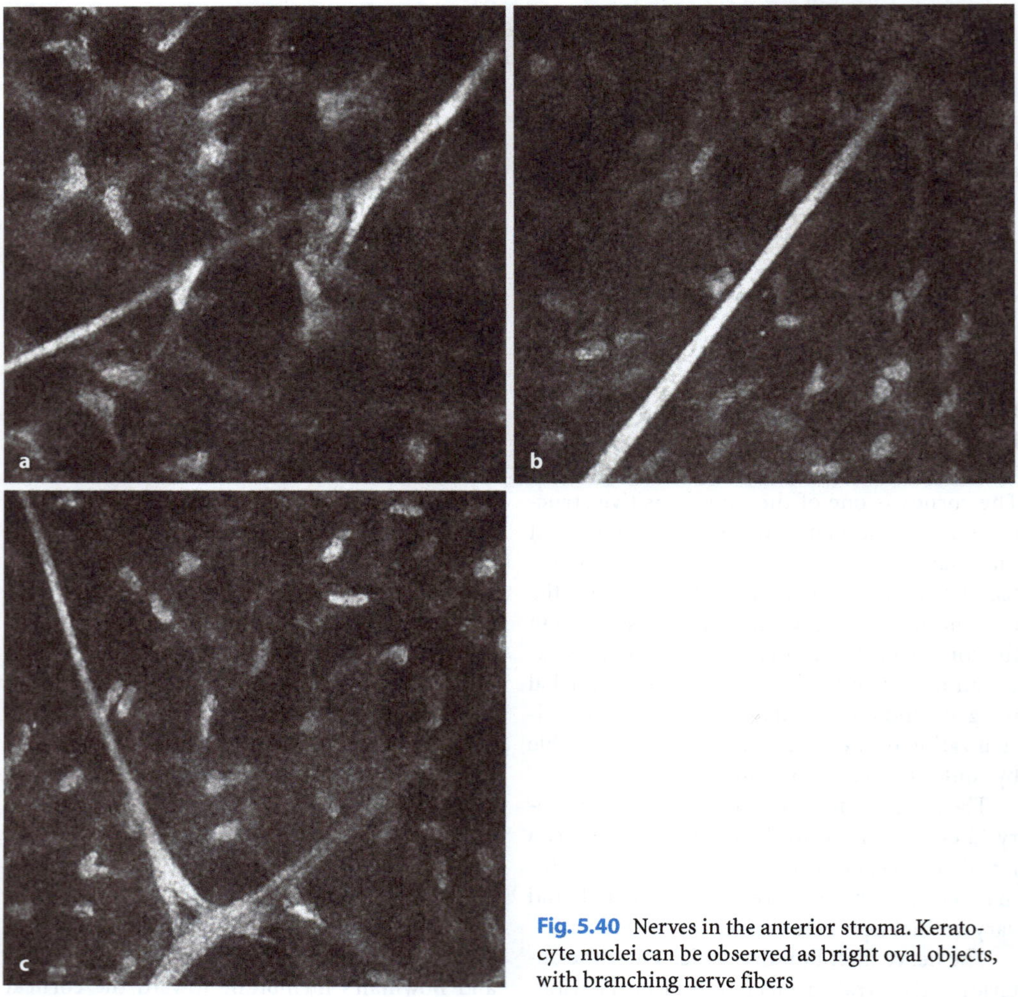

Fig. 5.40 Nerves in the anterior stroma. Keratocyte nuclei can be observed as bright oval objects, with branching nerve fibers

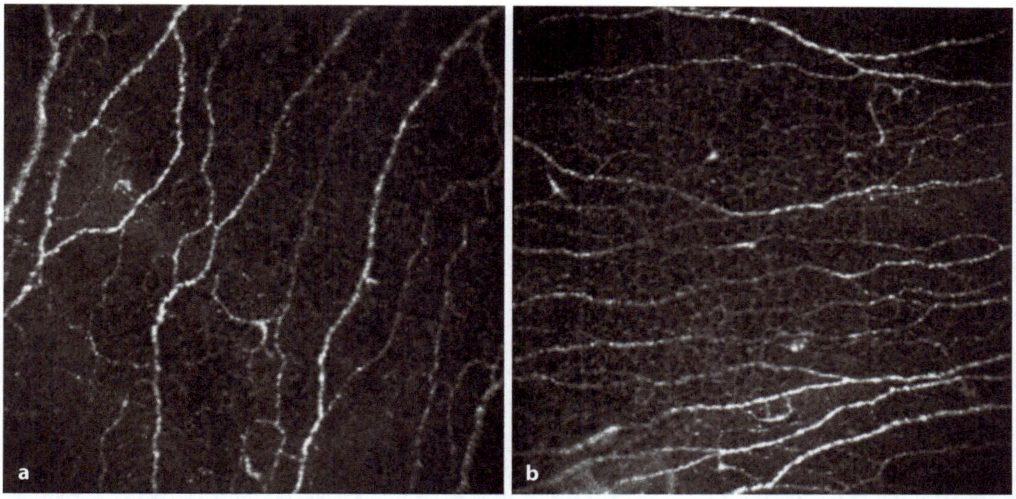

Fig. 5.41 Confocal laser scanning in vivo microscopy image of parallel nerve fiber bundles with bunches in the subepithelial plexus in a normal human cornea. **a** Vertical view. **b** Horizontal view

center and form the basal epithelial plexus (see Fig. 5.39). In so doing, they give off many small side branchlets directed both toward the corneal surface, where they end freely, and toward the center [60, 61]. The nerve fibers of the basal epithelial plexus mostly run parallel to each other and often form Y-shaped or T-shaped branches. Their predominantly granular, "string of pearl" structure is characteristic; more rarely they display a smooth surface. Unlike the stromal nerves, they are characterized by lesser reflectivity and frequently follow a meandering path. Occasionally, a thicker nerve fiber bundle will divide into two finer nerve fibers, which then reunite after a short distance into a single nerve fiber with the same thickness as before. The finer branchlets also form connections between larger nerve fibers (Figs. 5.40–5.44)

Fig. 5.42 Two-dimensional mapping of parallel nerve fiber bundles in the subepithelial plexus in a normal human cornea

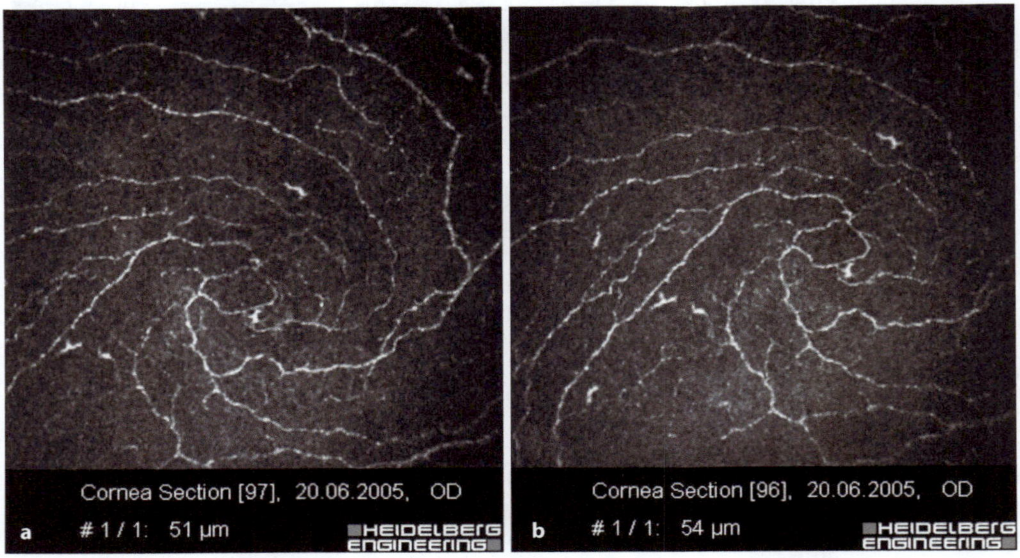

Fig. 5.43 Nerve structure of the cornea; individual image from a z-scan series. **a** Subepithelial nerve plexus in the cornea (z=51 µm). **b** Nerve plexus (z=53 µm)

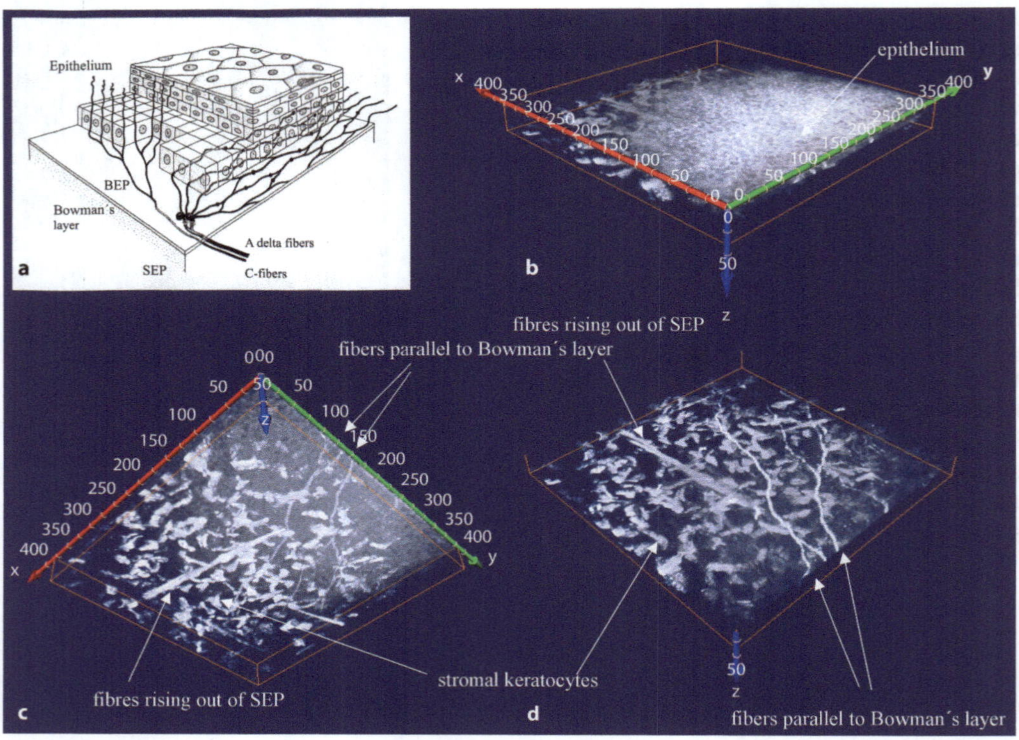

Fig. 5.44 Schematic illustration (**a**) and three-dimensional reconstruction (**b–d**) of the corneal epithelium with anterior stroma and nerves in a healthy human subject. **b** Anterior view. **c** Posterior view. **d** Anterior view with virtual removal of the epithelium Thin nerves run parallel with Bowman's membrane in the basal epithelial plexus, with thicker fibers originating from the subepithelial plexus

5.3.2 Pathological Findings

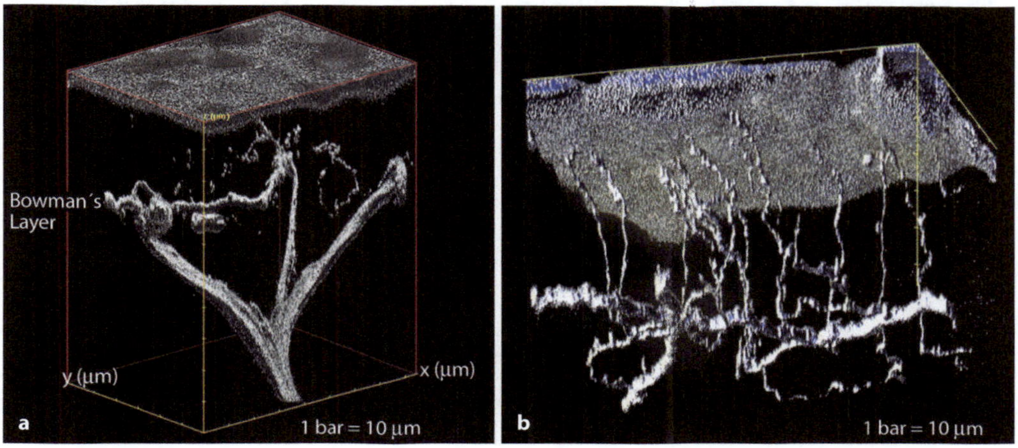

Fig. 5.45 Trephinates freshly obtained in cases of Fuchs corneal dystrophy, stained with calcein-AM/ethidium homodimer and evaluated by confocal microscopy. **a** Aδ-fiber rising out of the SEP, dividing trichotomously and kinking into the BEP; stepsize 0.5 μm, steps n=200, depth 10 μm, zoom 1.0, objective (water-immersion) ×60, average 32. **b** C-fibers sending off multiple branches that penetrate epithelial cell layers orthogonally, ending blindly underneath the superficial cells; stepsize 0.5 μm, steps n=132, depth 66 μm, zoom 0.82, objective (water-immersion) ×60, average 16

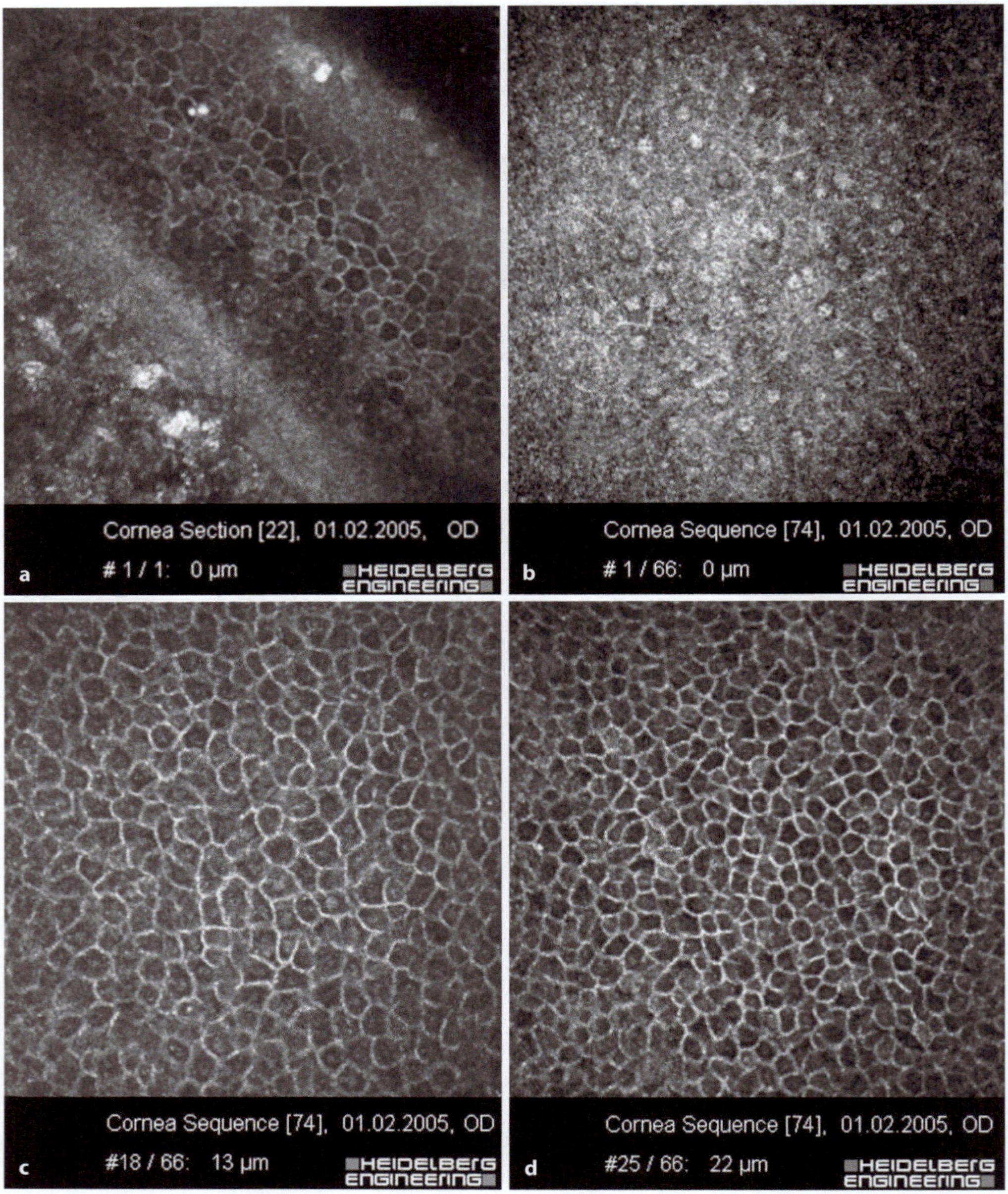

Fig. 5.46 a–h Status 4 weeks after corneal erosion. External *z*-scan through the corneal epithelium, no nerves can be visualized. **a** Oblique section through the epithelium/anterior stroma. **b–h** Surface-parallel sections through the epithelium/anterior stroma

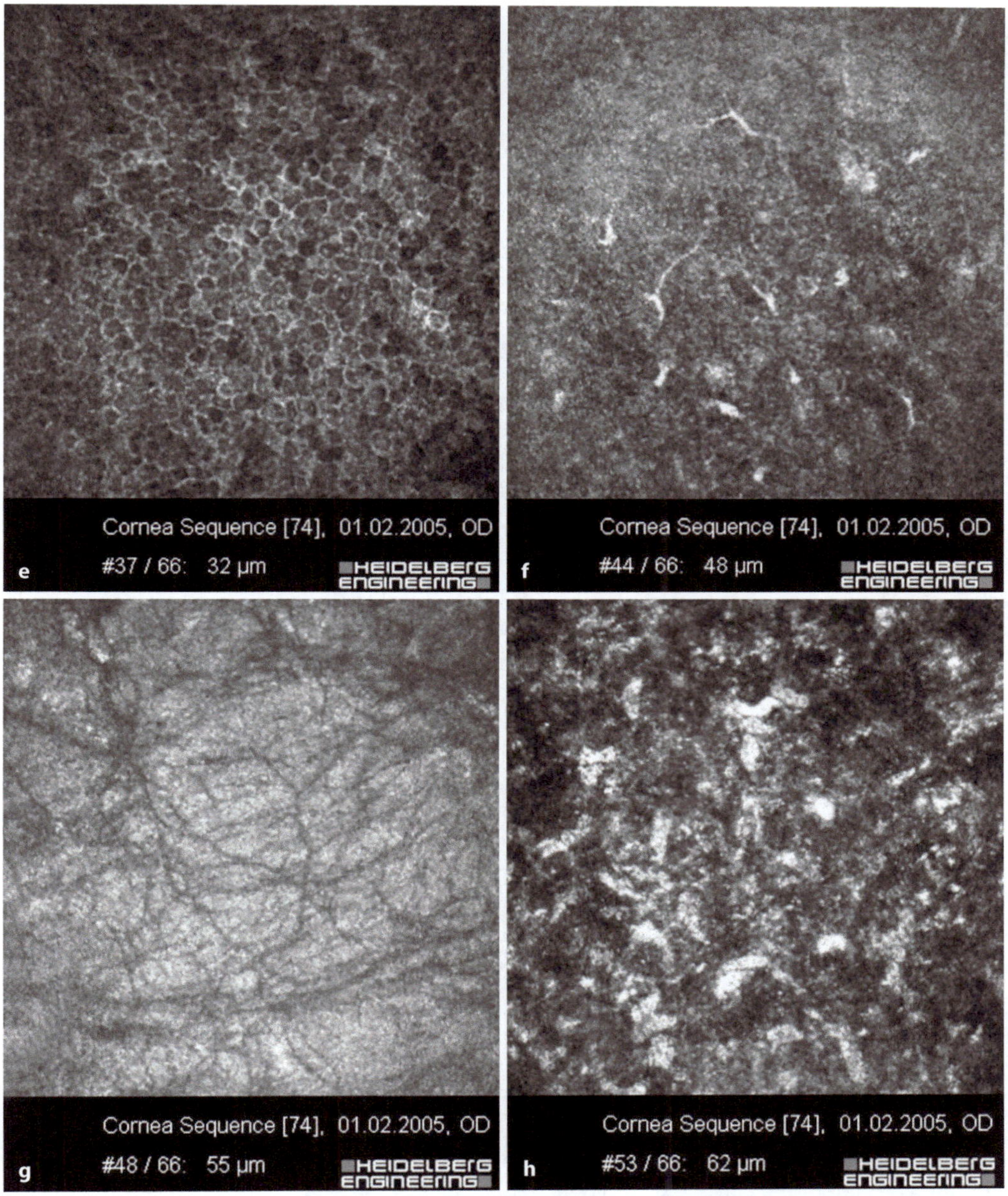

Fig. 5.46 (*continued*) **b–h** Surface-parallel sections through the epithelium/anterior stroma

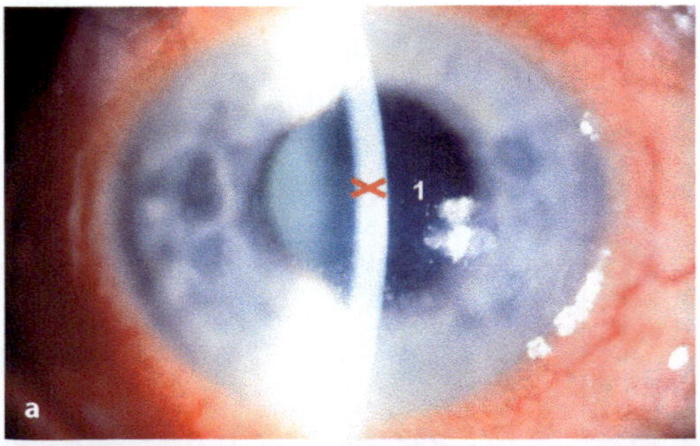

Fig. 5.47 Herpes keratitis. **a** Slit-lamp photograph of 56-year-old woman with diffuse stromal scars in central and paracentral areas, folding of Descemet's membrane and endothelial fitting; corneal sensation is decreased. **b, c** The same patient after 1 month. **d** Subepithelial nerve plexus shows unusual „wire netting" of nerve fibers and dendritic cells. **e, f** One month after successful therapy, no changes were detected in the epithelium or at the level of plexus, either in the central cornea (**e**) or in the periphery (**f**)

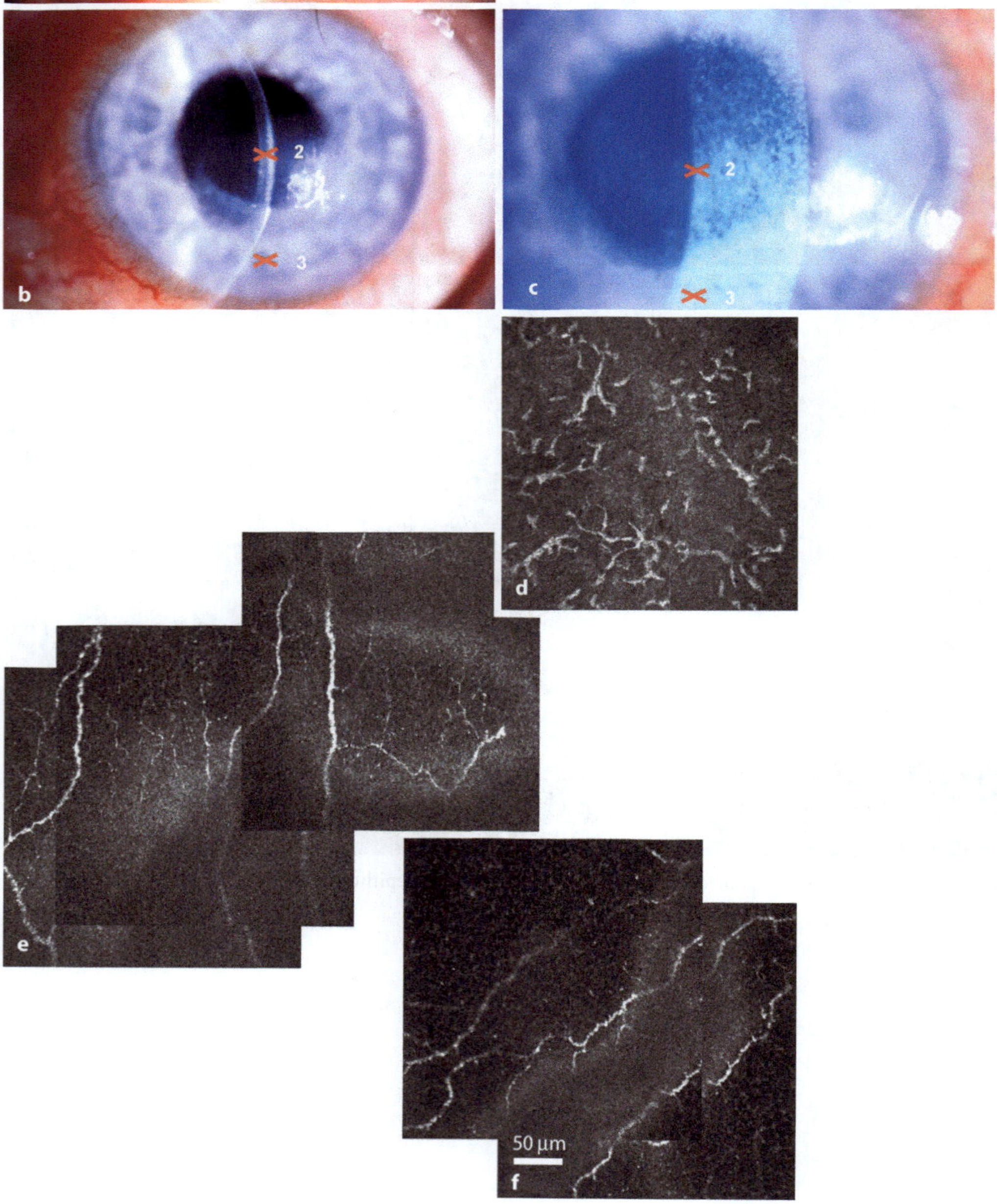

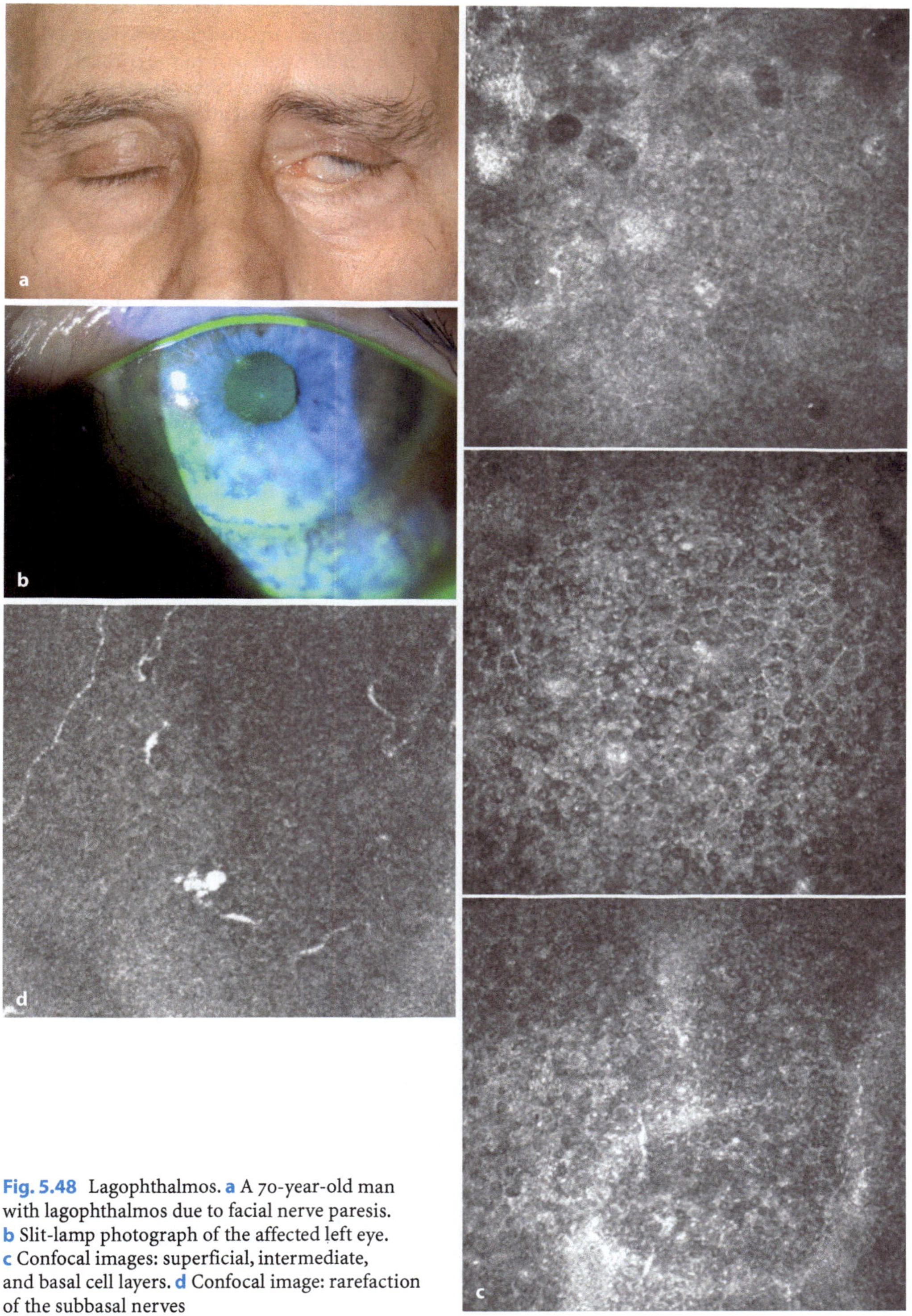

Fig. 5.48 Lagophthalmos. **a** A 70-year-old man with lagophthalmos due to facial nerve paresis. **b** Slit-lamp photograph of the affected left eye. **c** Confocal images: superficial, intermediate, and basal cell layers. **d** Confocal image: rarefaction of the subbasal nerves

5.3.2.1 Corneal Innervation After Refractive Surgery

The physiological structure of the corneal nerves is appreciably disturbed in the course of refractive surgery. This is considered to be the reason why keratoconjunctivitis sicca is the most common complication following LASIK, being encountered for a certain time in virtually all patients in varying degrees of severity.

In the undamaged cornea, the corneal nerves arising from the long ciliary nerves enter the middle stroma at 3 o'clock and 9 o'clock and ramify horizontally and vertically in their further course toward the middle of the cornea, ultimately forming a dense and branched subepithelial nerve plexus underneath Bowman's

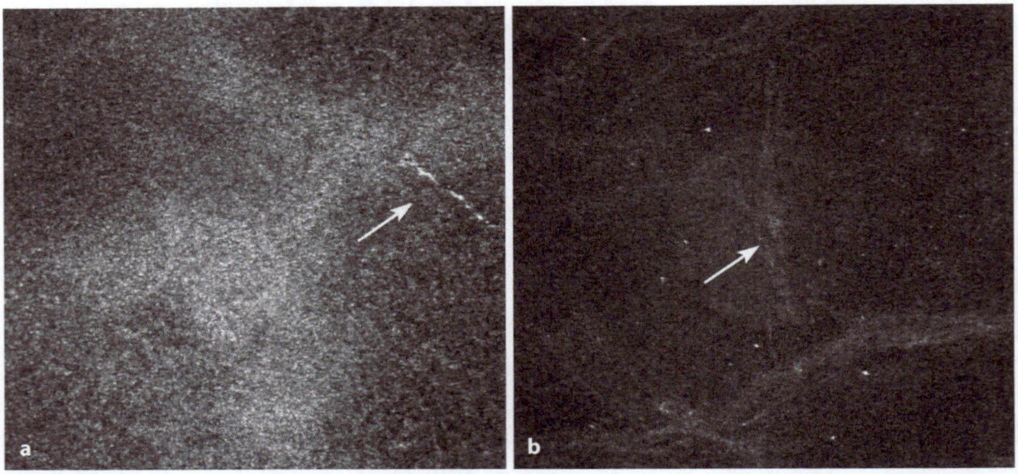

Fig. 5.49 Innervation after refractive surgery. Rarefaction of the nerve structures a few hours after laser-assisted in situ keratomileusis; rare and only thin subbasal nerve fibers. **a** Nerve fibers near subbasal layer. **b** Nerve fibers near Bowman's membrane

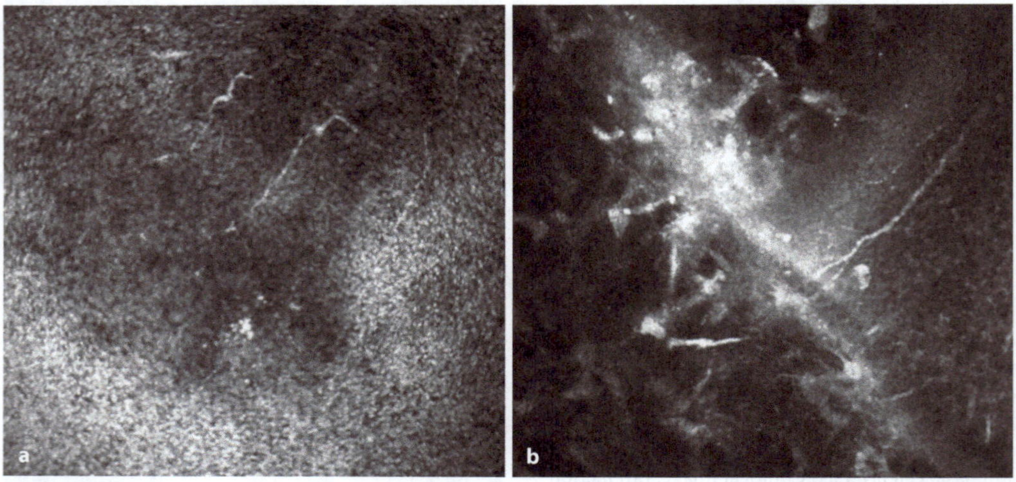

Fig. 5.50 Innervation after refractive surgery. **a** Two months after laser-assisted in situ keratomileusis (LASIK): small regenerating nerve fibers in central cornea. **b** Nine months after LASIK: new nerve fiber growing from the edge of flap region to center

membrane. From there, the nerve fibers travel vertically as far as the subbasal nerve plexus, before finally ascending into the upper epithelial layers [61].

Photorefractive keratectomy (PRK) involves ablation of the densely innervated epithelium, the branched subepithelial nerve plexus, and the anterior stroma. In the LASIK procedure, the microkeratome separates the subbasal nerve fiber bundles and the nerves of the anterior stroma along the flap region.

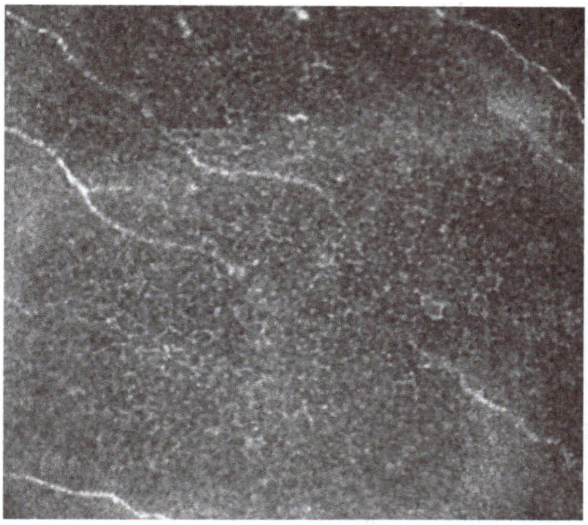

Fig. 5.51 Innervation after refractive surgery. Two years after laser-assisted in situ keratomileusis, there is still rarefaction of the subepithelial nerves, with short and unbranched nerve fibers

Degeneration of nerve structures in the flap vicinity can be detected on confocal microscopy just a few hours after LASIK, as reflected in rarefaction of the subbasal and stromal nerves (Fig. 5.49).

According to Donnenfeld et al., transection of both arms of the corneal nerve plexus following creation of a superior-hinge flap leads to more pronounced loss of corneal sensation and more pronounced dry eye signs and symptoms than with a nasal-hinge flap [13]. The same phenomenon has been reported with a narrow nasal hinge flap compared with a wider hinge flap in which more nerve fibers are left undamaged [14].

The onset of corneal nerve regeneration appears to occur early. Thus, Linna et al. were able to demonstrate gentle nerve regeneration 1 week after LASIK. Central corneal sensation was restored after 6 months. However, even after 6 months, short unconnected subbasal nerve fibers were still found, in contrast with long interconnected nerve fibers in a control population [44] (Fig. 5.50). Rarefied subbasal nerve fibers of this kind can still be detected 2 years after LASIK (Fig. 5.51). In addition, morphological changes have also been found in the area of the anterior stromal nerves or in the flap region [42] (Fig. 5.52). Using refractive corneal surgery as an example, confocal microscopy is thus able to provide a direct comparison between innervation detectable on microscopy and the sensitivity or symptom severity of the dry eye.

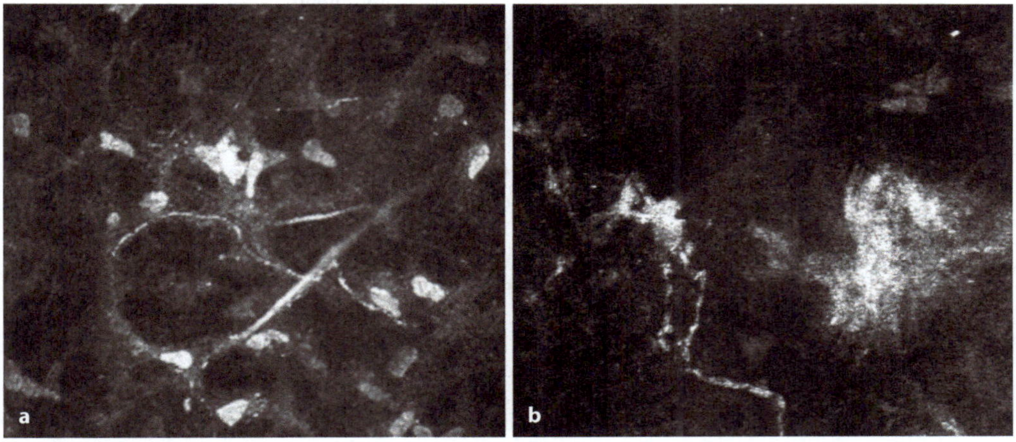

Fig. 5.52 Innervation after refractive surgery. Abnormal stromal nerves after laser-assisted in situ keratomileusis: curved pattern and very thin nerve fibers

5.3.2.2 Reinnervation After Penetrating Keratoplasty

Corneal innervation is the cornerstone of the normally functioning cornea. Although reinnervation processes have formerly been discussed in the literature, no suitable model was previously available. Different investigators [50, 71, 72, 76, 90] have used a Cochet–Bonnet esthesiometer to estimate graft innervation. Most studies have failed to differentiate between the peripheral and central sensitivity of the corneal graft. The results of these studies are summarized in Table 5.1.

Tervo et al. have demonstrated histochemically that complete regeneration does not occur either in the subepithelial nerve plexus or in the stromal nerves during the 3 years following penetrating keratoplasty. Moreover, a grafted cornea obtained 29 years after surgery did not show normal corneal sensitivity: The subepithelial nerve plexus presented with branched nerve fibers, but only a few stromal nerve trunks had regenerated [85].

Richter et al. [70] investigated corneal grafts using a Microphthal confocal microscope and compared the results with those obtained by esthesiometry. In the peripheral graft, the first atypically curved nerve fibers were detected in the middle stroma at 2 months and under Bowman's membrane at 3 months after surgery. In the central cornea, the first stromal nerve trunks were detected at 7 months after kera-

Table 5.1 Studies of corneal graft innervation

Study	Follow-up	Number of grafts investigated	Sensitivity (Cochet–Bonnet esthesiometry)
Ruben and Colebrook, 1979 [72]	7 months to 10 years	48	Incomplete sensitivity after 3 years
Skriver, 1978 [76]	12 months	45	Close to normal after 12 months
Mathers et al., 1988 [50]	1 month to 10 years	91	- With the exception of herpes simplex keratitis, the level of sensitivity did not depend on the diagnosis - Corneal sensitivity returned progressively from the periphery toward the center of the graft - The rate of return of sensitivity averaged 0.029 mm/month
Tugal Tutkun et al., 1993 [90] sensitivity	2 weeks to 15 years	71	- 36 grafts were completely anesthetic - 35 grafts had some level of sensitivity - Of these 35, only one graft had normal central
Richter et al., 1996 [70]	3 years	46	Graft center: - At 1 month, completely anesthetic - At 6 months, incomplete sensitivity - At 24 months, one-third of grafts had normal sensitivity, and one-seventh were completely anesthetic Graft periphery: - At 12 months, 50% of the grafts were completely anesthetic - At 24 months, 25% of the grafts were completely anesthetic - Best innervation in keratoconus or scars

toplasty, while the first bunches of subepithelial nerve plexus were noted only after 24 months. To date, the results of confocal microscopy appear to correspond to esthesiometry findings.

Confocal microscopy provides insights into the morphological and functional aspects of corneal reinnervation after surgery. The sample images of corneal innervation after grafting illustrate the structures of the subepithelial nerve plexus and stromal nerves at about 24 months after surgery (Fig. 5.53).

Figure 5.54a–c shows hypertrophy of corneal nerves in a patient with neurofibromatosis.

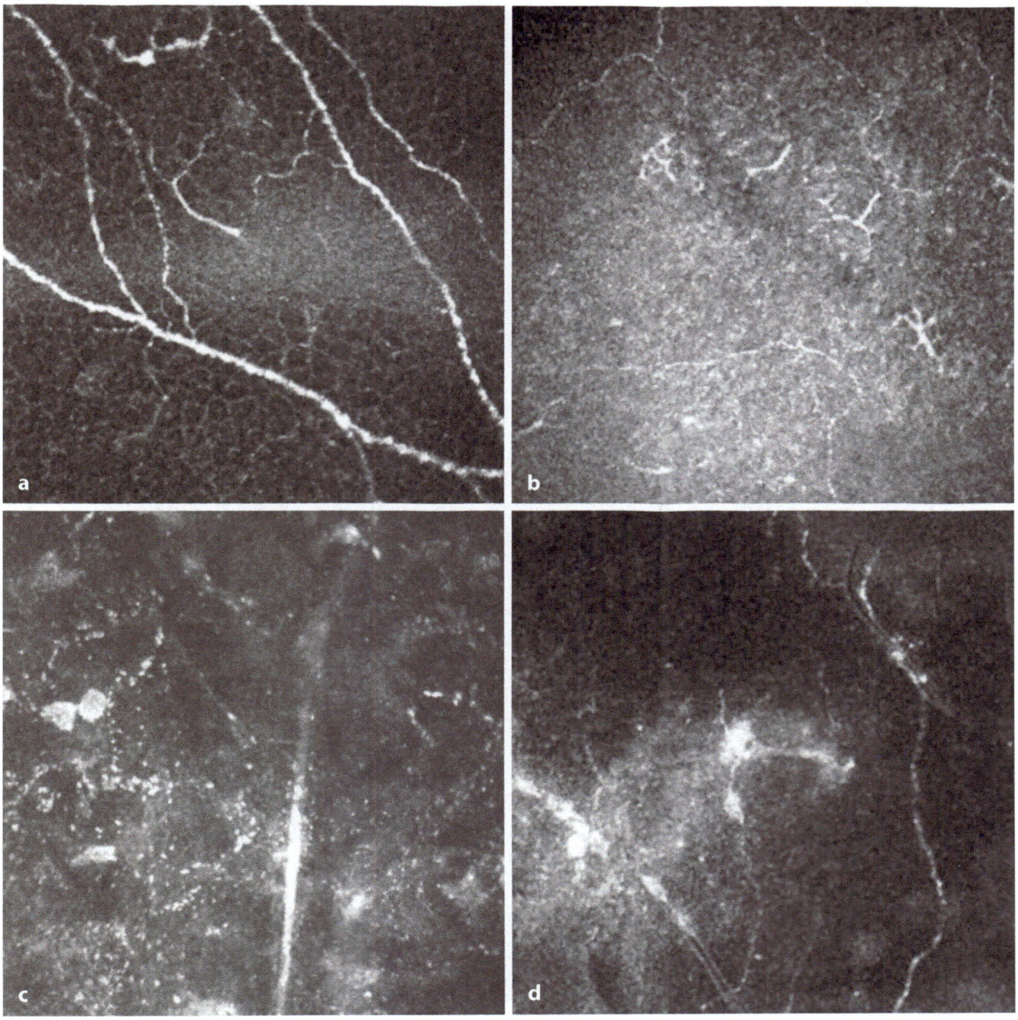

Fig. 5.53 Reinnervation after penetrating keratoplasty. **a** Graft center: complete reinnervation of subepithelial nerve plexus 24 months after surgery. **b** Graft center: incomplete reinnervation of subepithelial nerve plexus; curved nerve bunches 25 months after surgery; dendritic cells. **c** Graft center: nerve trunk in middle stroma 24 months after surgery. **d** Graft periphery: incomplete reinnervation of subepithelial nerve plexus, hyperreflective scattering at the level of Bowman's membrane 24 months after surgery

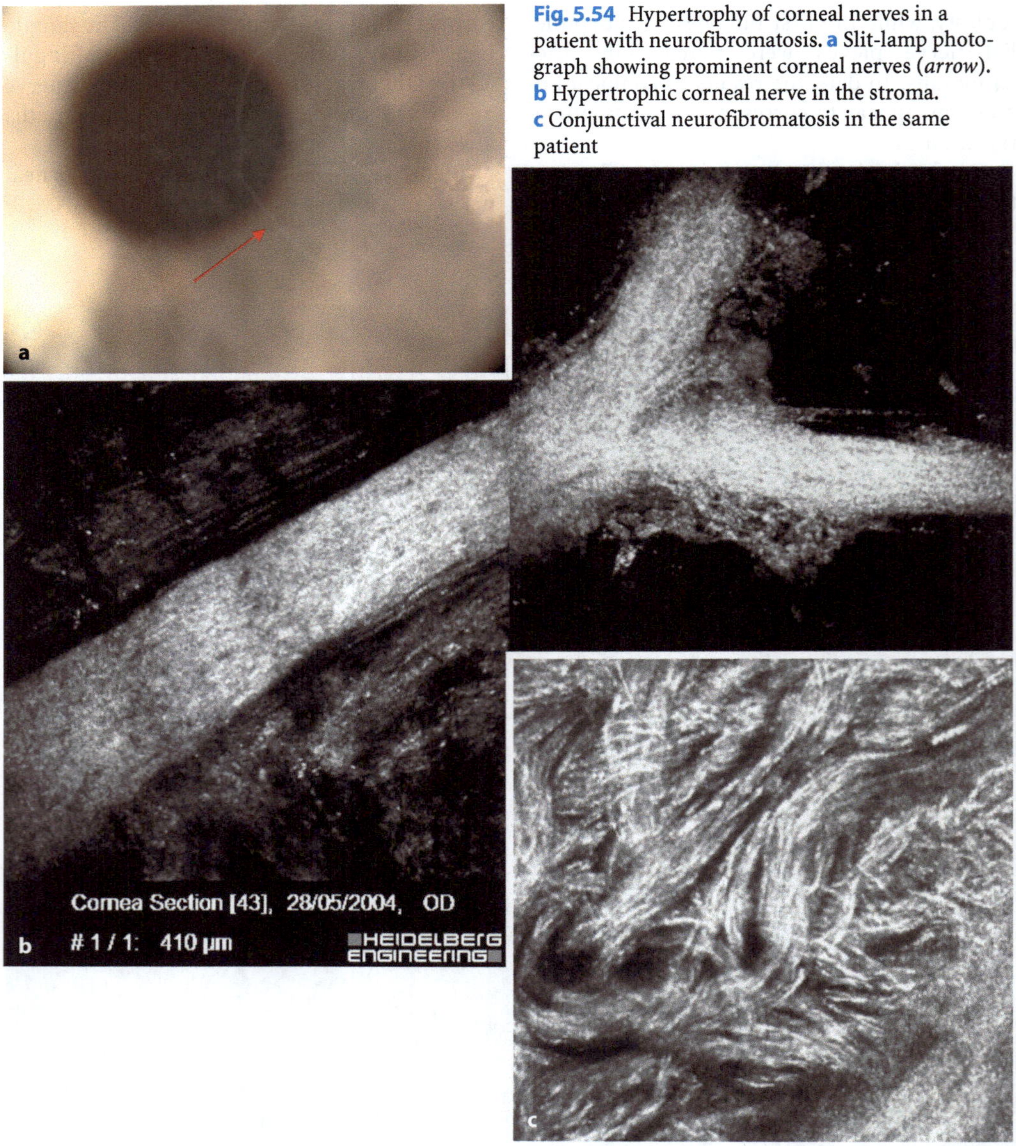

Fig. 5.54 Hypertrophy of corneal nerves in a patient with neurofibromatosis. **a** Slit-lamp photograph showing prominent corneal nerves (*arrow*). **b** Hypertrophic corneal nerve in the stroma. **c** Conjunctival neurofibromatosis in the same patient

5.4 Bowman's Membrane

5.4.1 Normal Anatomy

The anterior limiting membrane has an amorphous appearance. Its location can be established from the nerves of the basal epithelial plexus, which ramify there (Fig. 5.55a–c).

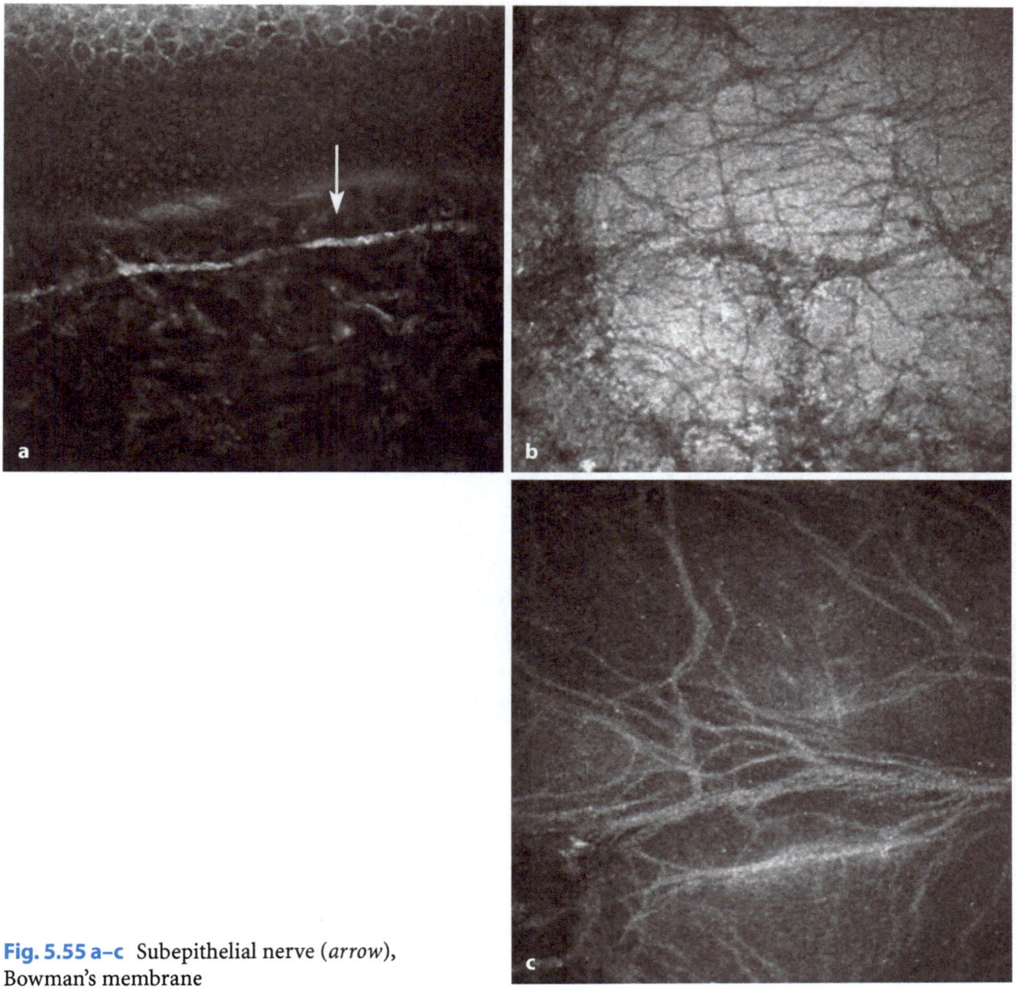

Fig. 5.55 a–c Subepithelial nerve (*arrow*), Bowman's membrane

5.4.2 Pathological Findings

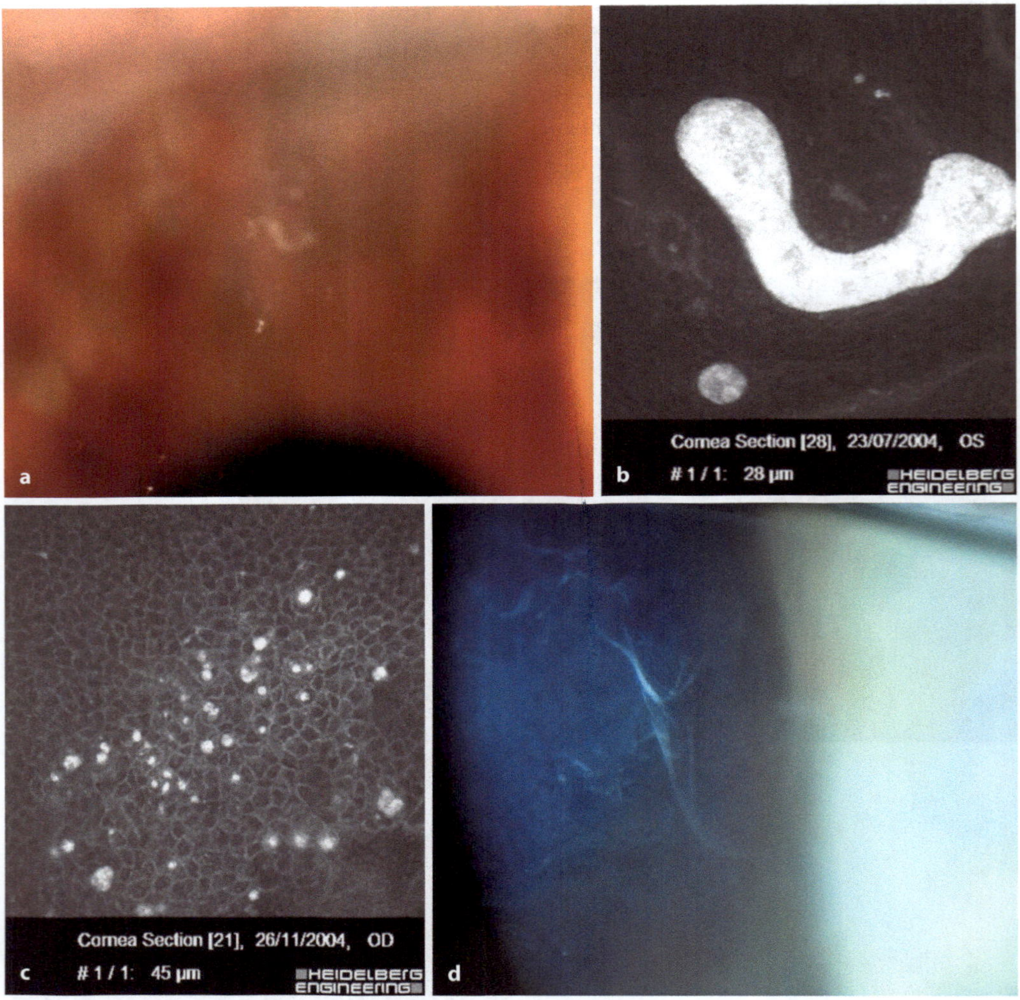

Fig. 5.56 a–f Dots and maplike lesions. **a** Slit-lamp photograph of dots. **b, c** Round structures corresponding to microcysts containing hyperreflective material. **d** Slit-lamp photograph of maplike lesions

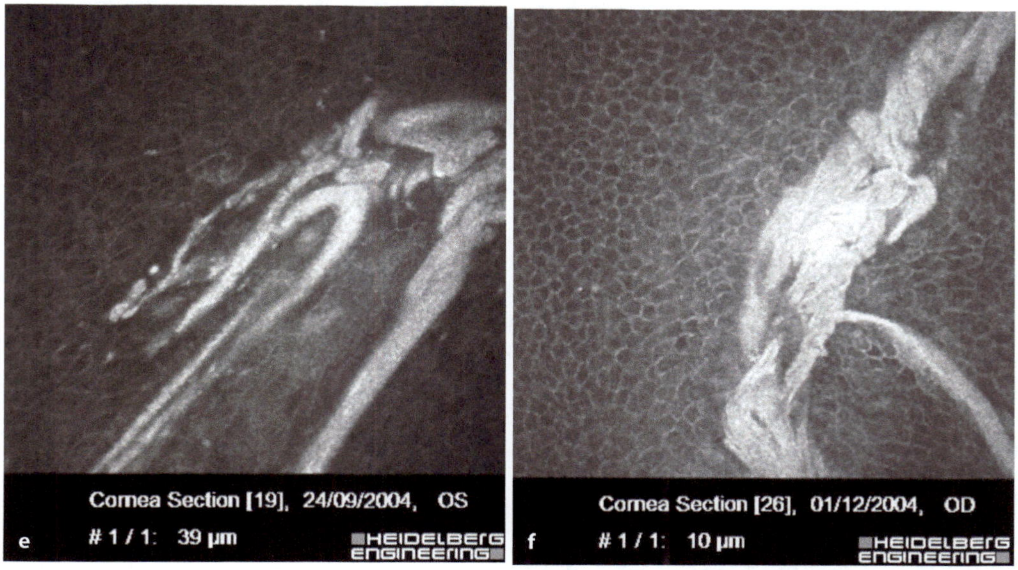

Fig. 5.56 a–f (*continued*) **e, f** Abnormal basement membrane within the corneal epithelium. This abnormal tissue terminates in bizarre shapes.

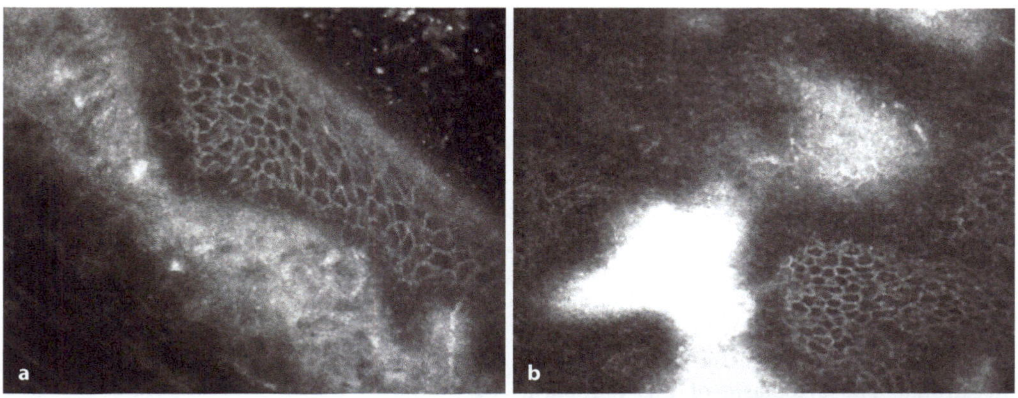

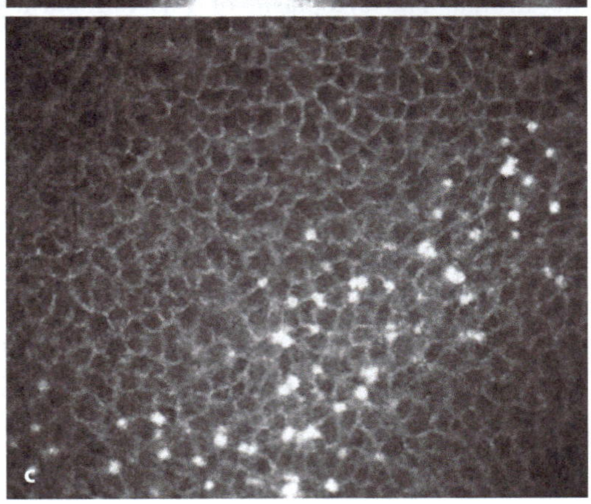

Fig. 5.57 Reis–Bückler corneal dystrophy. **a** Oblique section showing reflective and irregular material at the level of Bowman's layer and the anterior stroma. **b** Highly reflective tissue at the level of Bowman's layer. **c** Deposition of fine reflective material in the basal cell layer

Fig. 5.58 Bowman's membrane 4 weeks after ammonia burn injury and complete epithelial wound healing with extreme development of a fiber matrix structure

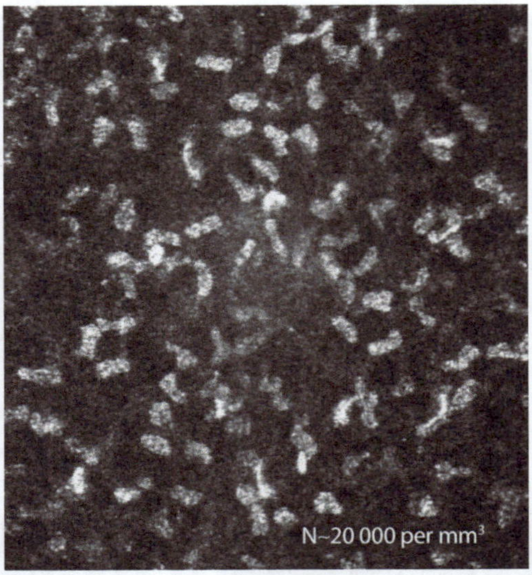

Fig. 5.59 Anterior stroma. In the corneal stroma only the keratocyte nuclei are visualized. Cell nucleus density is highest in the anterior stroma; the cell nuclei shown are approximately 15 µm in diameter

5.5 Stroma

5.5.1 Normal Anatomy

Apart from neural structures, only the highly reflective, sharply demarcated cell nuclei of the keratocytes are visualized on examination of the stroma. The cytoplasm of this fibroblast subpopulation and the collagen fibers produced by them are not visible. Keratocyte nucleus density is higher in the anterior stroma close to Bowman's membrane than in the central and deep stroma (Figs. 5.59 and 5.60). Keratocyte density is highest in the anterior stroma, clearly declines toward the central stroma, and increases again slightly in the region immediately before Descemet's membrane.

Fig. 5.61 depicts posterior stroma, and Fig. 5.62 compares normal and activated stroma.

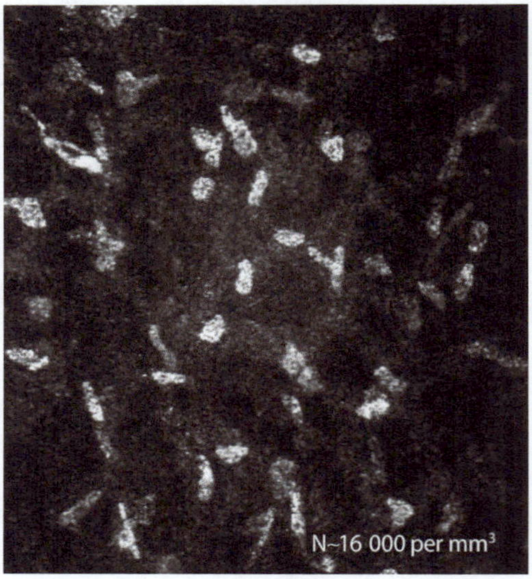

Fig. 5.60 Central stroma. Clearly demarcated, highly reflective, oval keratocyte nuclei in the central stroma; cell nucleus density is lowest in the central stroma

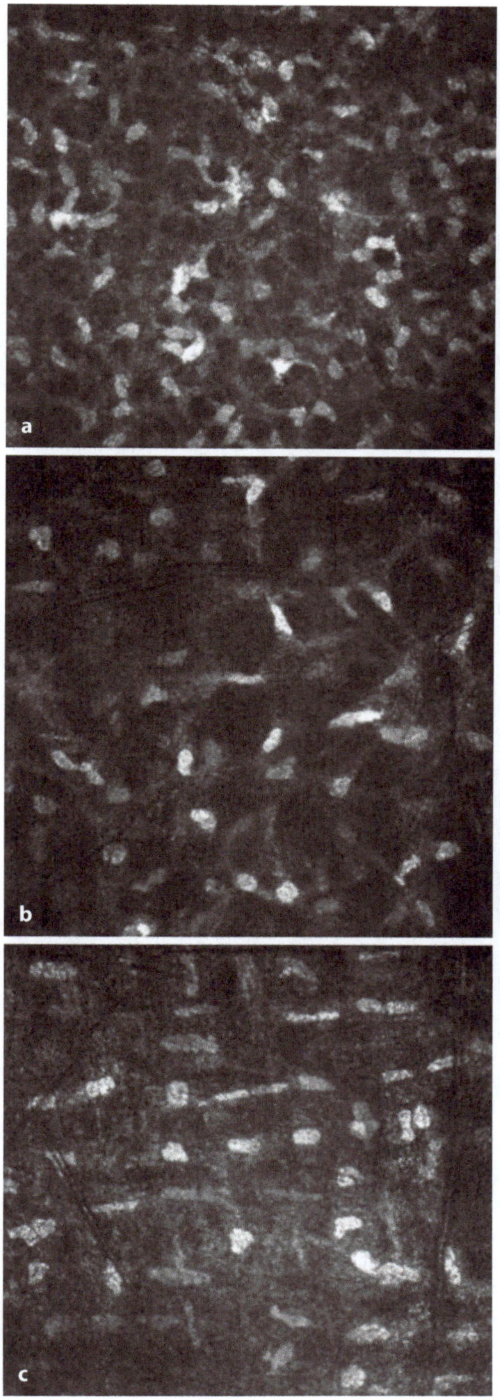

Fig. 5.61 Posterior stroma. Clearly demarcated, highly reflective, oval nuclei, which appear larger and are more numerous than in the central stroma but less numerous than in the anterior stroma. Anterior (**a**), central (**b**), and posterior stroma (**c**)

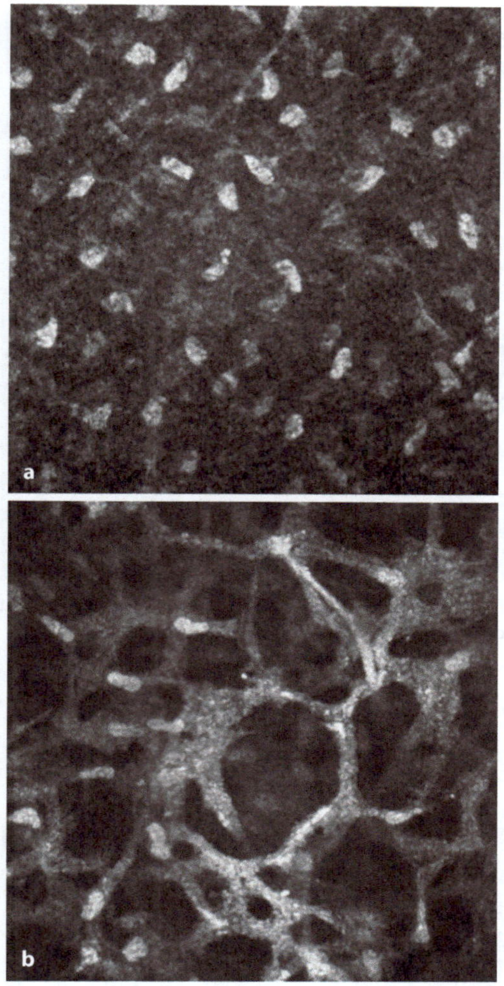

Fig. 5.62 Central stroma. **a** Normal keratocytes. **b** Activated keratocytes

5.5.2 Pathological Findings

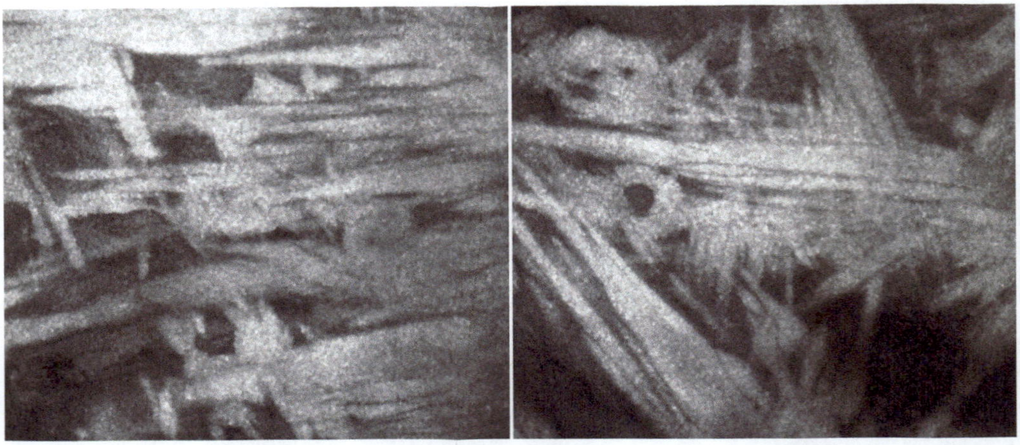

Fig. 5.63 Schnyder's crystalline dystrophy. Accumulation of reflective material consistent with subepithelial crystalline deposits

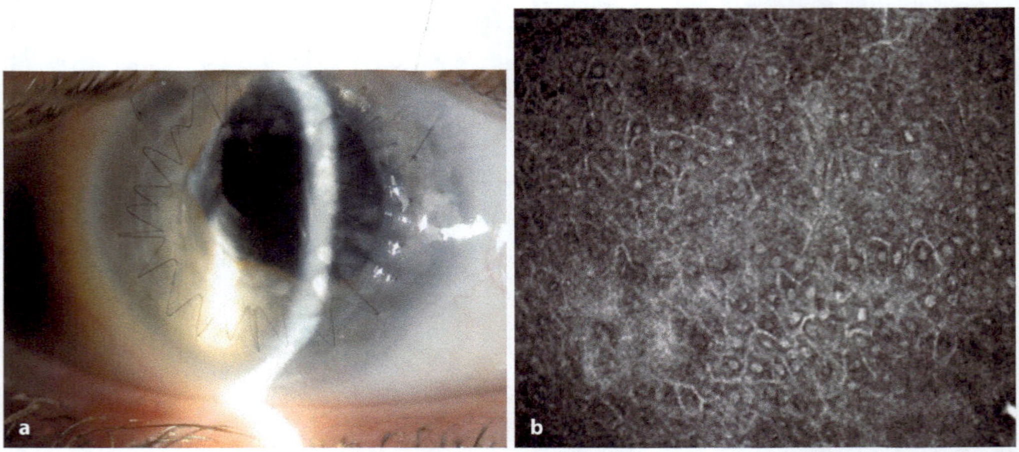

Fig. 5.64 a–f Penetrating keratoplasty. a Slit-lamp photograph of 72-year-old woman after penetrating keratoplasty. The graft is well located, clear, and fully epithelialized, with small areas of fibrination near the endocorneal suture. b–f Normal epithelial structure with a number of Langerhans cells (arrow) at the level of the basal cells and subepithelial nerve plexus. Scar areas (d) of basal epithelium in fibrination zones. Nerve fibers are absent both in the epithelium and in the stroma of the graft

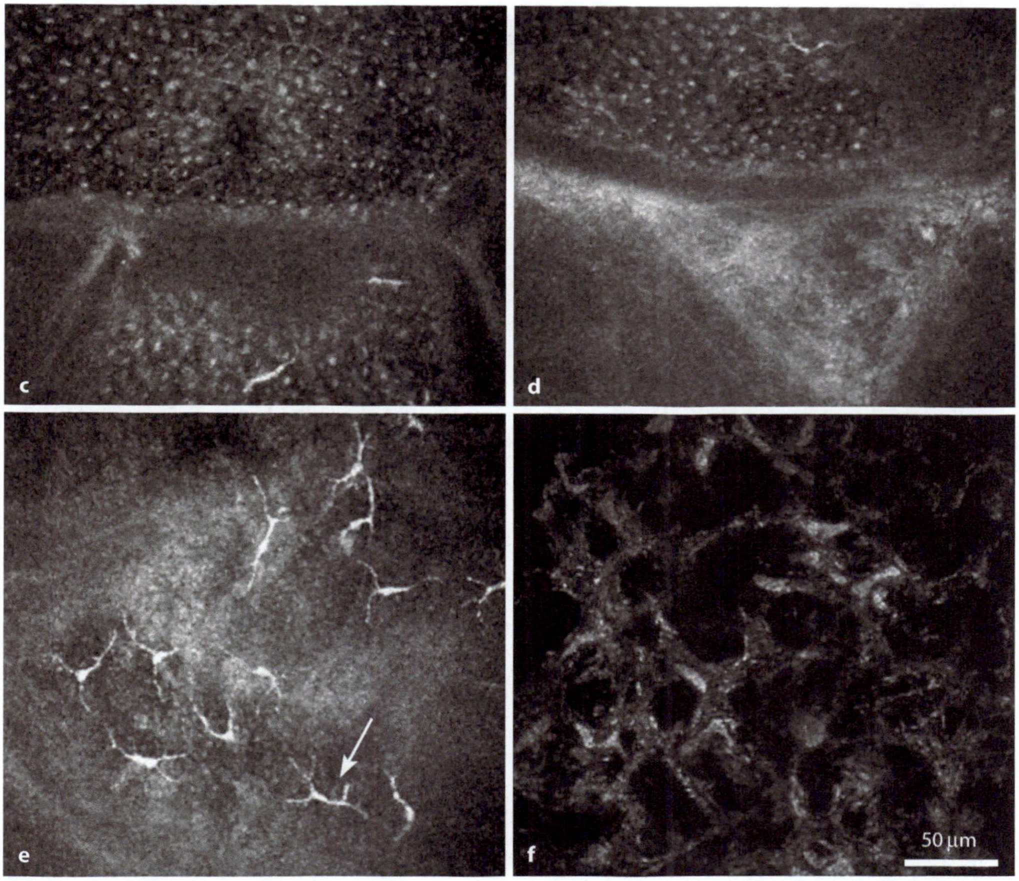

Fig. 5.64 (*continued*) **b–f** Normal epithelial structure with a number of Langerhans cells (arrow) at the level of the basal cells and subepithelial nerve plexus (**e**). Scar areas (**d**) of basal epithelium in fibrination zones. Nerve fibers are absent both in the epithelium and in the stroma of the graft

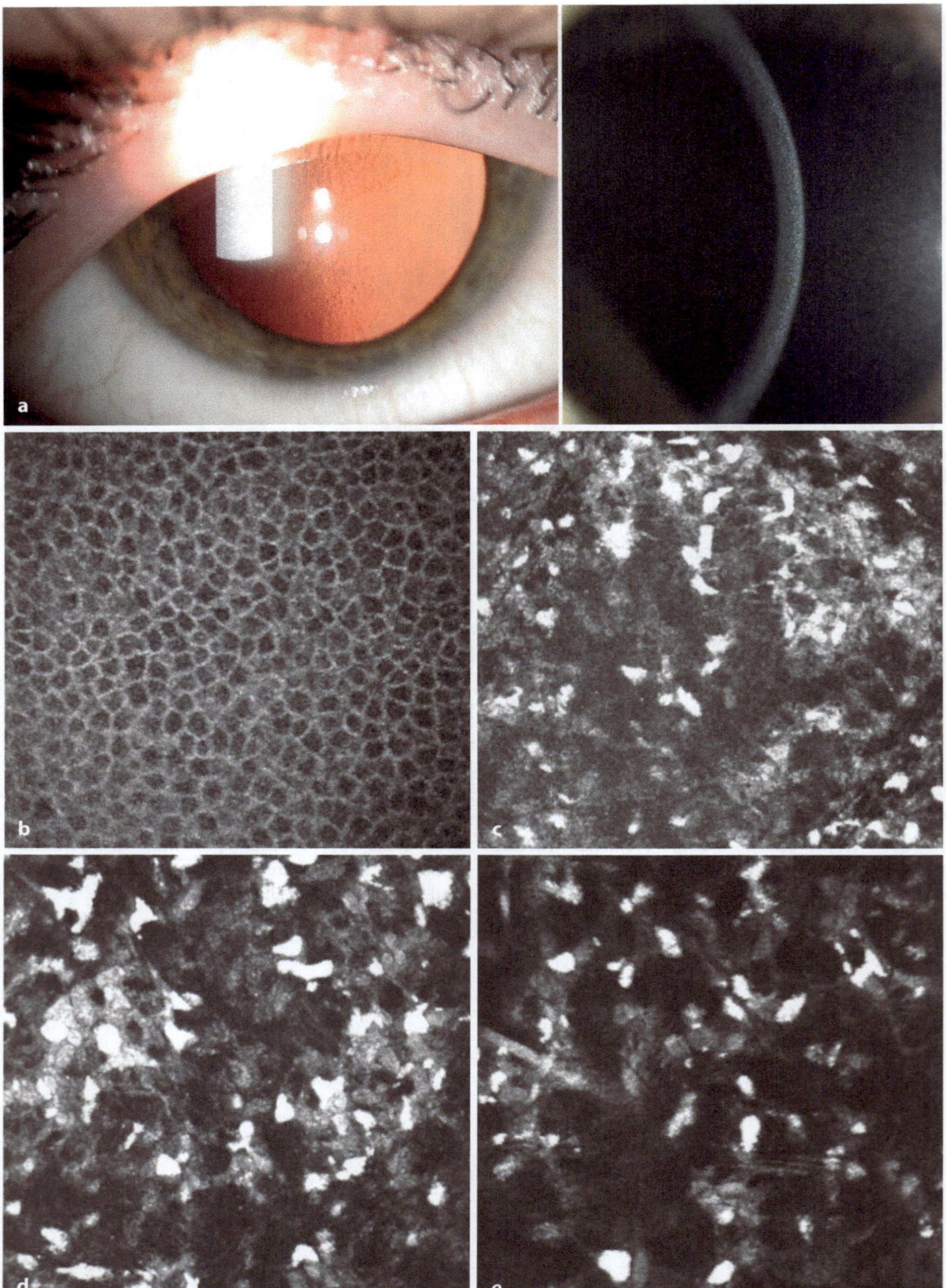

Fig. 5.65 a–f Cystinosis. a Slit-lamp photographs of an 18-year-old woman with dazzling; visual acuity 20/20. b–f Confocal images from the same patient: normal epithelium, 14 µm (b); highly reflective structures in the stroma at depths of 60 µm, 160 µm, and 540 µm, precise differentiation from keratocytes not possible (c–e)

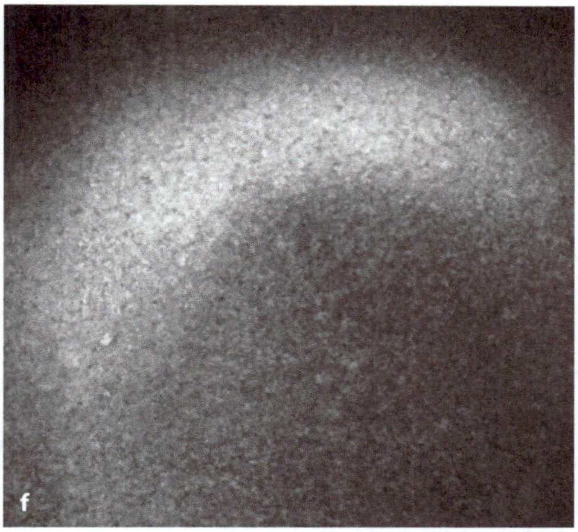

Fig. 5.65 (*continued*) normal endothelium (**f**)

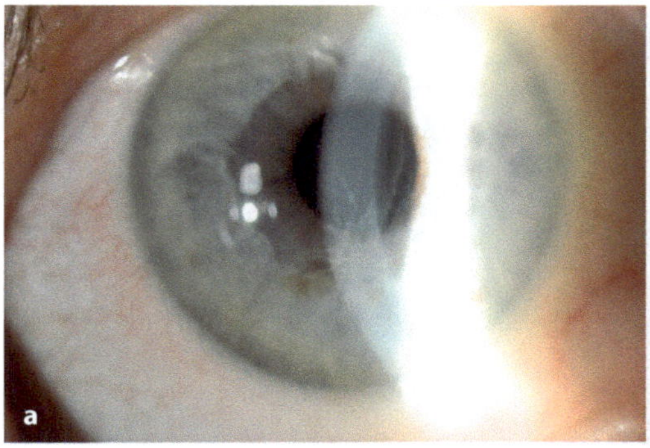

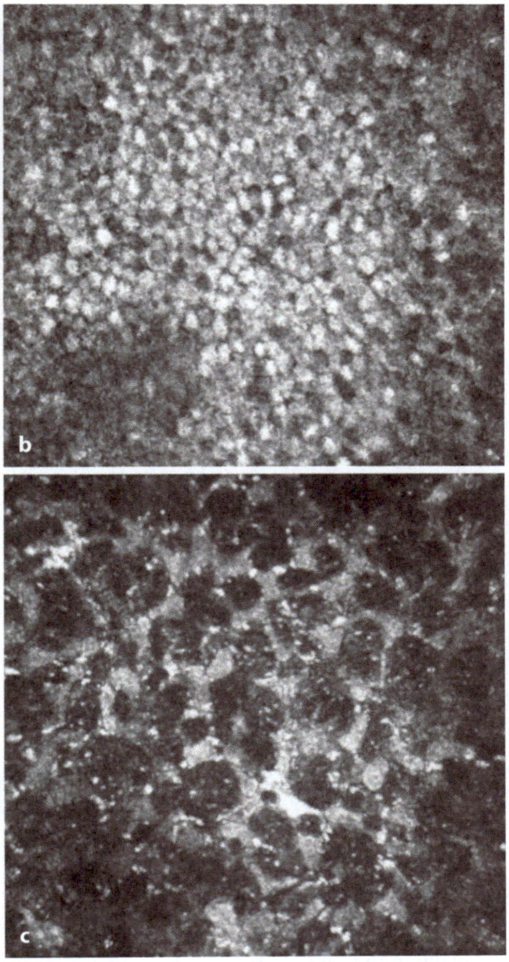

Fig. 5.66 Fabry's disease. **a** Slit-lamp photograph of a 40-year-old man. **b, c** Confocal images from the same patient: highly reflective round structures in the basal epithelium (z=28 μm; **b**) and the very anterior stroma (z=77 μm; c)

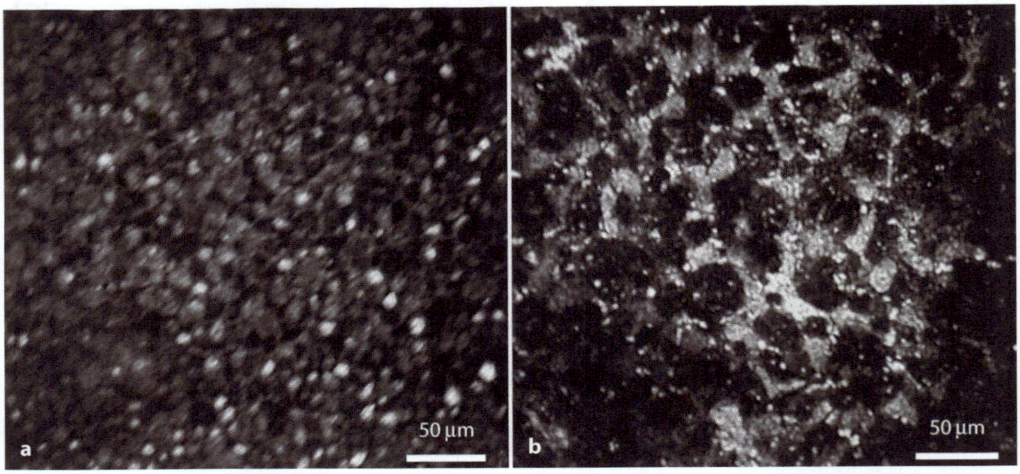

Fig. 5.67 Confocal images from a patient after amiodarone therapy: consistent finding of highly reflective round structures in the basal epithelium (z=20 µm)

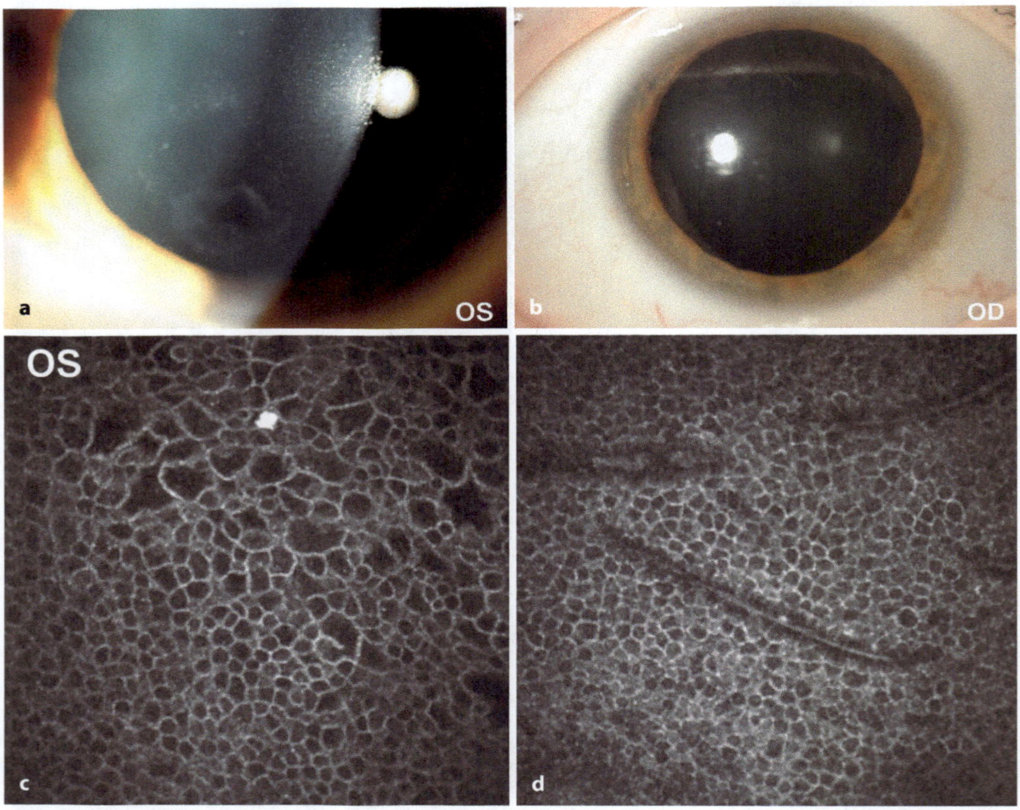

Fig. 5.68 a–h Interstitial keratitis. **a, b** Slit-lamp photographs of a 34-year-old man after bilateral interstitial keratitis. No signs of inflammation. On the left side, scar tissue is visible at the periphery and minimally in the central area (**a**); the cornea of the right eye is clear (**b**). **c–h** Confocal images of the left eye show structural changes at the level of lower intermediate-basal cells with formation of microscars and opacities (**c–e**) with an inconspicuous stroma (**f**).

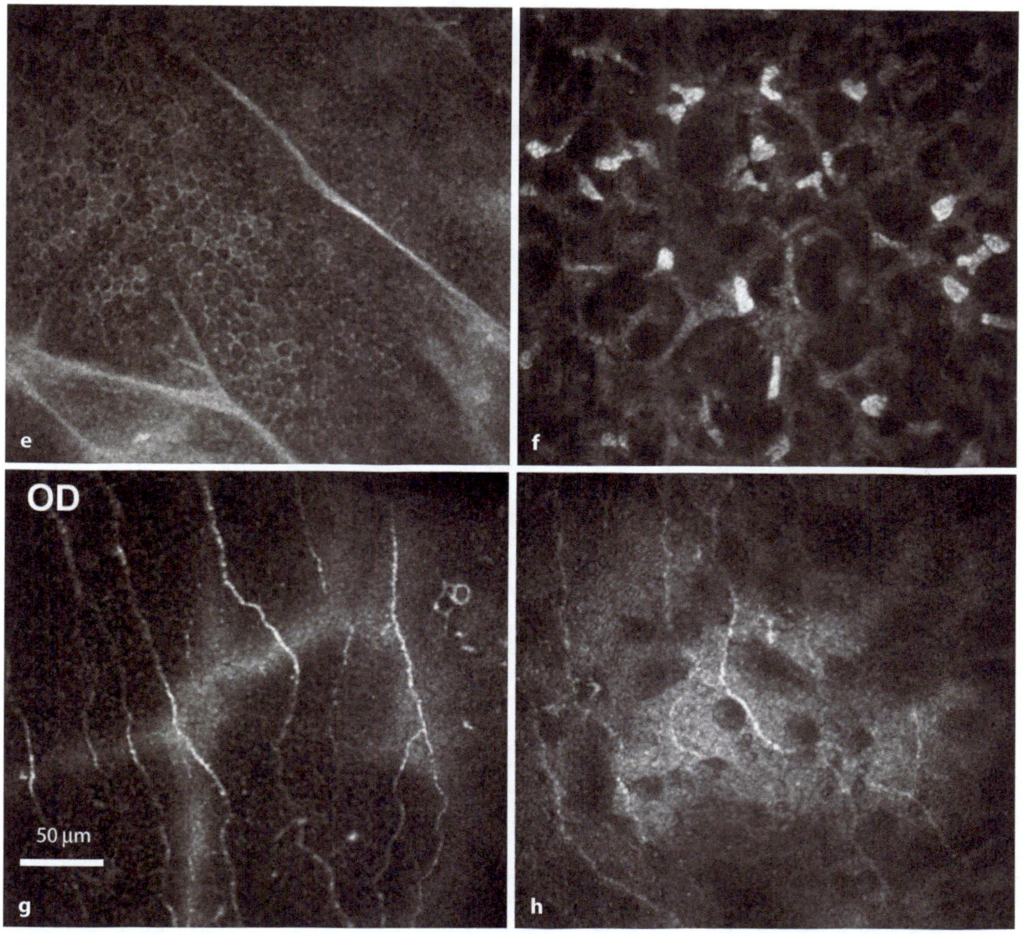

Fig. 5.68 (*continued*) The right side shows normal structure in the center (**g**) with peripheral dystrophic areas at the level of the subepithelial nerve plexus and Bowman's membrane (**h**)

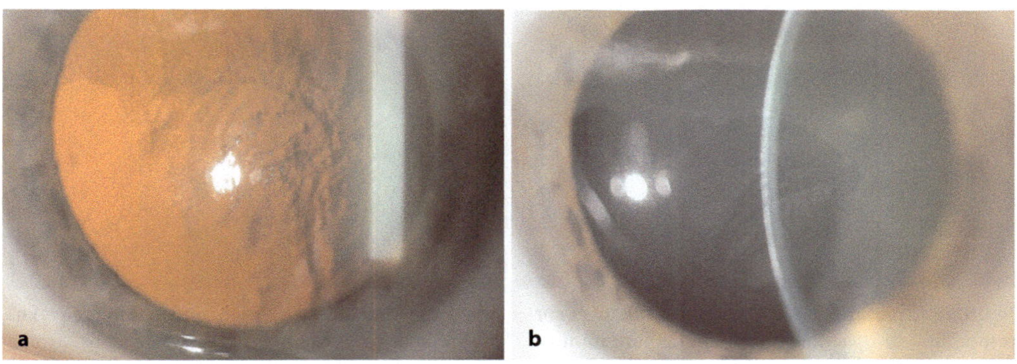

Fig. 5.69 a–h Interstitial keratitis. (**a, b**) Slit-lamp photographs of 61-year-old woman after interstitial keratitis. Scar tissues are visible in the anterior stroma on retro-illumination

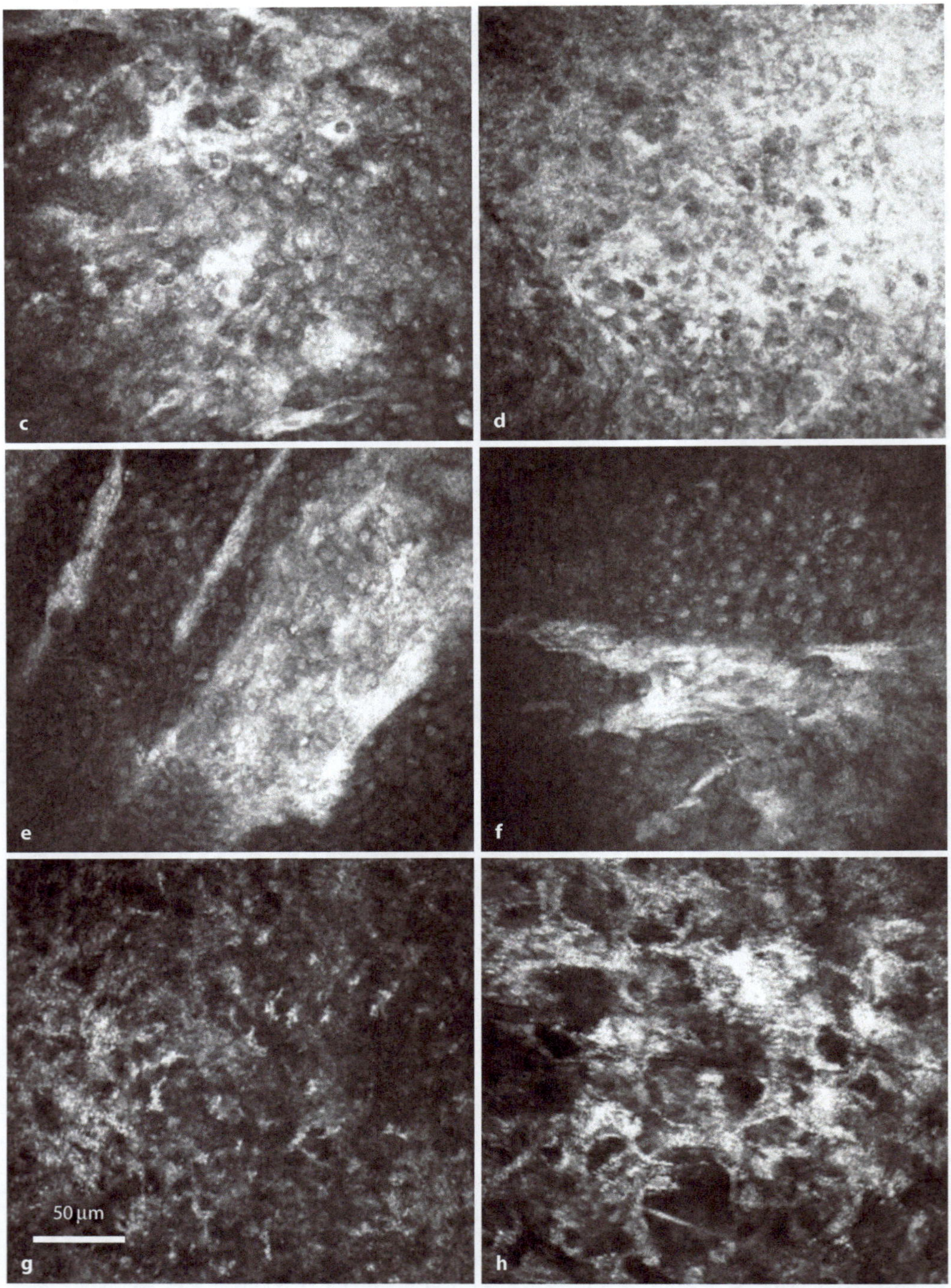

Fig. 5.69 (*continued*) **c–h** Pathological changes corresponding to opacities in the epithelium; minimal degree of corneal edema in the superficial and upper intermediate cells (**c, d**); highly reflective acellular structures at the level of the deeper intermediate and basal cells and absence of subepithelial nerve plexus in the area concerned (**e, f, g**). The anterior stroma displays an irregular collagen fiber structure with an atypical keratocyte pattern (**h**)

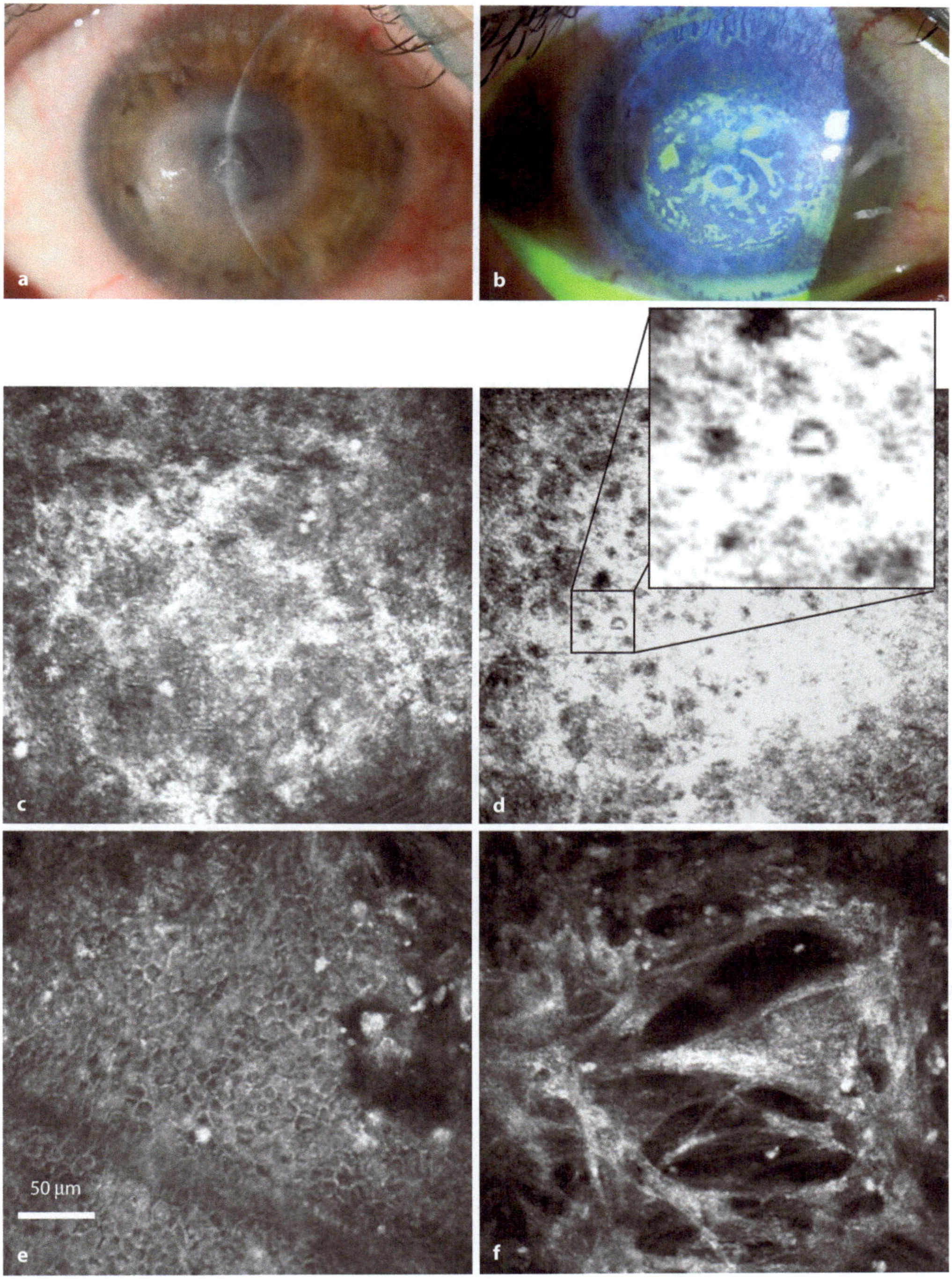

Fig. 5.70 *Acanthamoeba* keratitis. **a, b** Slit-lamp photographs from a 42-year-old woman with *Acanthamoeba* keratitis with epithelial and stromal ring infiltrate formation after dendritic epitheliopathy; corneal sensitivity is decreased. PCR (herpes zoster) and corneal culturing (for pathogens including *Acanthamoeba*) was negative. **c–f** Corneal microcysts (cystic stage of life cycle, round and up to 10 µm with double wall) are visible at the level of the deeper intermediate-basal cells (**c**) and in the anterior stroma (d). Same areas of the cornea after 3 months of specific therapy: no signs of cysts either in epithelium (**e**) or in stroma (**f**)

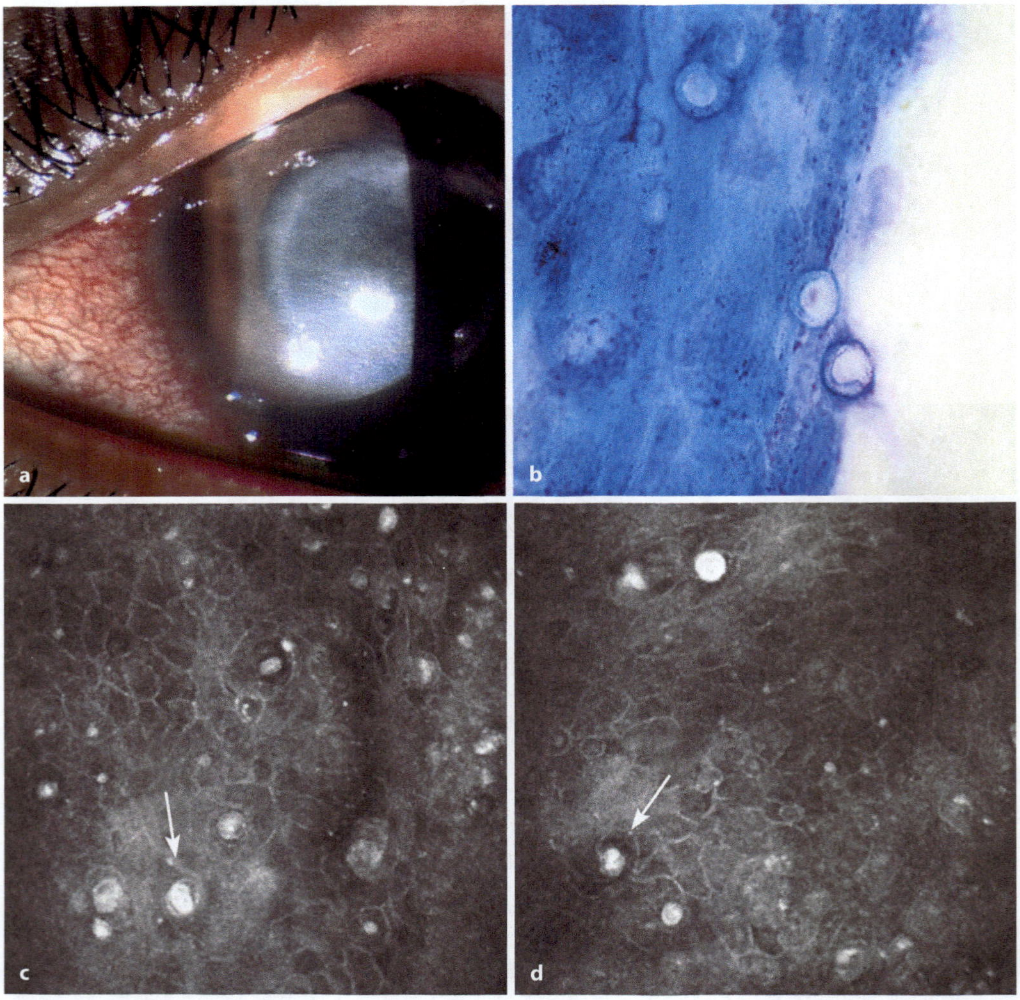

Fig. 5.71 *Acanthamoeba* keratitis. **a** Slit-lamp photograph of the right eye of a 22-year-old woman with *Acanthamoeba* keratitis. **b** *Acanthamoeba* cysts on cytological analysis. **c, d** *Acanthamoeba* cysts observed by confocal in vivo microscopy

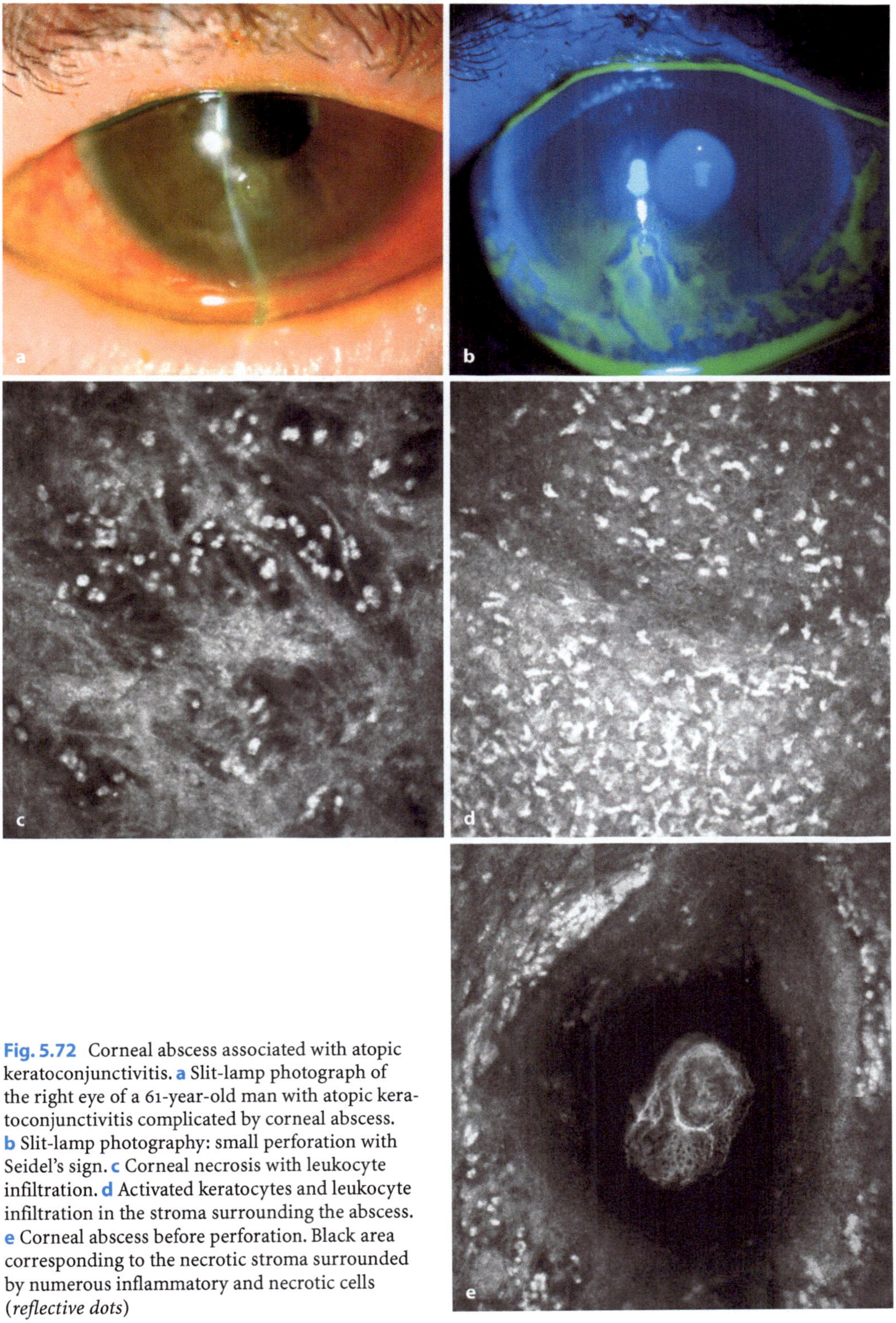

Fig. 5.72 Corneal abscess associated with atopic keratoconjunctivitis. **a** Slit-lamp photograph of the right eye of a 61-year-old man with atopic keratoconjunctivitis complicated by corneal abscess. **b** Slit-lamp photography: small perforation with Seidel's sign. **c** Corneal necrosis with leukocyte infiltration. **d** Activated keratocytes and leukocyte infiltration in the stroma surrounding the abscess. **e** Corneal abscess before perforation. Black area corresponding to the necrotic stroma surrounded by numerous inflammatory and necrotic cells (*reflective dots*)

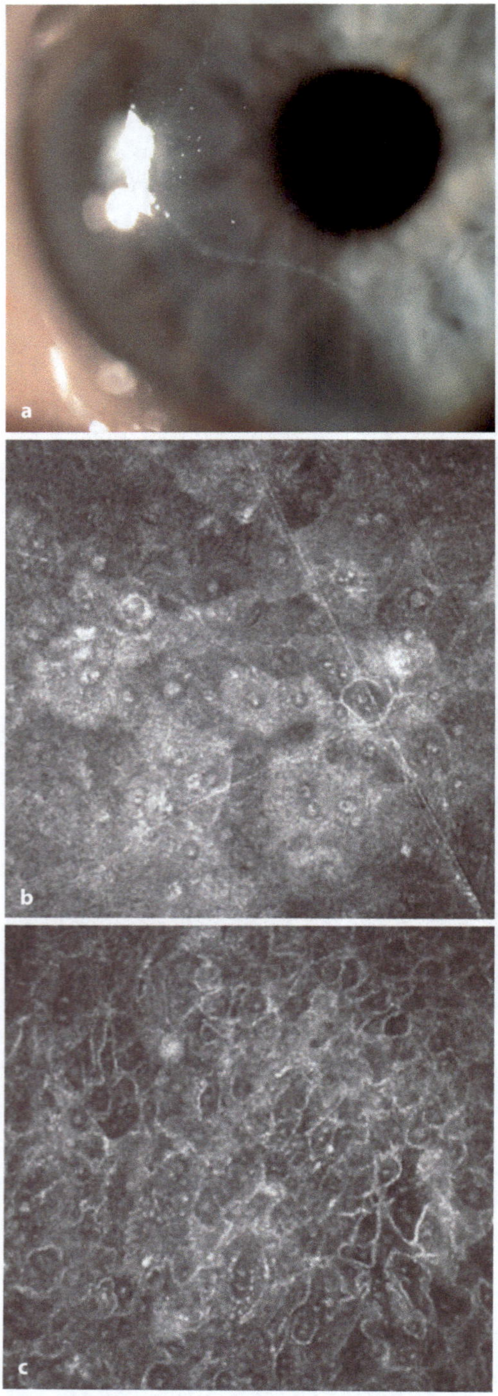

Fig. 5.73 Viral keratitis. a Slit-lamp photograph of the right eye of a 44-year-old man with viral keratitis, with immune infiltrates surrounding the keratitis. b, c Epithelial cell abnormalities, reflective nuclei, and binucleated cells

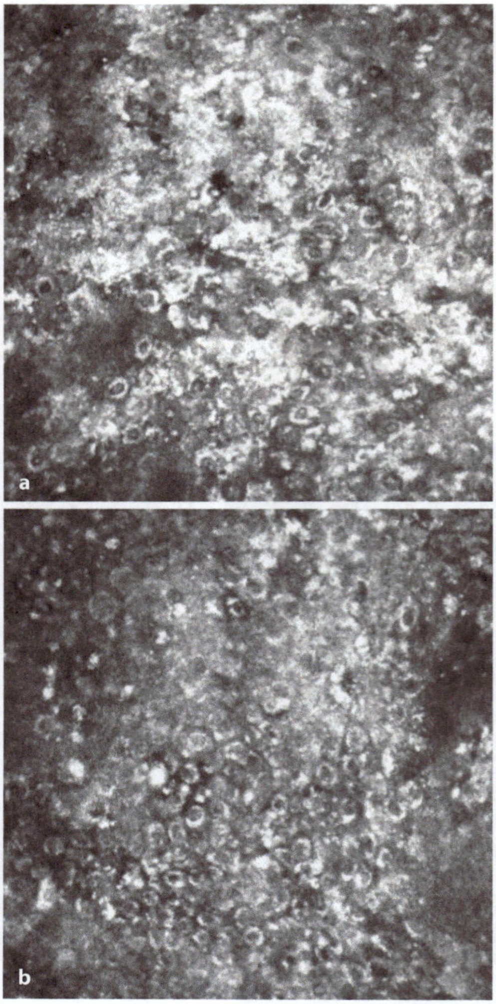

Fig. 5.74 Corneal wound healing: corneal ulcer. Multiple cellular abnormalities at the ulcer edges: hyperreflective cells, reflective nuclei, and polymorphism

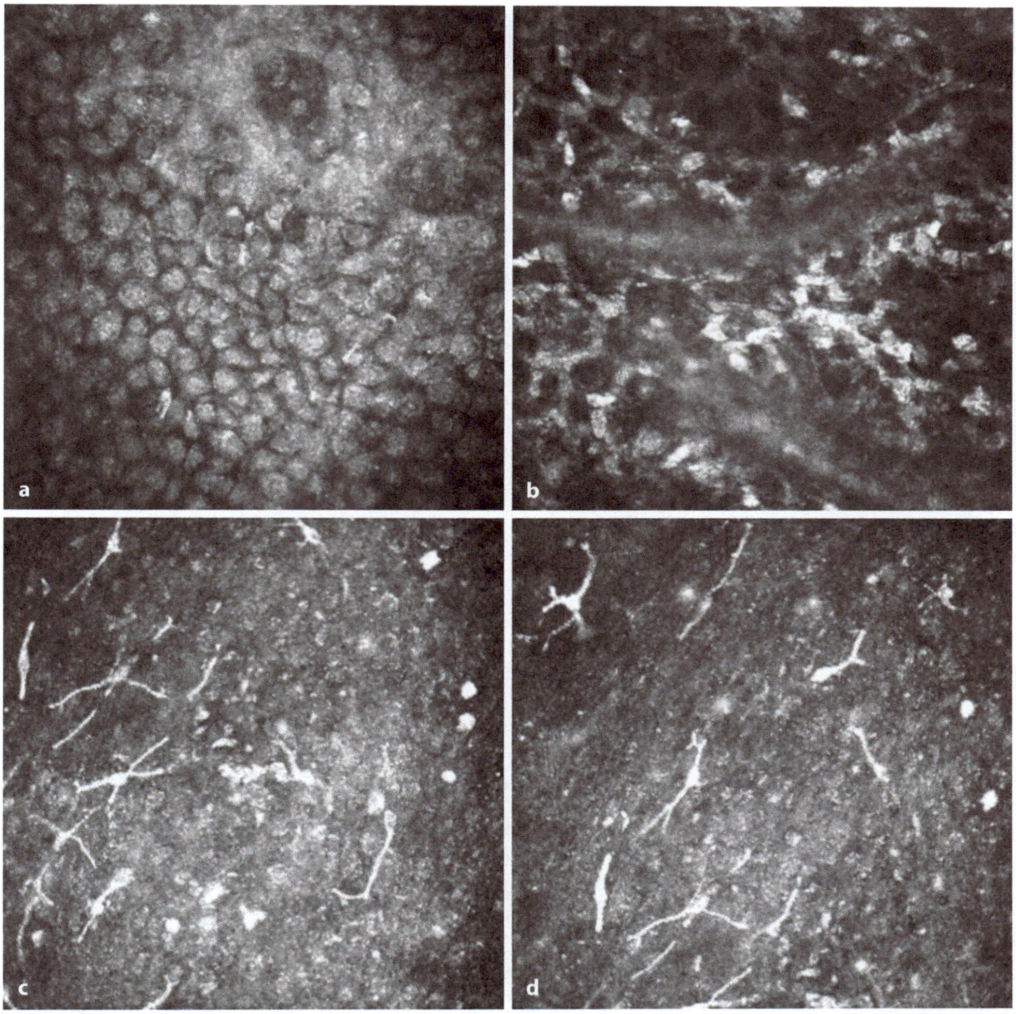

Fig. 5.75 Corneal wound healing: recently healed ulcer. a Activated corneal epithelium with reflective cells. b Numerous activated keratocytes in the anterior stroma below Bowman's layer. c, d Multiple inflammatory cells: dendritic cells

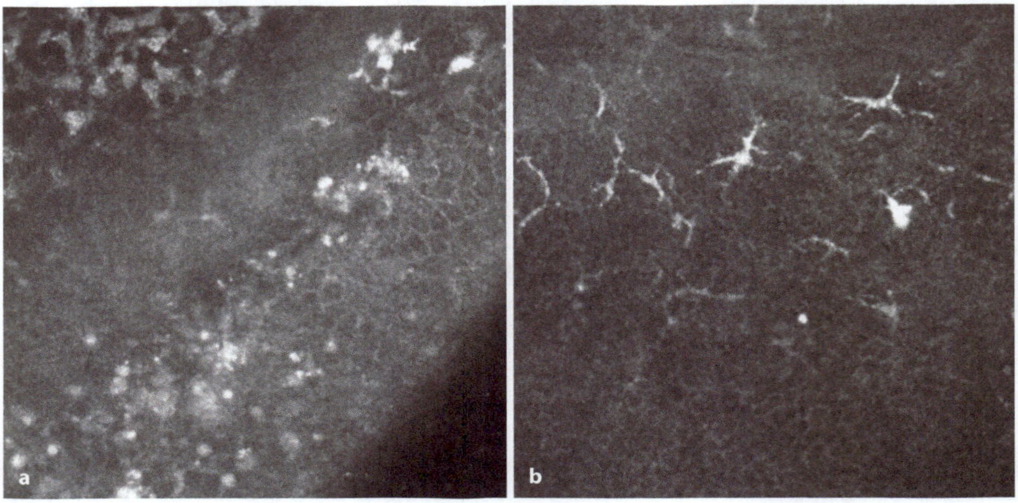

Fig. 5.76 Penetrating keratoplasty. a Oblique section showing the corneal graft: irregular and hyperreflective epithelium and activated stroma. b Dendritic cells at the edge of the corneal graft

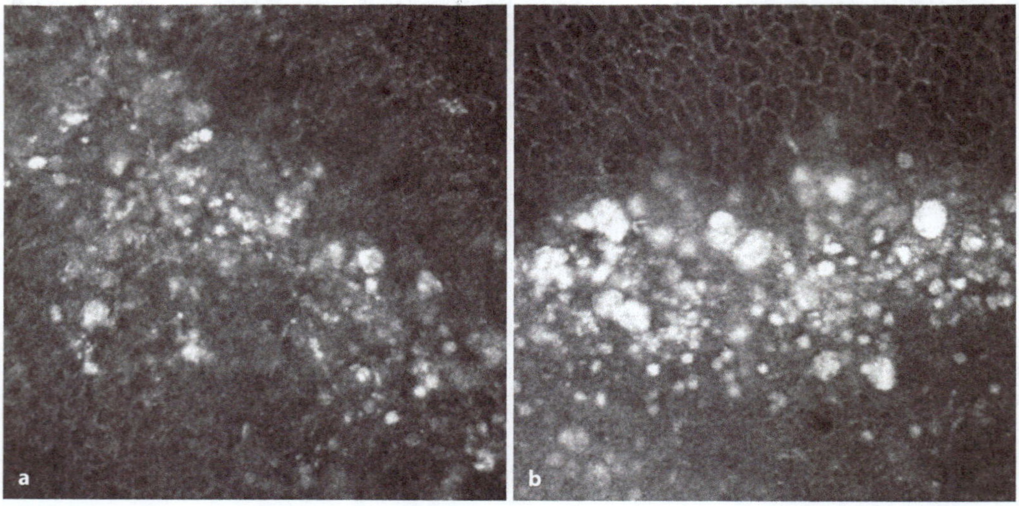

Fig. 5.77 Penetrating keratoplasty. Immune ring at the edge of a corneal graft in a case of epithelial rejection, with precipitates (*reflective dots*) in the corneal epithelium

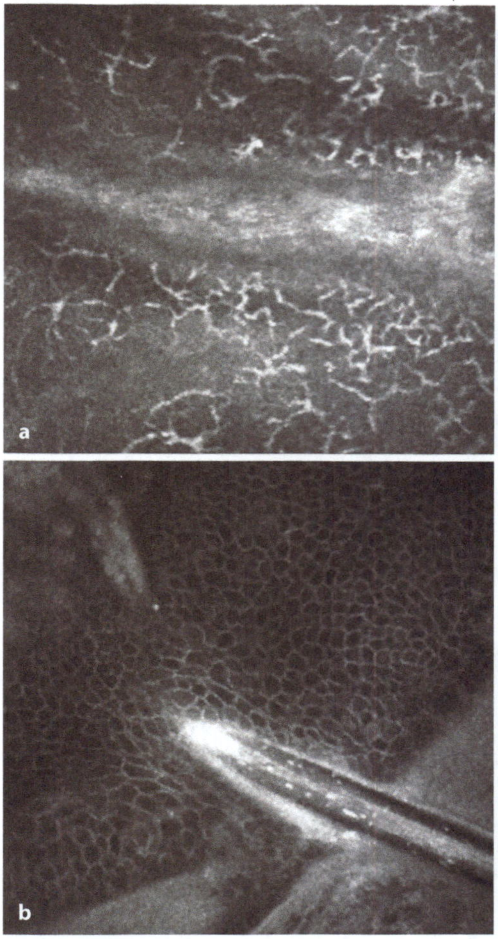

Fig. 5.78 Penetrating keratoplasty. **a** Periphery of the corneal graft: inflammatory cells on both sides. **b** Corneal suture penetrating the basal corneal epithelium

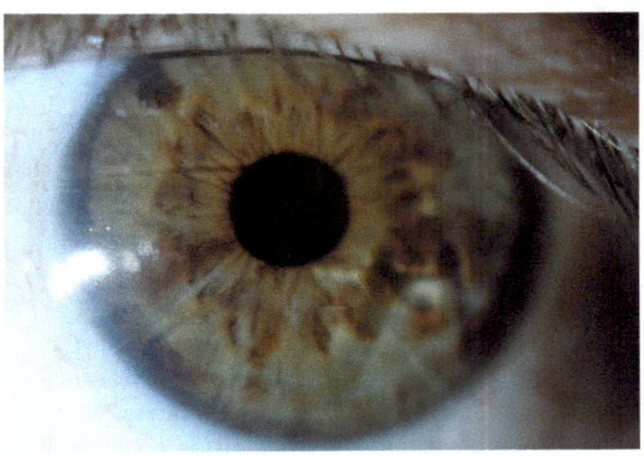

Fig. 5.79 Refractive surgery: slit-lamp photograph of a radial keratotomy performed 15 years previously

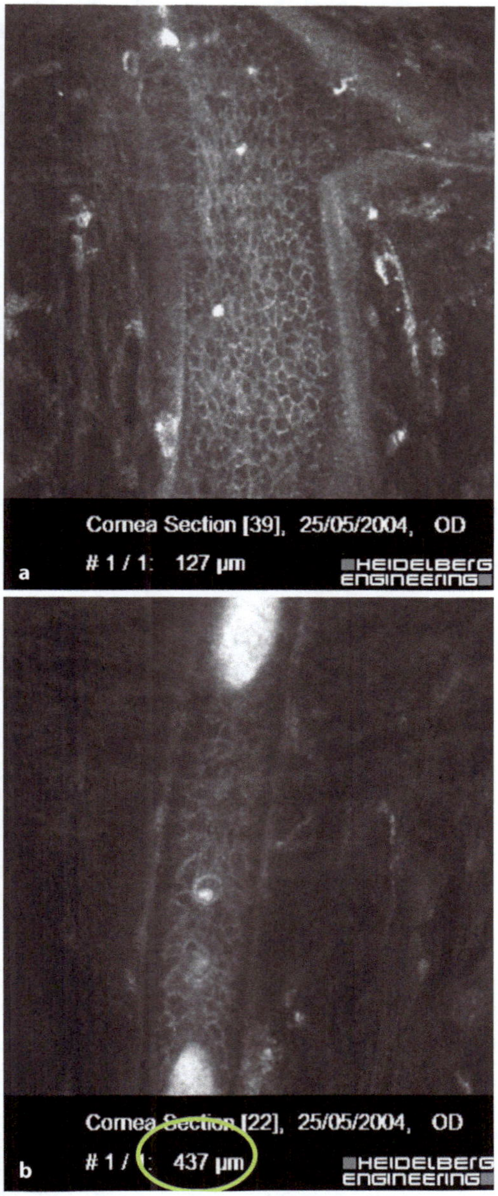

Fig. 5.80 Same patient as in Fig. 5.79: confocal microscopy images of a furrow. Note the presence of corneal epithelial cells located very deeply in the bed of the incision

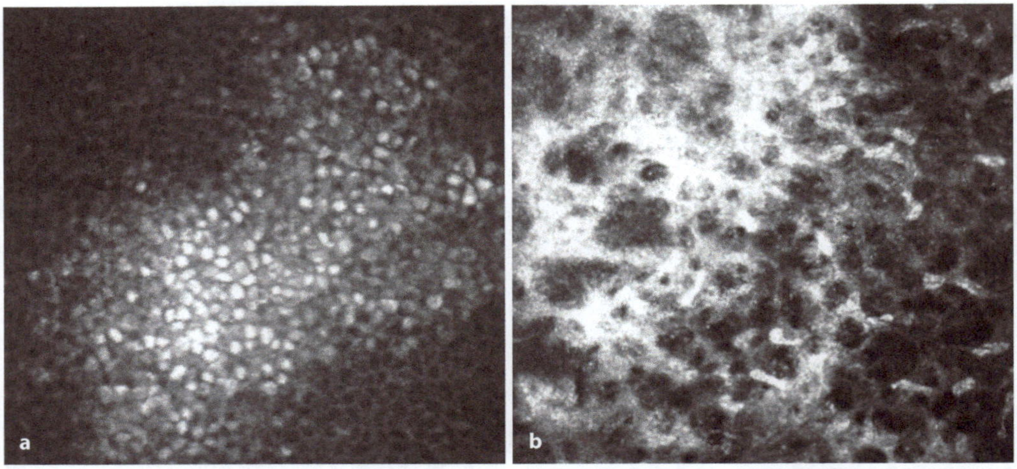

Fig. 5.81 Refractive surgery: photorefractive keratectomy. a Corneal haze 3 years after photorefractive keratectomy. Activated corneal epithelial cells are abnormally hyperreflective. b At a deeper level, fibrosis of the epithelial basement membrane, activated keratocytes, and hyperreflective deposits

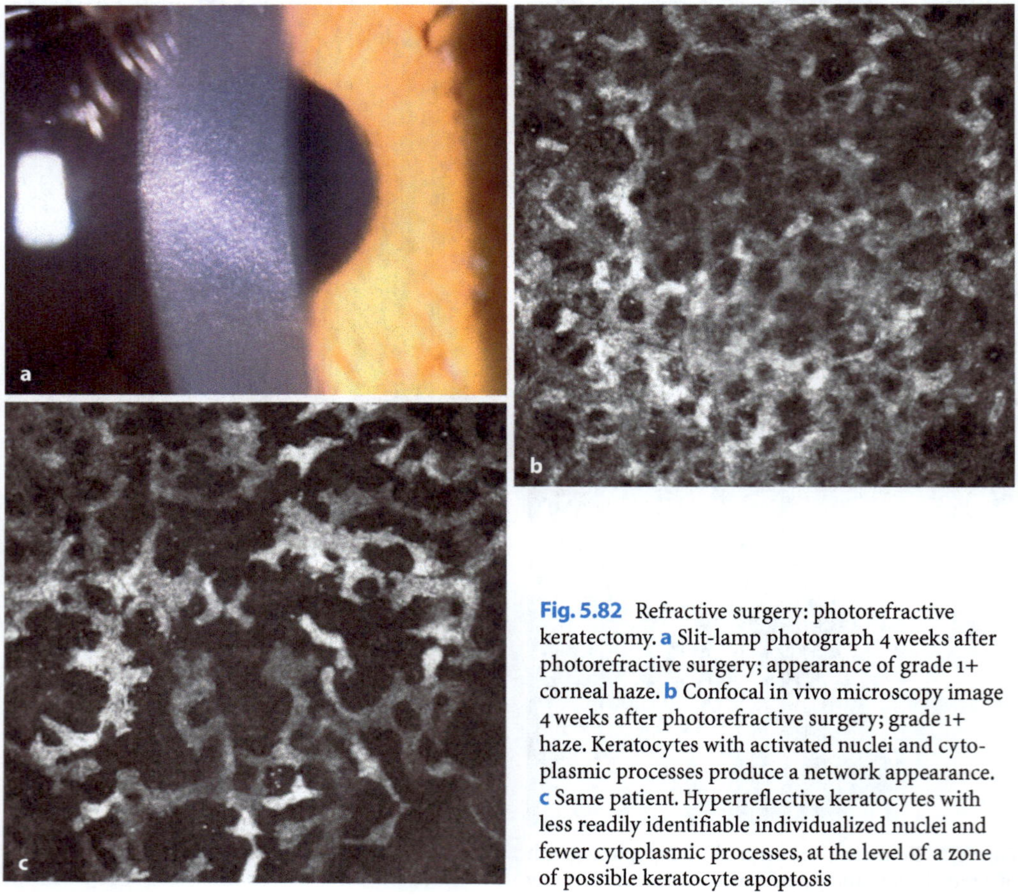

Fig. 5.82 Refractive surgery: photorefractive keratectomy. a Slit-lamp photograph 4 weeks after photorefractive surgery; appearance of grade 1+ corneal haze. b Confocal in vivo microscopy image 4 weeks after photorefractive surgery; grade 1+ haze. Keratocytes with activated nuclei and cytoplasmic processes produce a network appearance. c Same patient. Hyperreflective keratocytes with less readily identifiable individualized nuclei and fewer cytoplasmic processes, at the level of a zone of possible keratocyte apoptosis

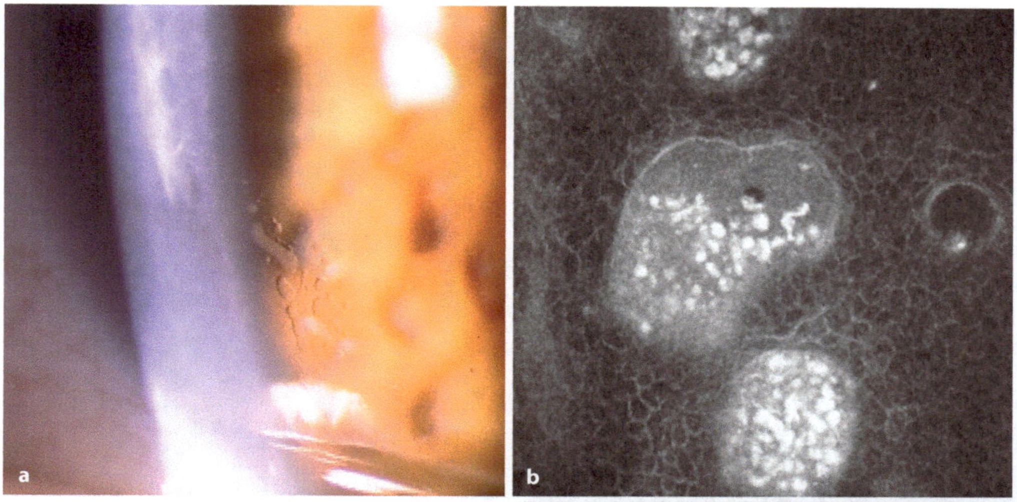

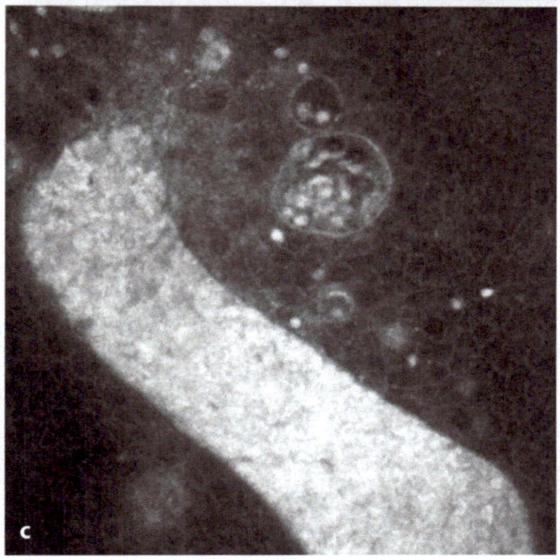

Fig. 5.83 Refractive surgery: laser-assisted in situ keratomileusis (LASIK). **a** Slit-lamp photograph of a 25-year-old woman after LASIK surgery. Epithelial invasion at the flap periphery. Fibrotic scar nasal to the flap periphery. **b** Epithelial invasion cysts containing still-identifiable epithelial cells. **c** Large epithelial invasion cysts containing dead epithelial cells surrounded by small cysts

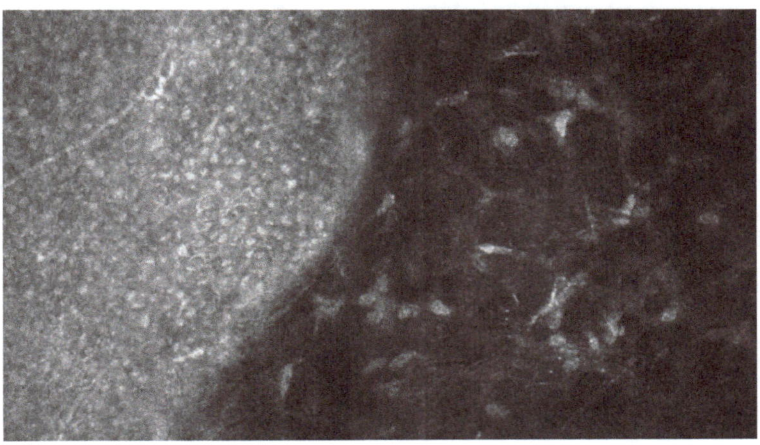

Fig. 5.84 Refractive surgery: laser-assisted in situ keratomileusis (LASIK). Confocal microscopy image of epithelial invasion at the interface with the stroma

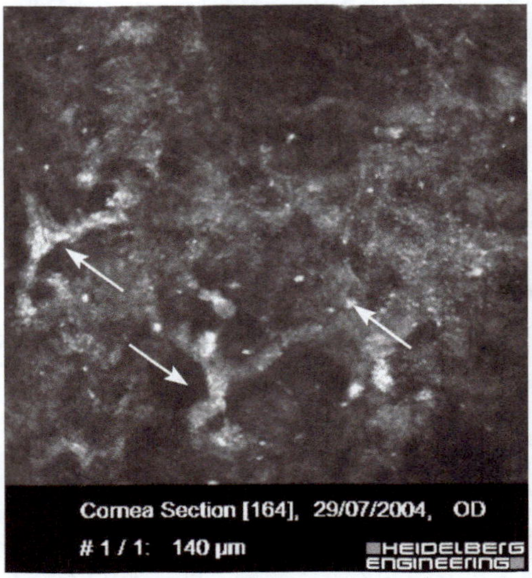

Fig. 5.85 Refractive surgery: laser-assisted in situ keratomileusis (LASIK). Confocal microscopy image of keratocyte activation 1 week after LASIK

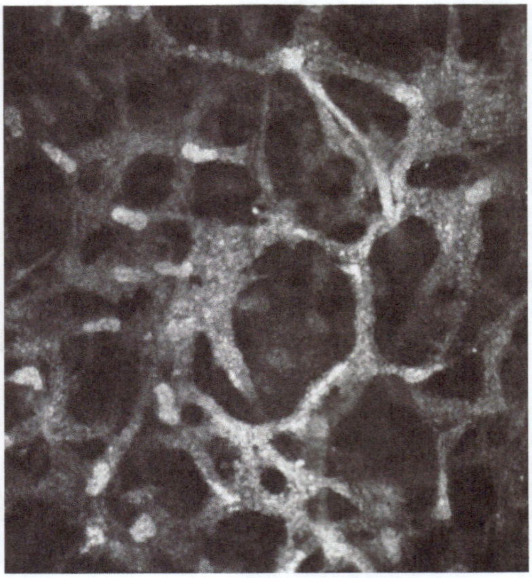

Fig. 5.86 Refractive surgery: laser-assisted in situ keratomileusis (LASIK). Confocal microscopy image of keratocyte activation 1 month after LASIK

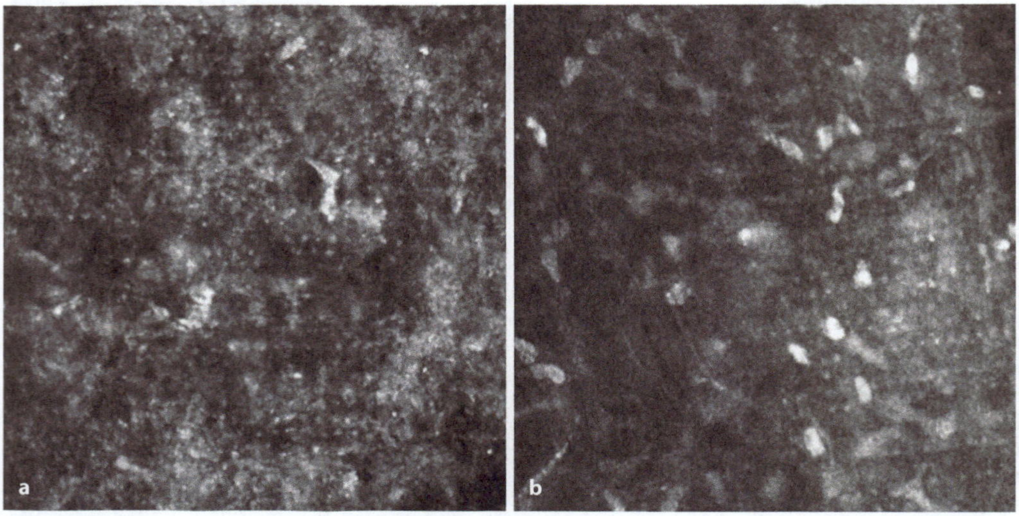

Fig. 5.87 Refractive surgery: laser-assisted in situ keratomileusis. **a, b** Early after flap creation with femtosecond laser (IntraLase) showing the existence of regularly disposed spots, probably corresponding to the microcysts created by laser impact. These spots disappear in the weeks following the procedure

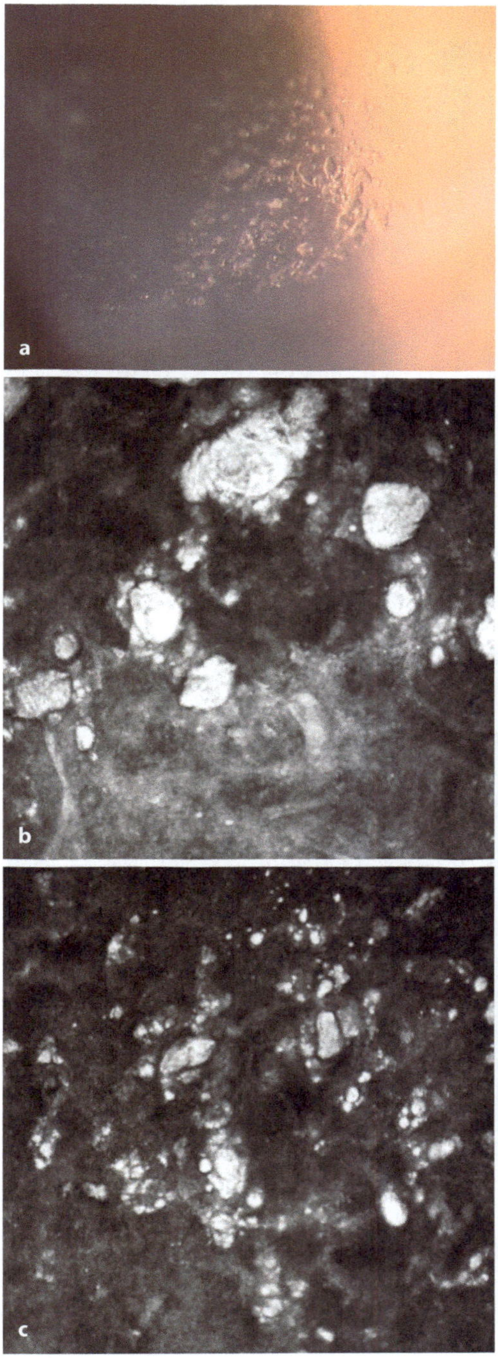

Fig. 5.88 Refractive surgery: laser-assisted in situ keratomileusis (LASIK). **a** Slit-lamp photograph of brightly reflective deposits at the LASIK interface. **b, c** Confocal in vivo microscopy image of these reflective deposits at the LASIK interface

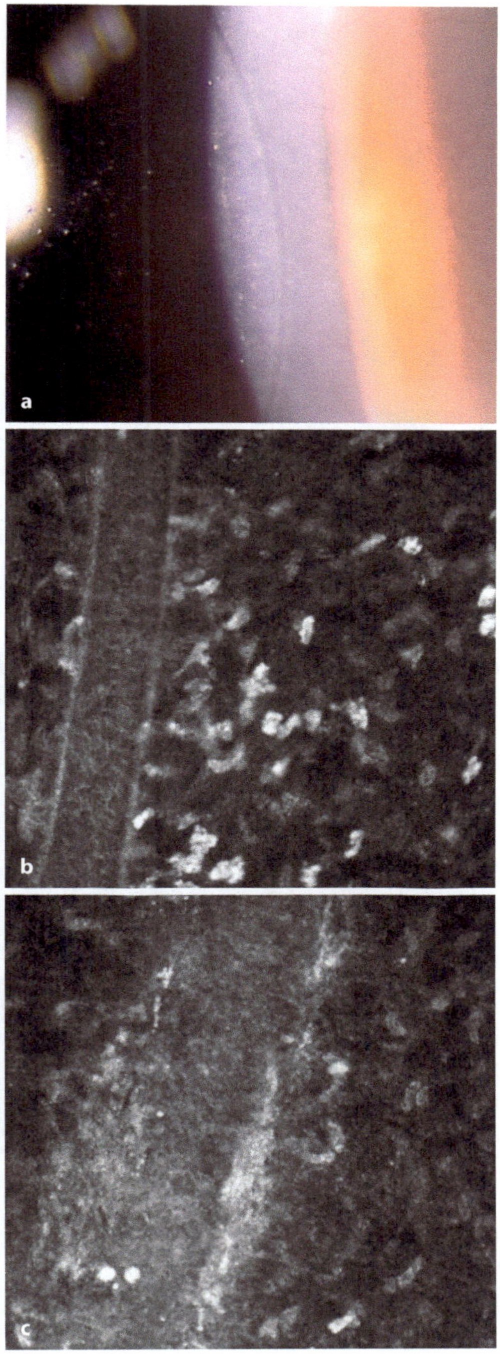

Fig. 5.89 Refractive surgery: laser-assisted in situ keratomileusis (LASIK). **a** Slit-lamp photograph of the flap margin after the IntraLase procedure. **b** Corresponding confocal in vivo microscopy image. **c** Confocal in vivo microscopy image showing the flap margin created by the mechanical microkeratome. Note the less clearly recognizable incision compared with IntraLase

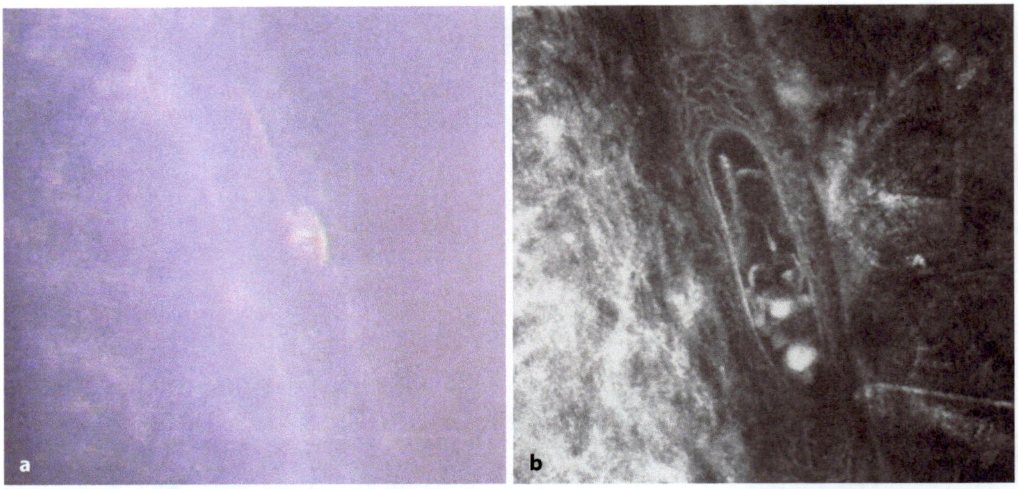

Fig. 5.90 Refractive surgery: laser-assisted in situ keratomileusis. **a** Slit-lamp photograph of an epithelial cyst induced by an inclusion at the flap periphery. **b** Corresponding confocal in vivo microscopy image

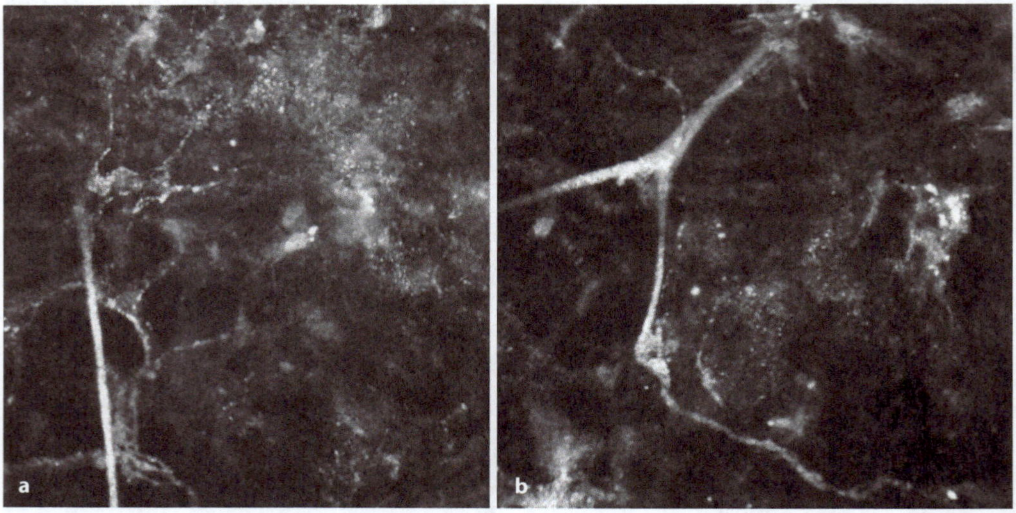

Fig. 5.91 Refractive surgery: laser-assisted in situ keratomileusis at 6 months. Corneal nerves are again visible ahead of the interface

Fig. 5.92 Refractive surgery: laser-assisted in situ keratomileusis. a Slit-lamp photograph of fibrosis at the flap margin. b Corresponding confocal in vivo microscopy image

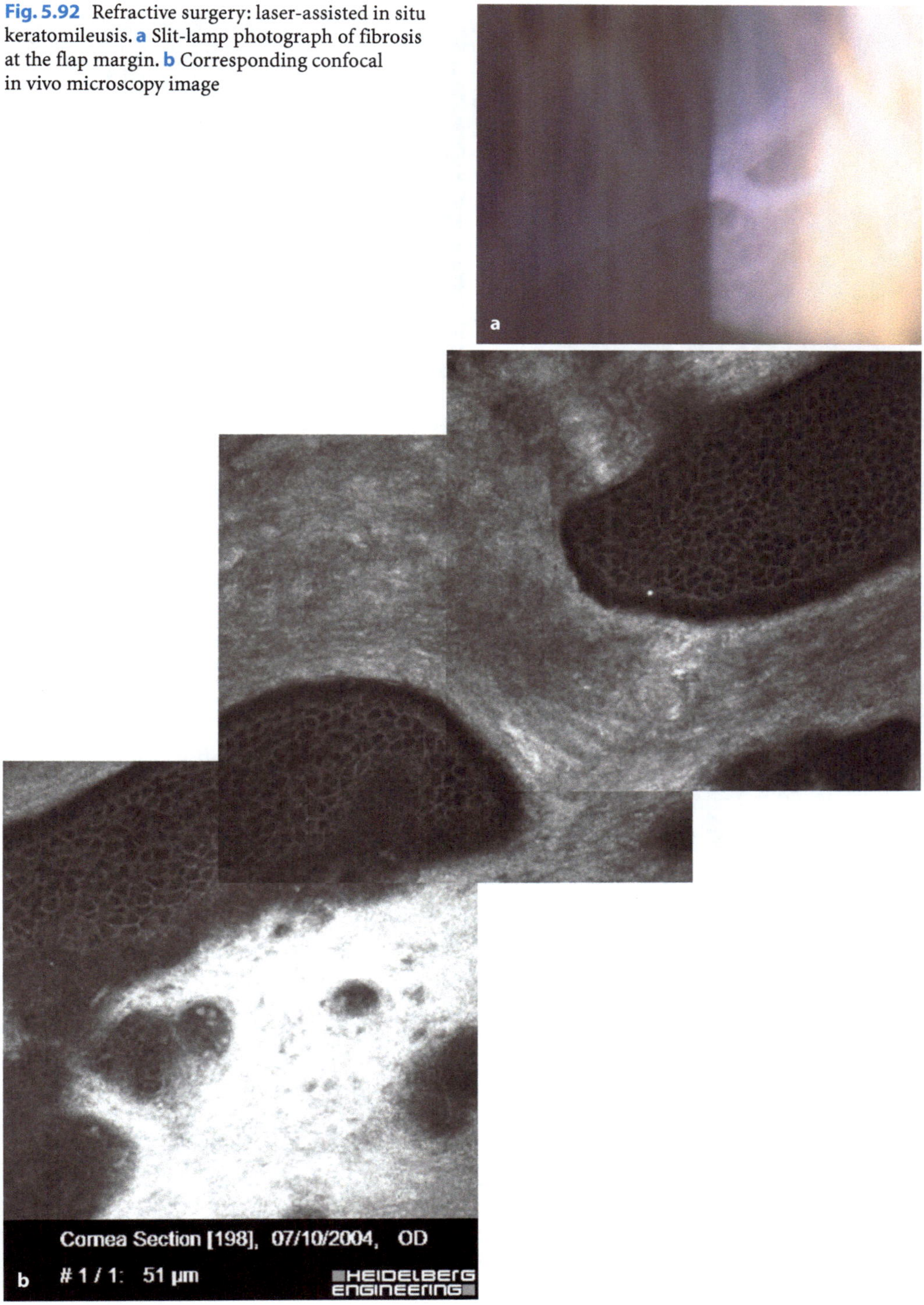

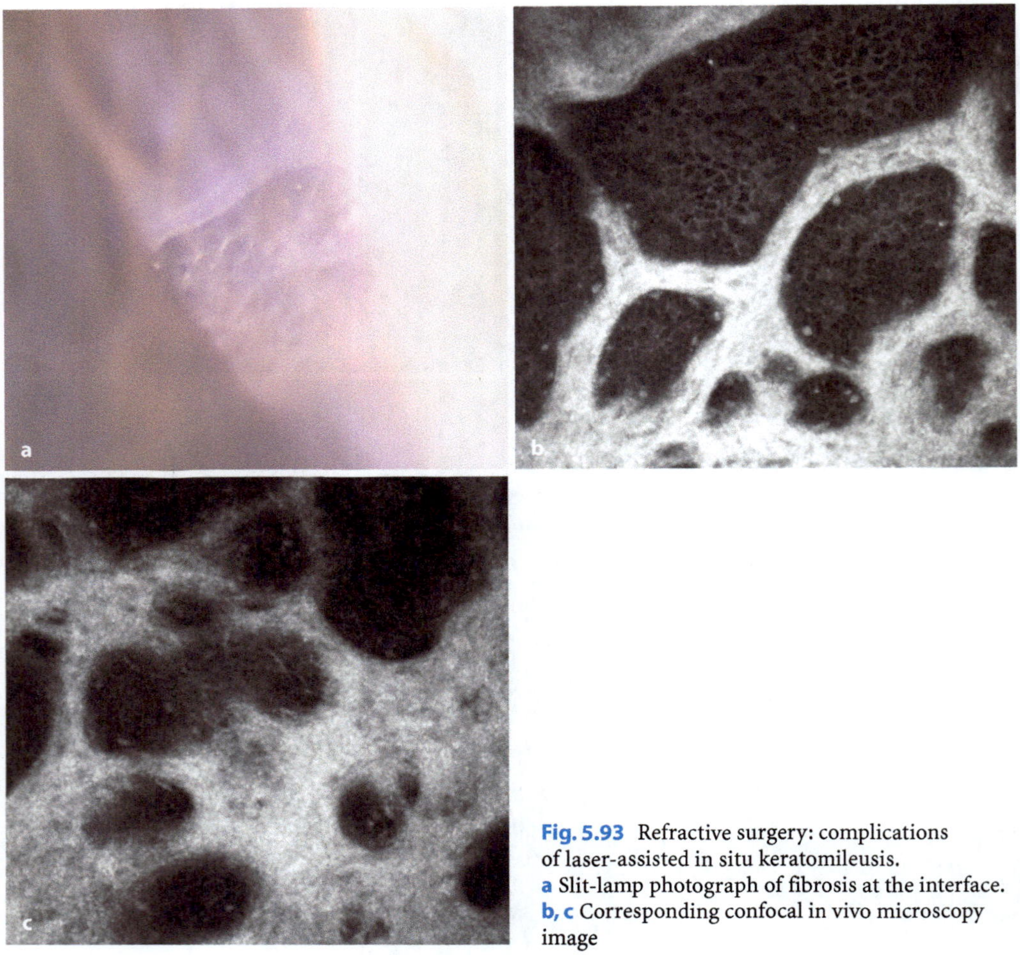

Fig. 5.93 Refractive surgery: complications of laser-assisted in situ keratomileusis. **a** Slit-lamp photograph of fibrosis at the interface. **b, c** Corresponding confocal in vivo microscopy image

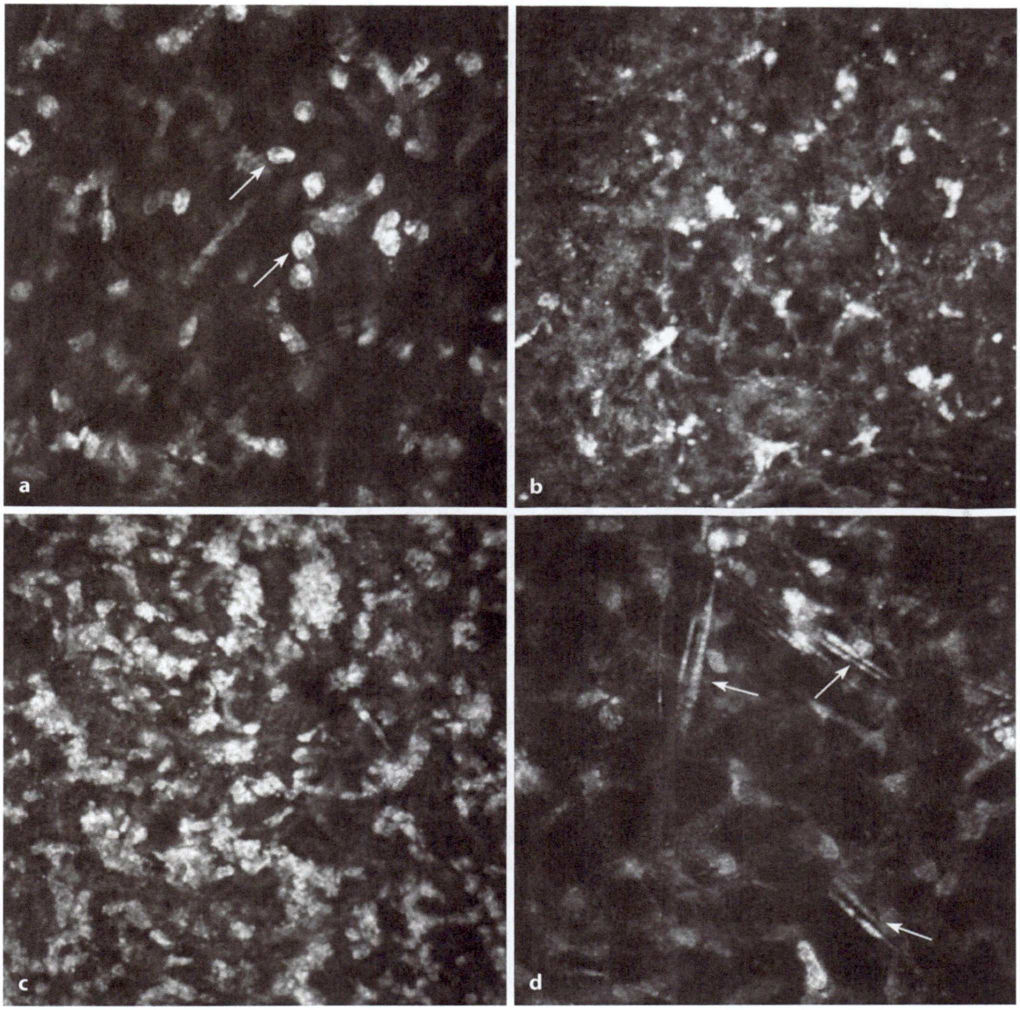

Fig. 5.94 Refractive surgery: complications of laser-assisted in situ keratomileusis. a Sands of Sahara syndrome on day 5. Reflective inflammatory cells. b Residual inflammatory cells at the interface on day 7. c Multiple reflective cells, possibly corresponding to neutrophils. d Numerous switchlike structures possibly corresponding to degradation products of neutrophils

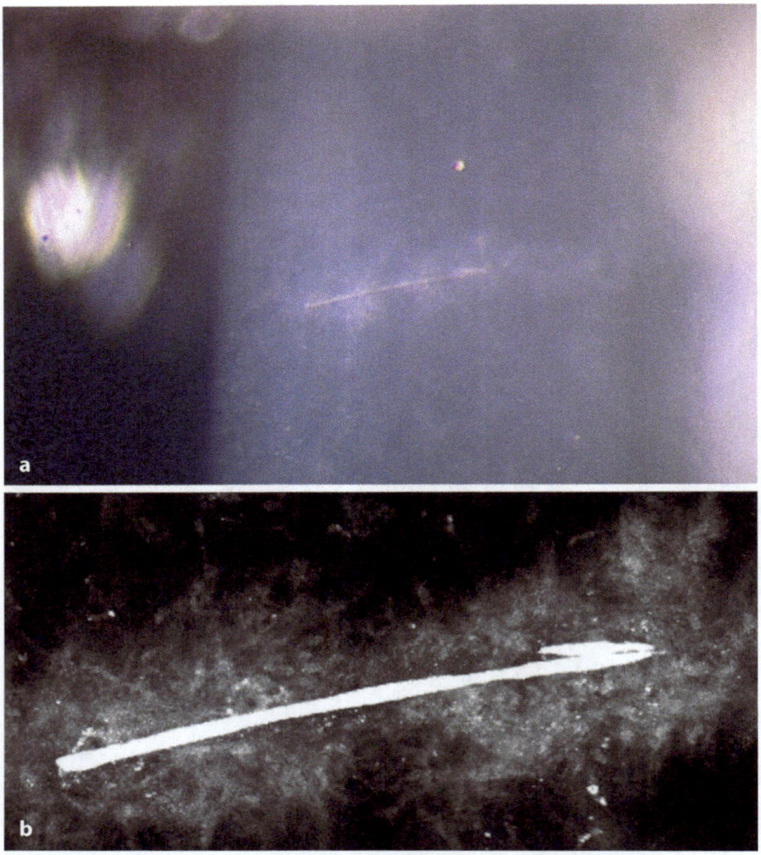

Fig. 5.95 Refractive surgery: complications of laser-assisted in situ keratomileusis).
a Slit-lamp photograph of a fiber from a microsponge at the interface.
b Corresponding confocal in-vivo microscopy image

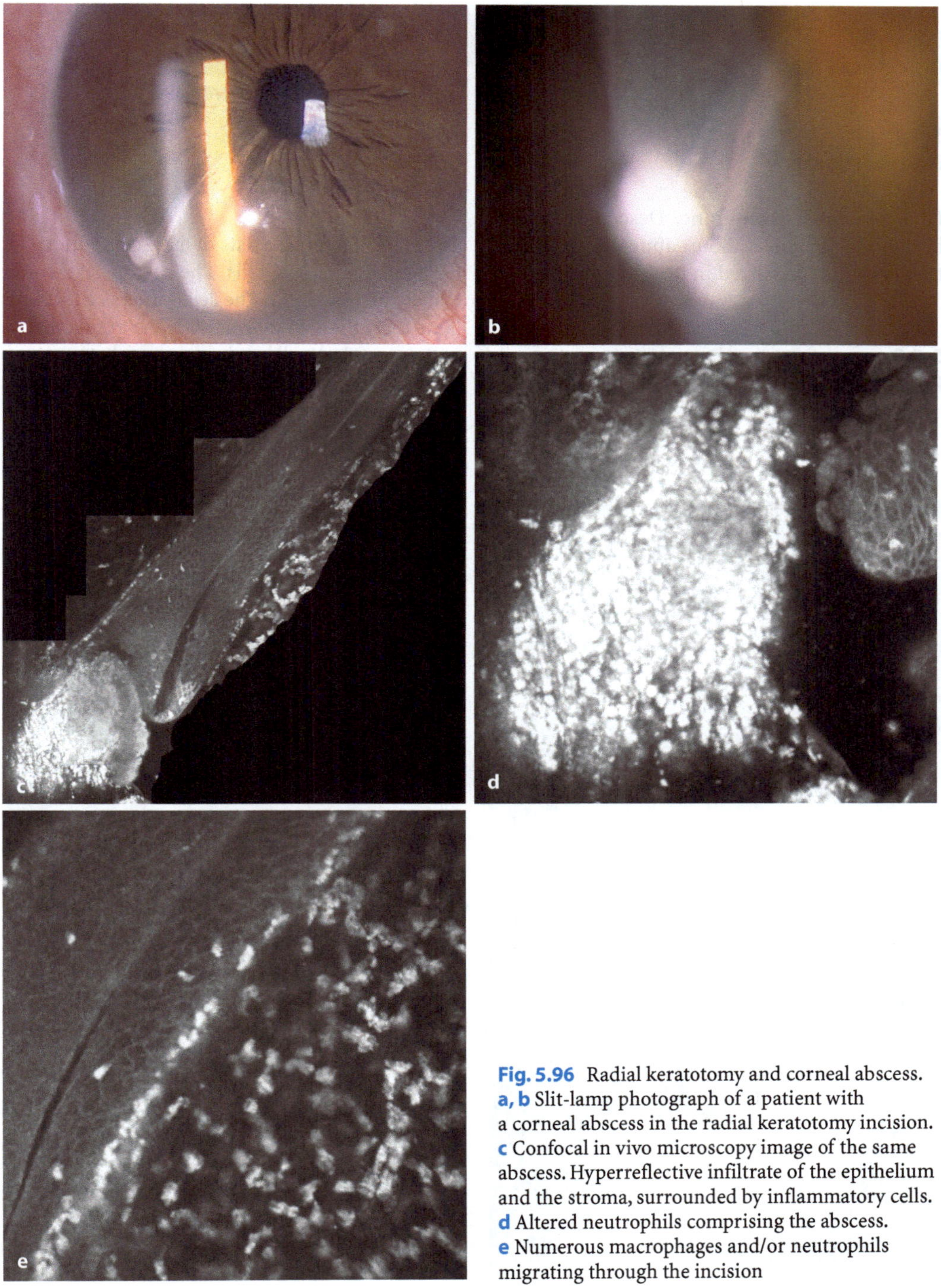

Fig. 5.96 Radial keratotomy and corneal abscess. a, b Slit-lamp photograph of a patient with a corneal abscess in the radial keratotomy incision. c Confocal in vivo microscopy image of the same abscess. Hyperreflective infiltrate of the epithelium and the stroma, surrounded by inflammatory cells. d Altered neutrophils comprising the abscess. e Numerous macrophages and/or neutrophils migrating through the incision

5.6 Descemet's Membrane

5.6.1 Normal Anatomy

Like Bowman's membrane, Descemet's membrane has an amorphous appearance and is therefore not visualized in healthy subjects.

5.6.2 Pathological Findings

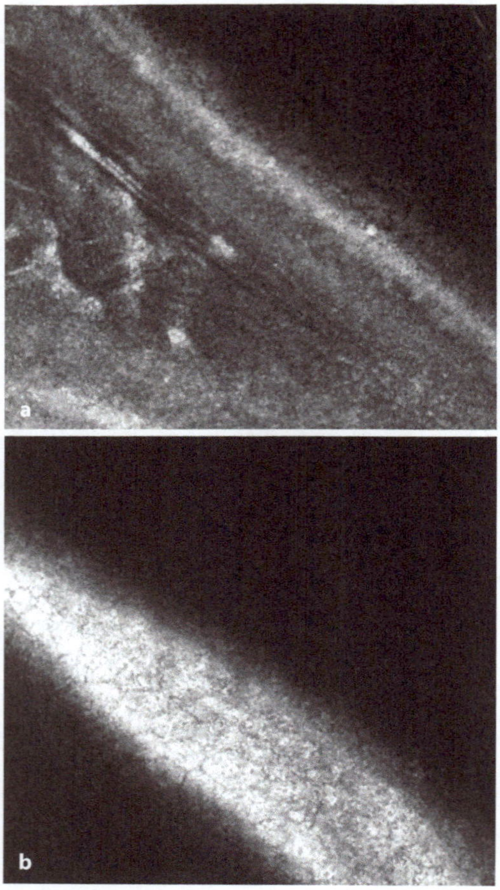

Fig. 5.97 Fold in Descemet's membrane. **a** Due to the oblique section, the changing structure is visible between the posterior stroma and the endothelium. **b** The borders of the endothelial cells are fading

5.7 Endothelial Cells and Trabecular Structures

5.7.1 Normal Anatomy

The endothelium consists of a regular pattern of hexagonal reflective cells. The cell nuclei cannot usually be visualized. The cell borders reflect less light than the cytoplasm, with the result that a network of dark cell borders appears between areas of bright cytoplasm. Endothelial cell density can be determined by counting (Figs. 5.98–5.102).

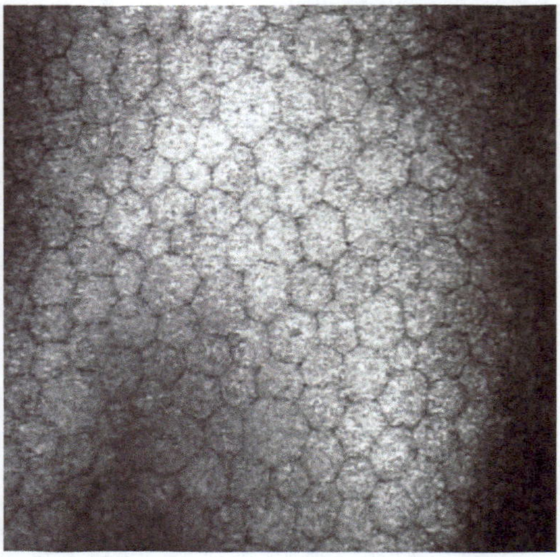

Fig. 5.98 Endothelium. A monolayer of regularly arranged hexagonal cells completely covering the posterior surface of the cornea. Unlike the basal cells, these cells have a brightly reflecting cytoplasm and dark cell borders. The cell nucleus is not visible

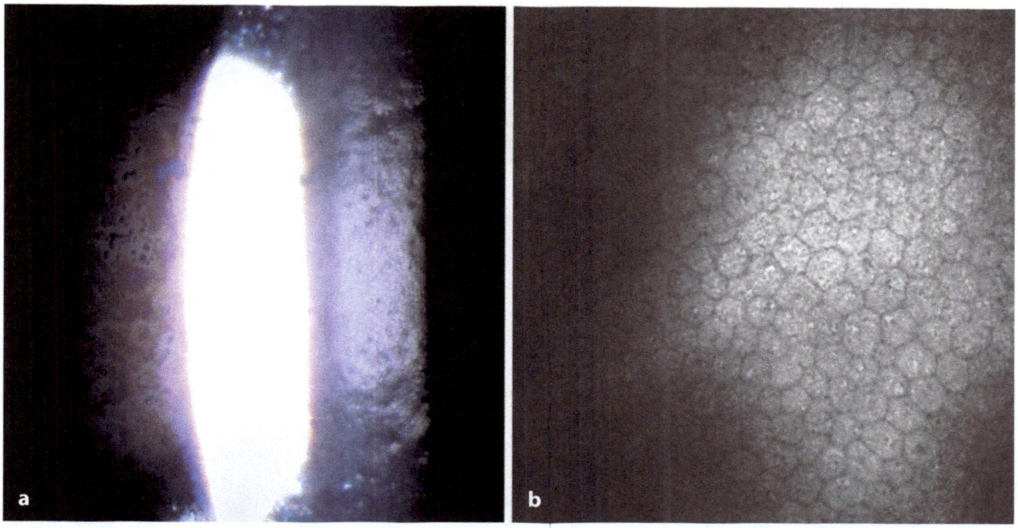

Fig. 5.99 Endothelium. **a** Slit-lamp photograph of normal endothelial cells. **b** Confocal in vivo microscopy image of normal endothelial cells

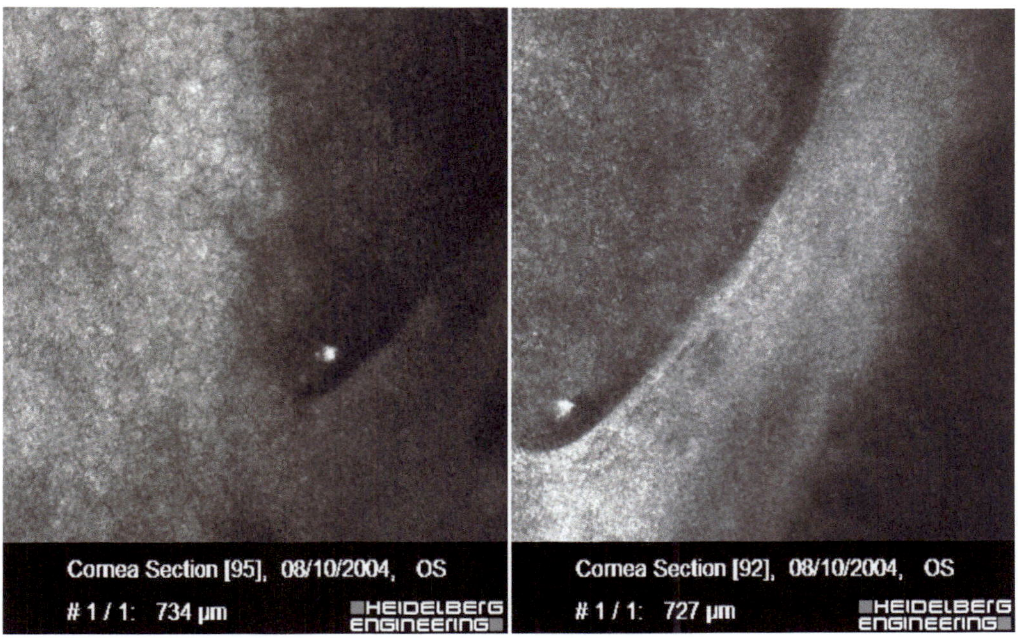

Fig. 5.100 Trabeculum. Schwalbe's ring at the periphery of the endothelium

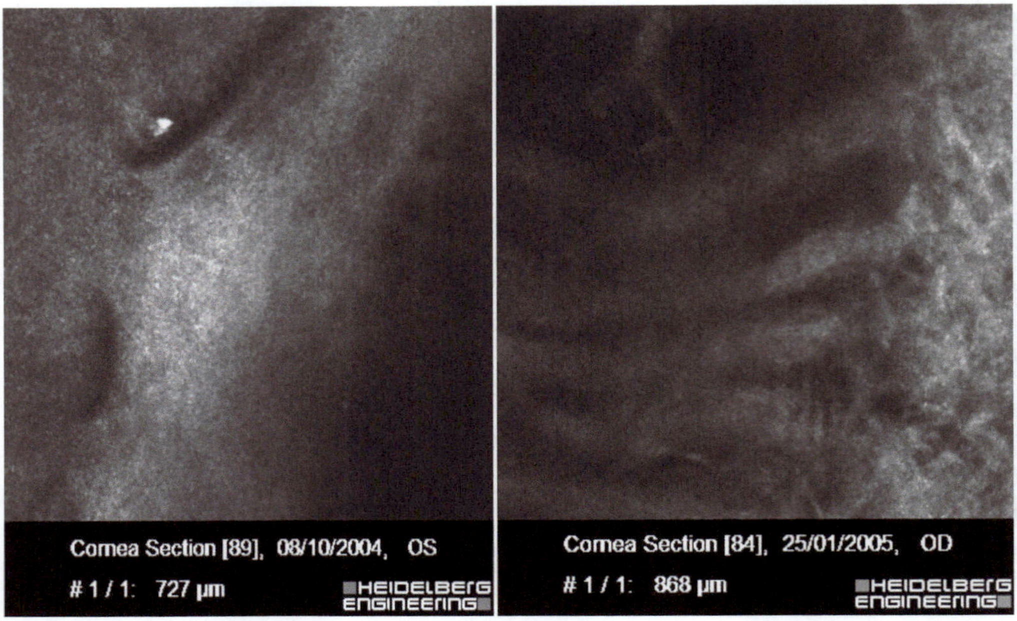

Fig. 5.101 Trabeculum. Schwalbe's ring and trabecular meshwork

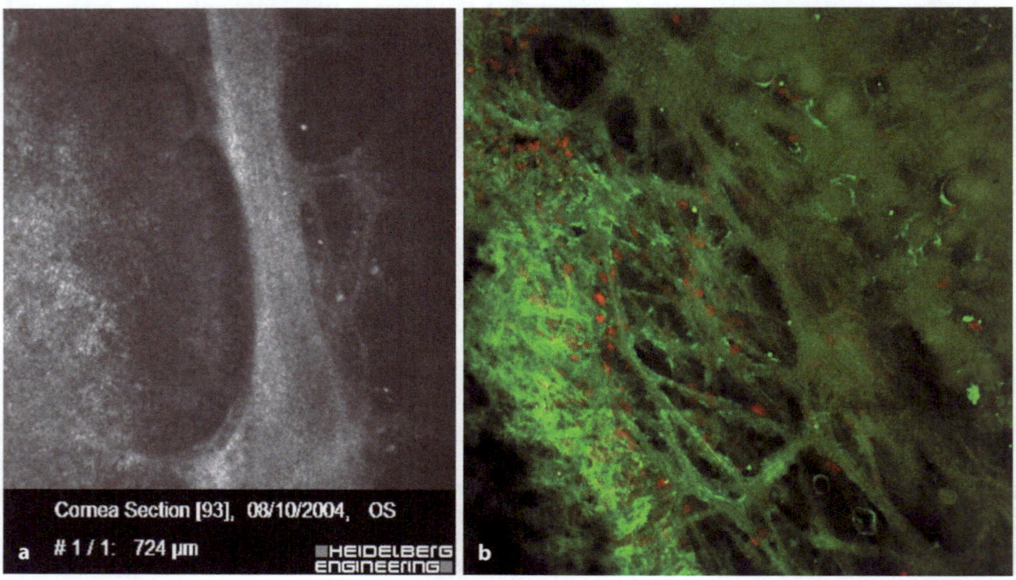

Fig. 5.102 Trabeculum. **a** Confocal in vivo microscopy image of Schwalbe's ring and trabecular meshwork. **b** Ex vivo histologic image of the same area

5.7.2 Pathological Findings

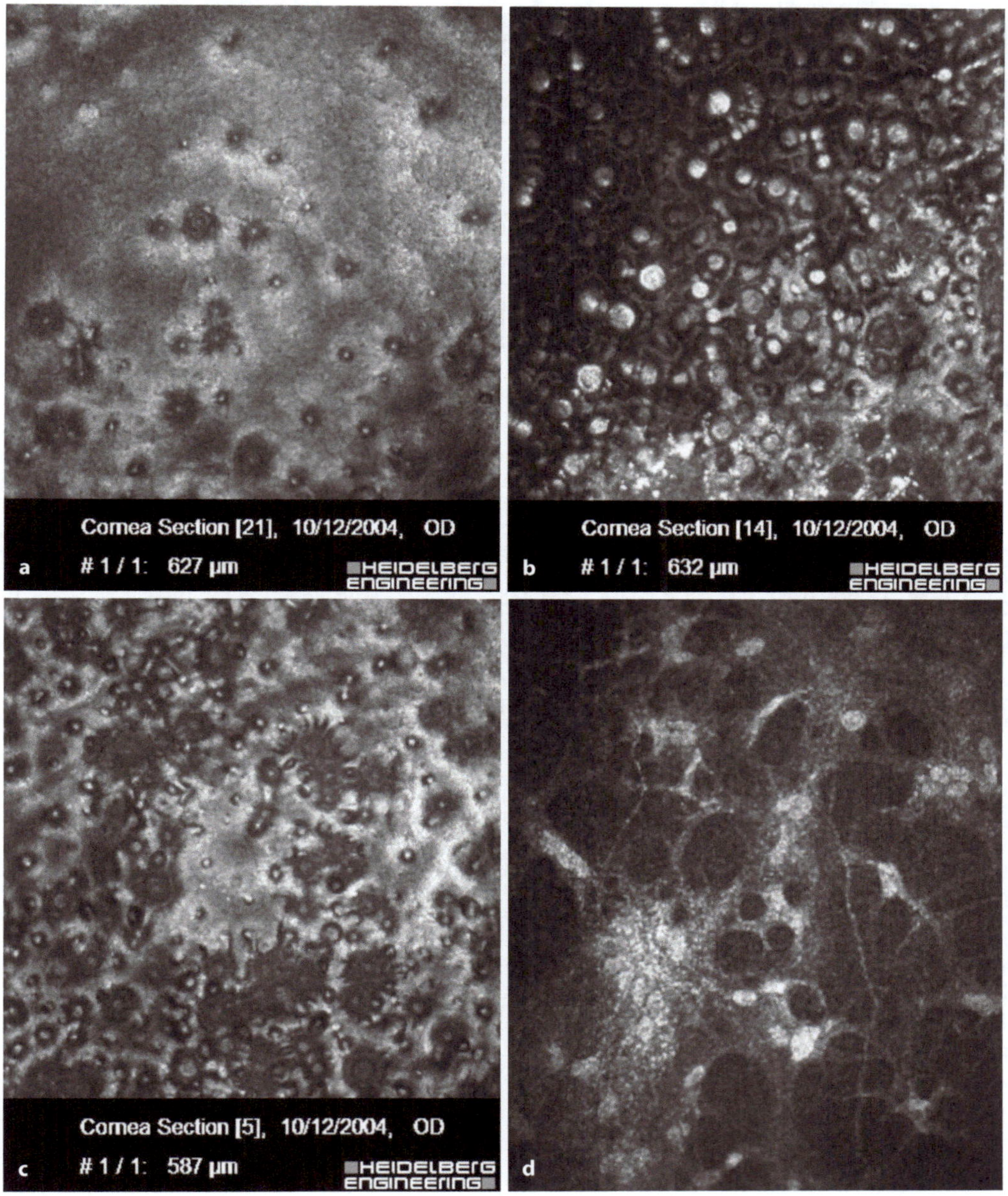

Fig. 5.103 a–f Endothelial dystrophy: cornea guttata. **a–c** Round hyporeflective structures (guttae) with occasional central reflective material at the level of the endothelium. **d–f** Stroma: fibrosis and activated keratocytes

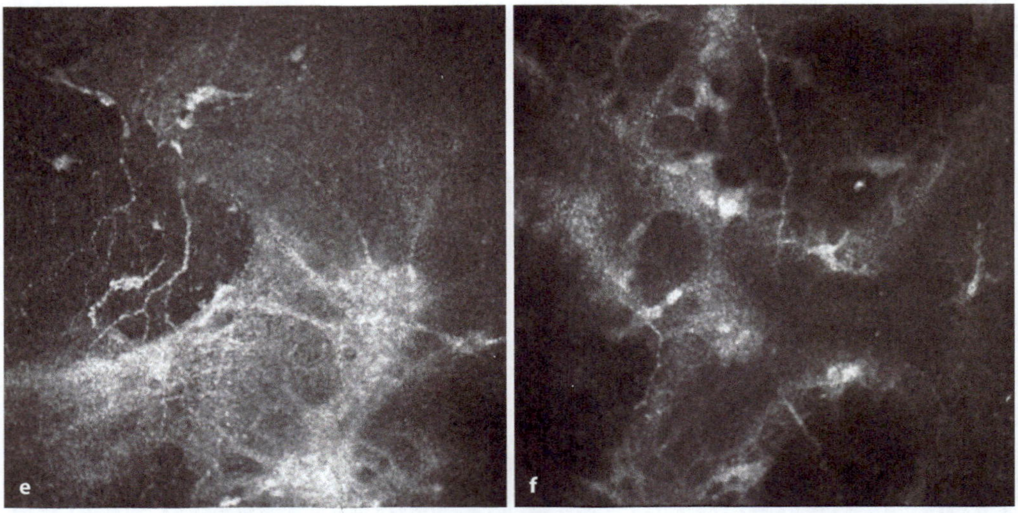

Fig. 5.103 (*continued*) Endothelial dystrophy: cornea guttata. **a–c** Round hyporeflective structures (guttae) with occasional central reflective material at the level of the endothelium. **d–f** Stroma: fibrosis and activated keratocytes

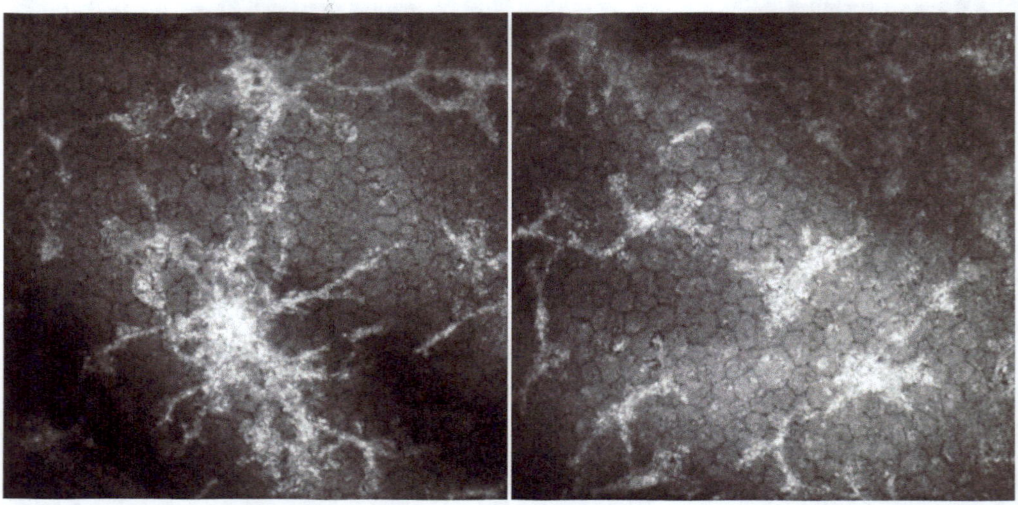

Fig. 5.104 Keratic precipitates

Fig. 5.105 Pigment dispersion. **a** Slit-lamp photograph of a 50-year-old man showing a Krukenberg's spindle. **b–e** Pigment dispersion: pigment deposits on the endothelium

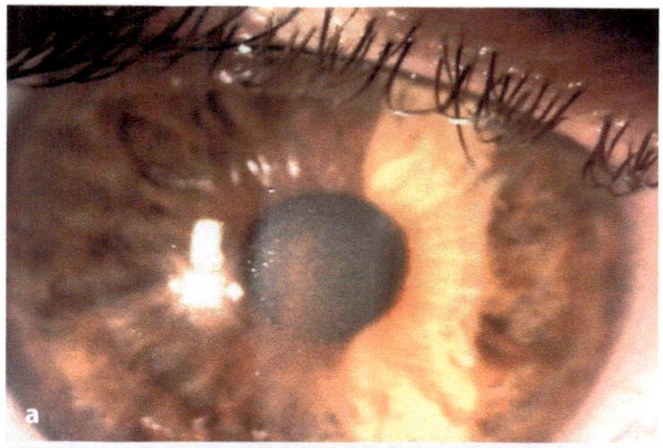

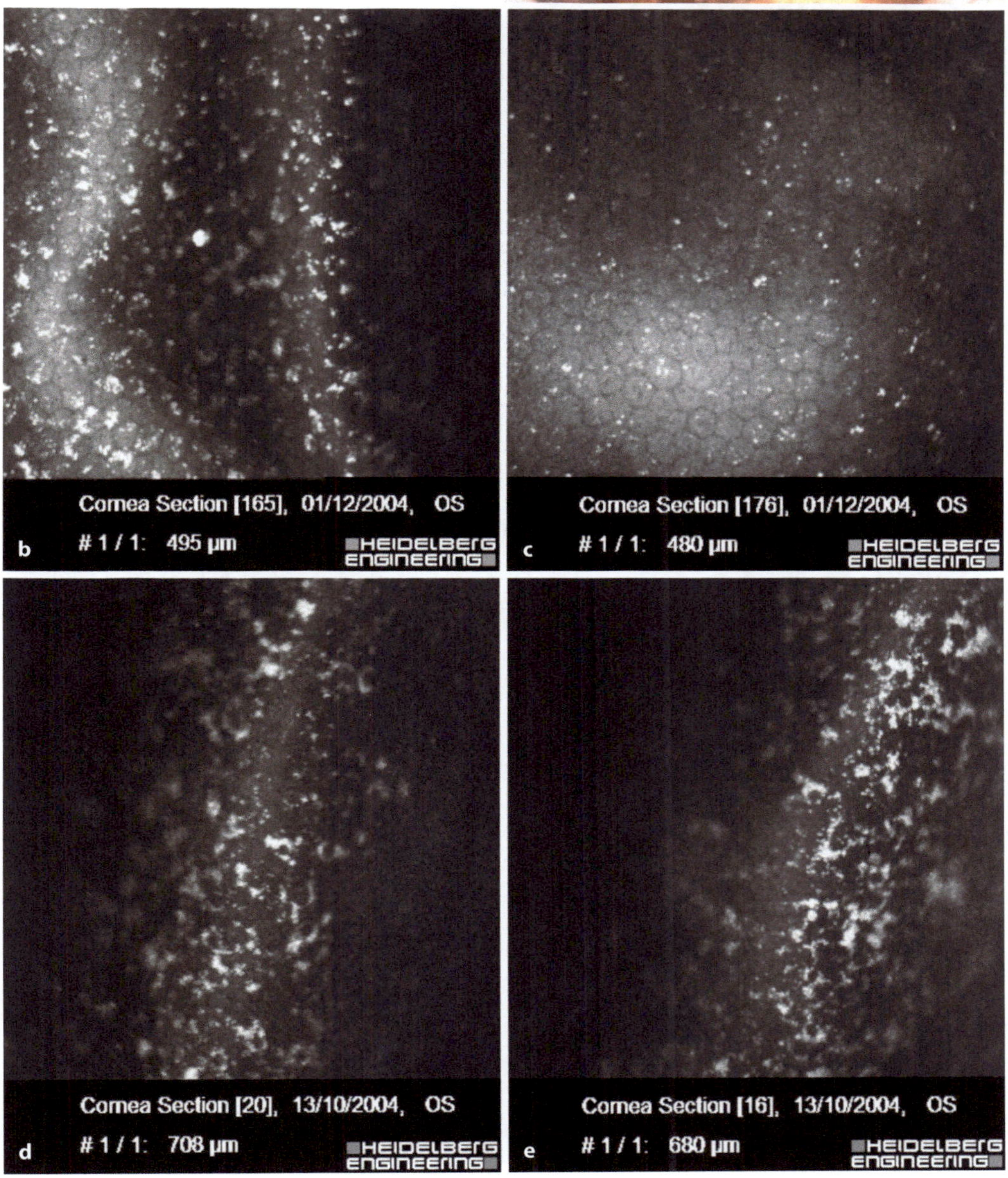

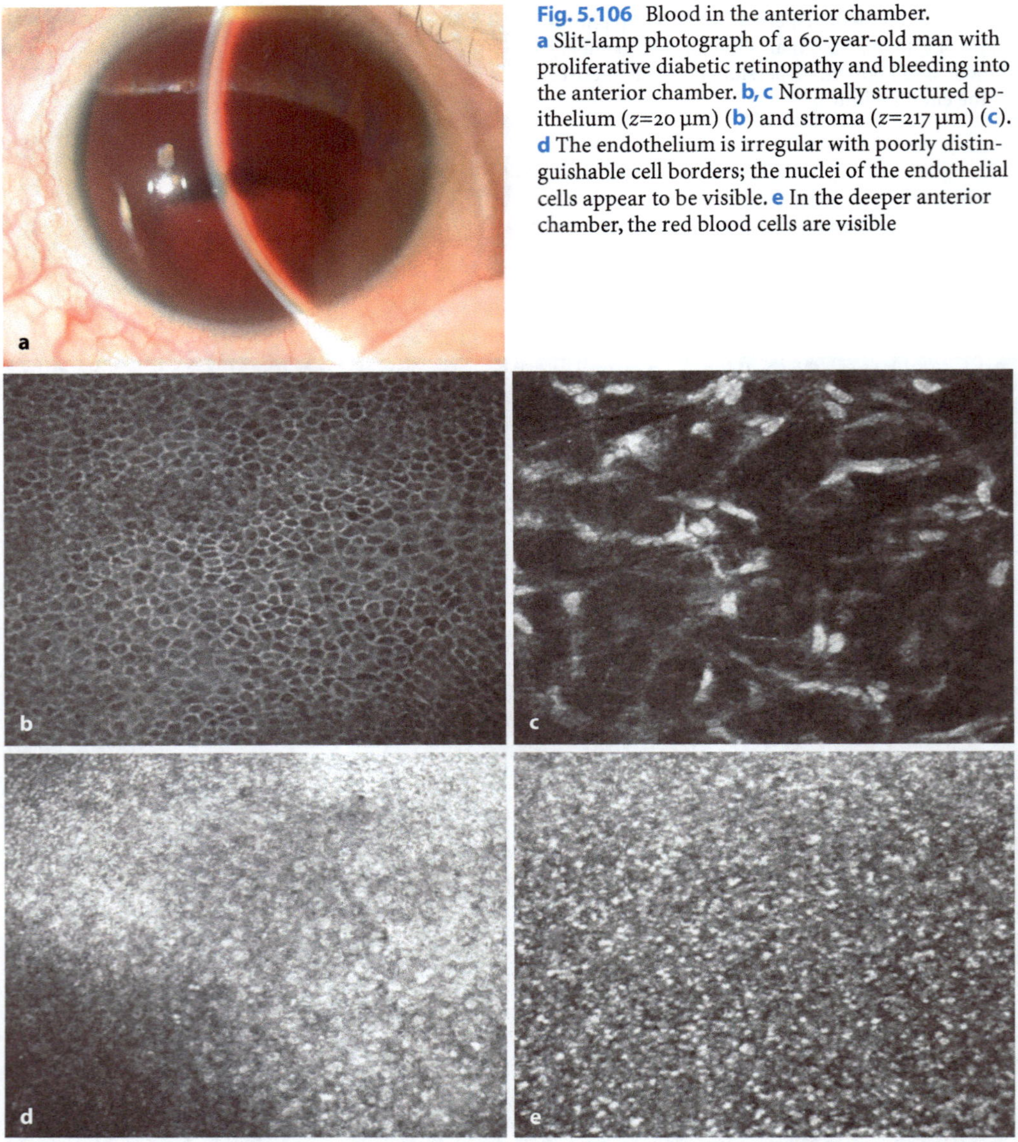

Fig. 5.106 Blood in the anterior chamber.
a Slit-lamp photograph of a 60-year-old man with proliferative diabetic retinopathy and bleeding into the anterior chamber. **b, c** Normally structured epithelium (z=20 µm) (**b**) and stroma (z=217 µm) (**c**). **d** The endothelium is irregular with poorly distinguishable cell borders; the nuclei of the endothelial cells appear to be visible. **e** In the deeper anterior chamber, the red blood cells are visible

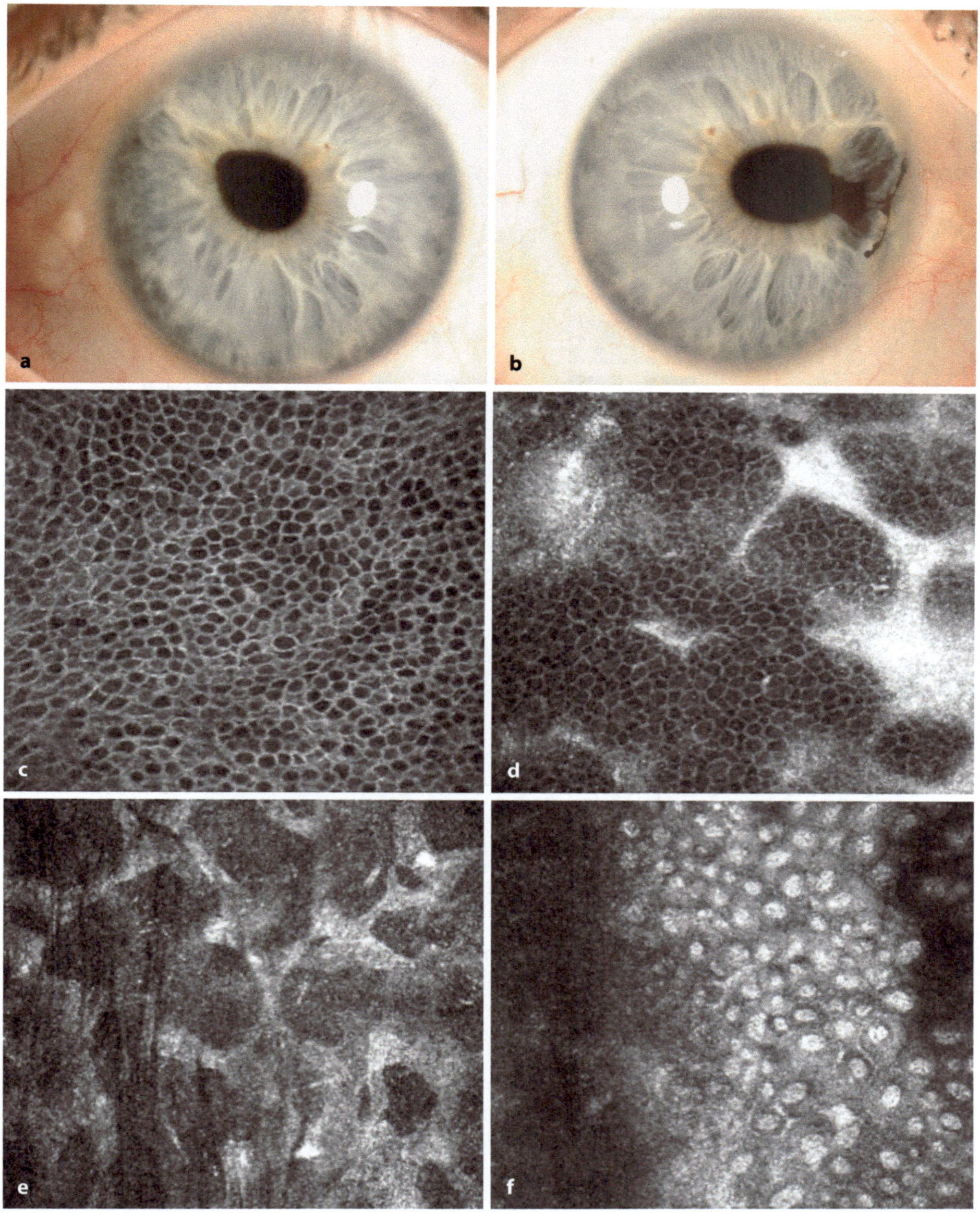

Fig. 5.107 Iridocorneal syndrome. **a, b** Slit-lamp photographs of right (**a**) and left (**b**) eye of a 35-year-old woman with iridocorneal syndrome. **c** Normal intermediate cells (z=28 µm). **d** Highly reflective fibers are visible between normally structured cells at the level of basal cell layer (z=49 µm). **e** In the stroma the cytoplasm is visible, possibly due to keratocyte activation (z=149 µm). **f** The endothelium displays a very abnormal structure, with rounded rather than hexagonal cells, with clearly visible, highly reflective nuclei (z=621 µm)

5.8
Intraocular Structures (Lens, Iris)

5.8.1
Normal Anatomy

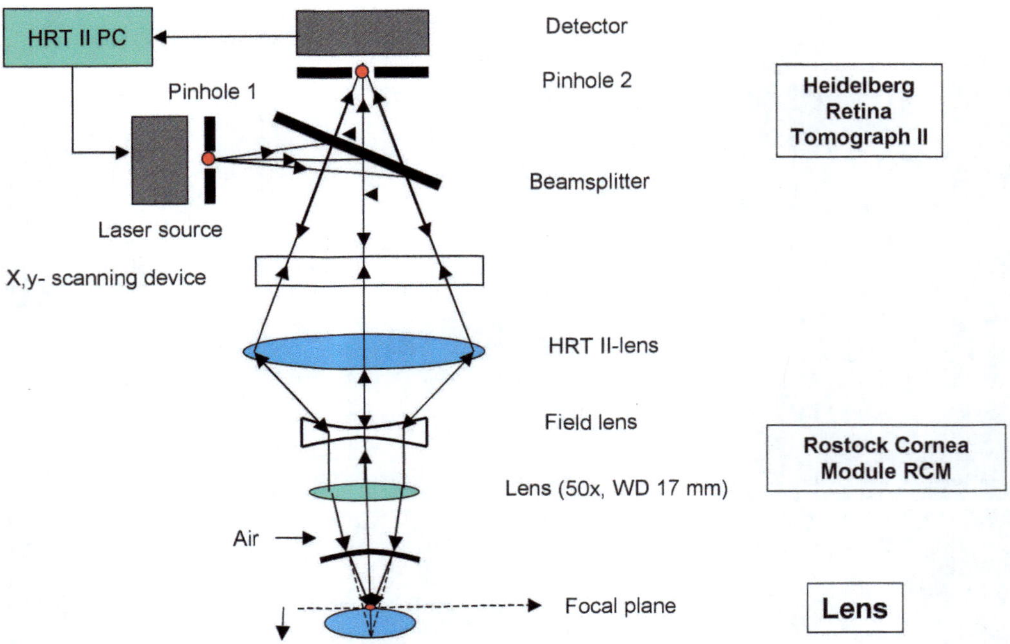

Fig. 5.108 Noncontact confocal laser scanning in vivo microscopy of the lens. Optical principle

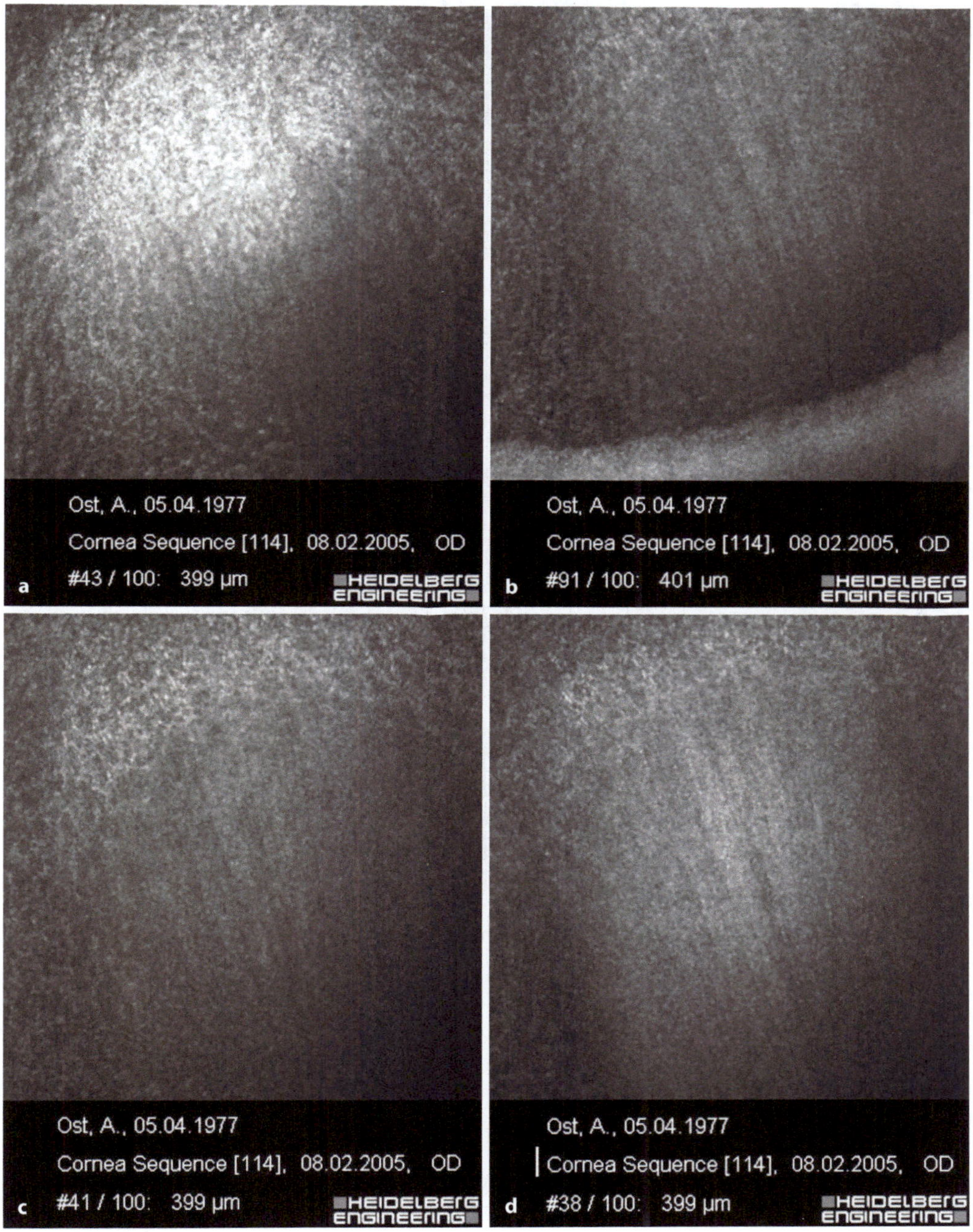

Fig. 5.109 Noncontact confocal microscopy (Nikon ×50; tube: 20 mm). **a** Lens epithelium. **b–d** Fibers

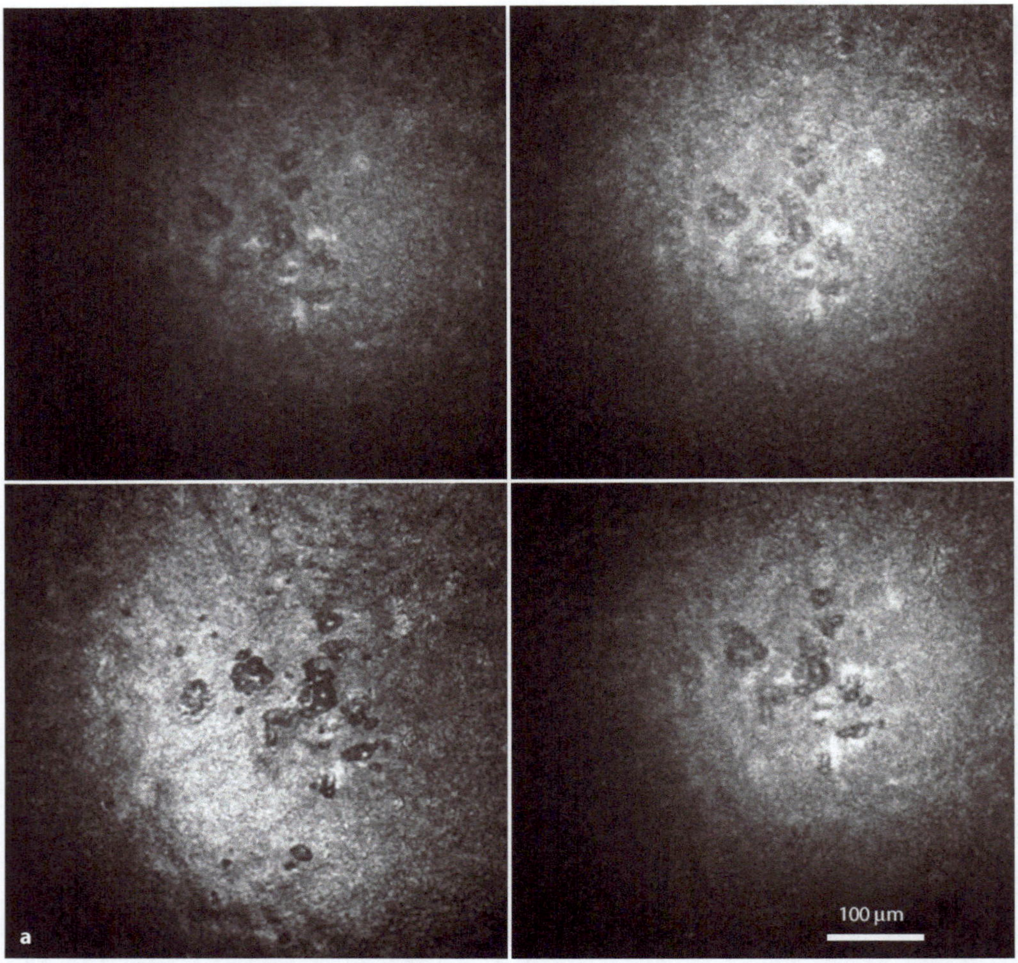

Fig. 5.110 a–c *Z*-scan (in vivo) through the human lens. **a** Lens epithelium

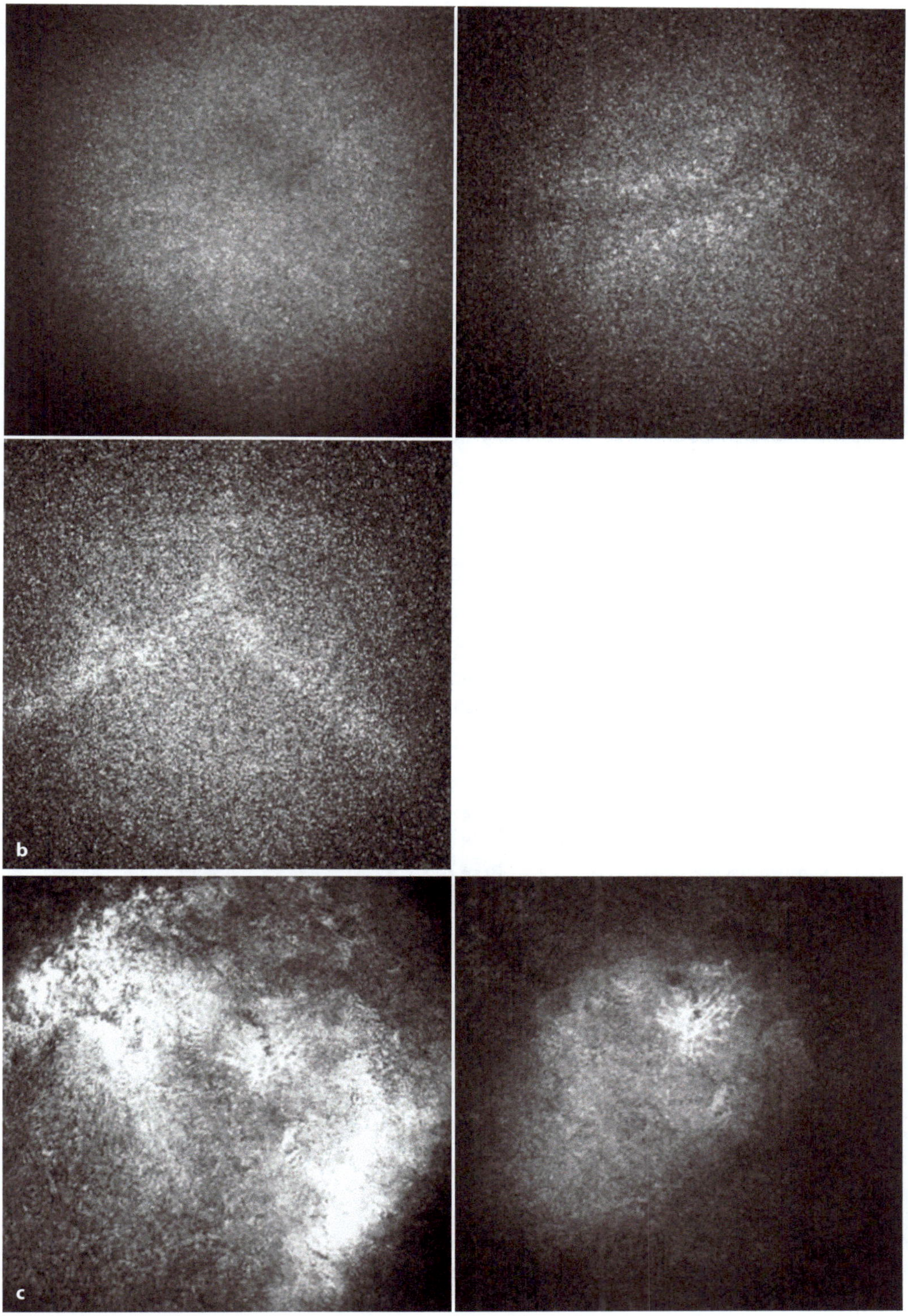

Fig. 5.110 (*continued*) **b** Lens nucleus with fibers. **c** Reverse side of the lens

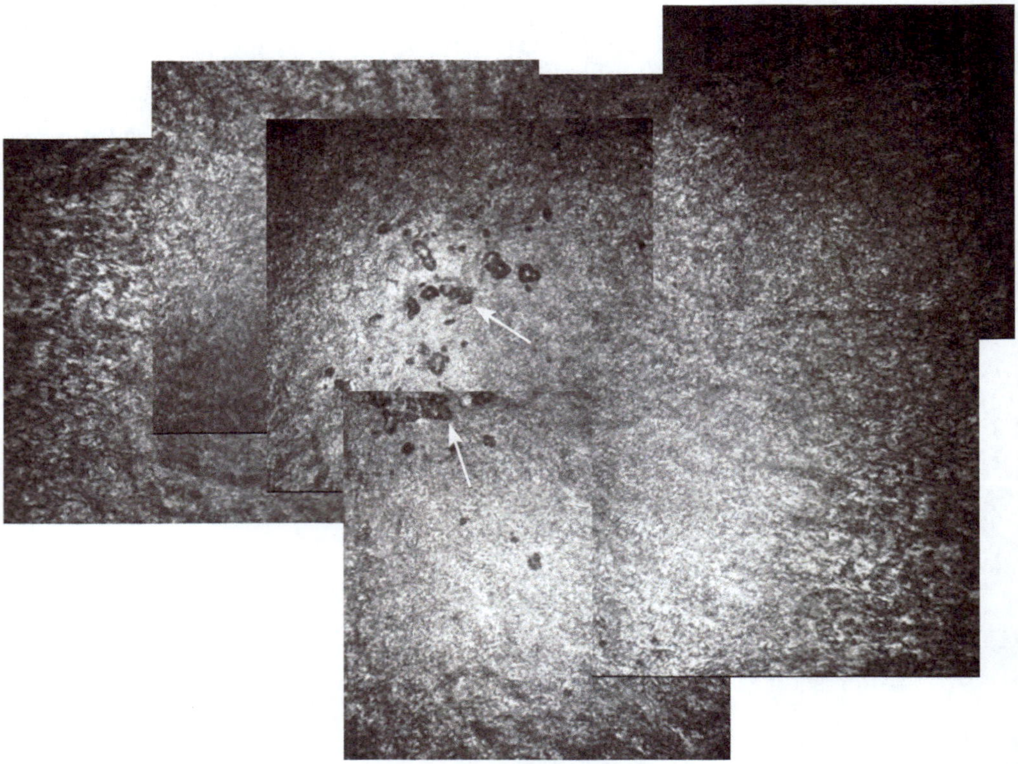

Fig. 5.111 Front side of the lens epithelium with small defects (*arrows*)

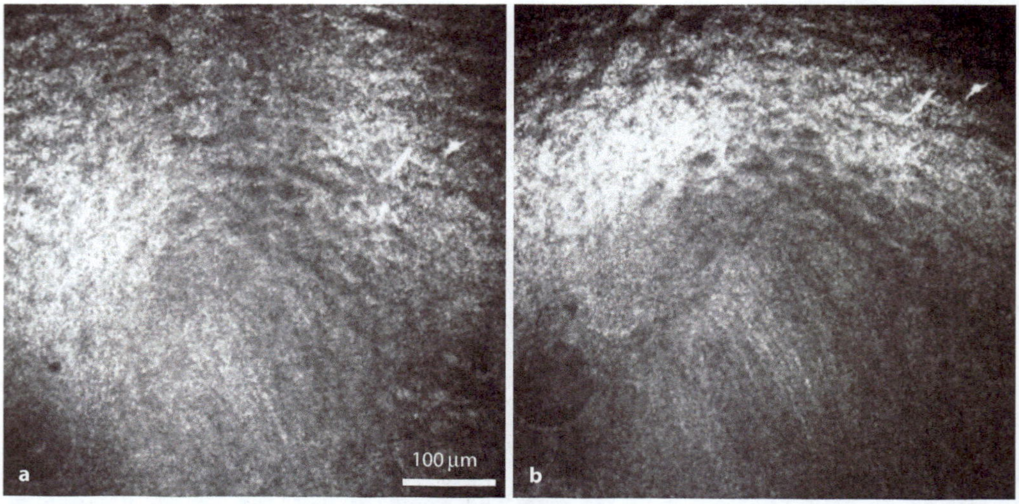

Fig. 5.112 a–f Microstructures of the lens. a Lens epithelium. b–e Lens fibers

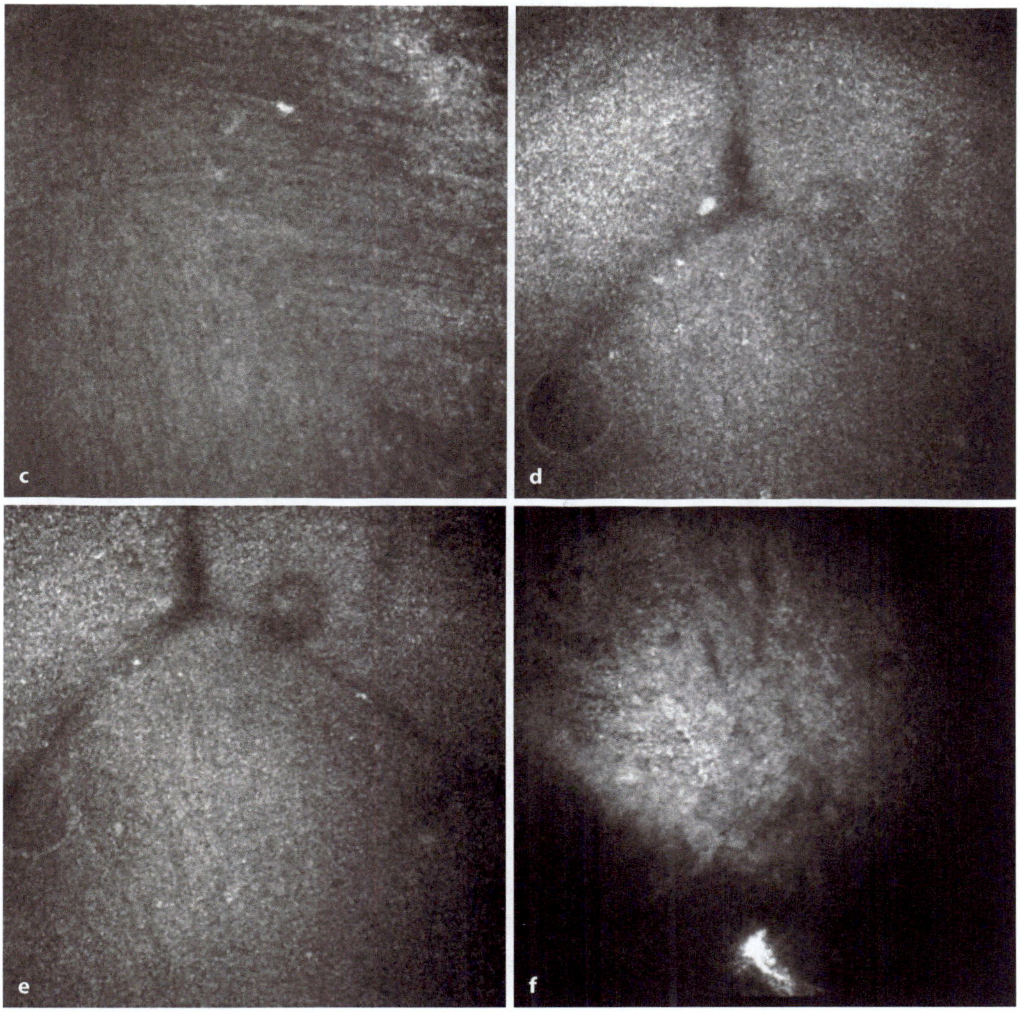

Fig. 5.112 (*continued*) **b–e** Lens fibers. **f** Cell structures on the posterior capsular bag

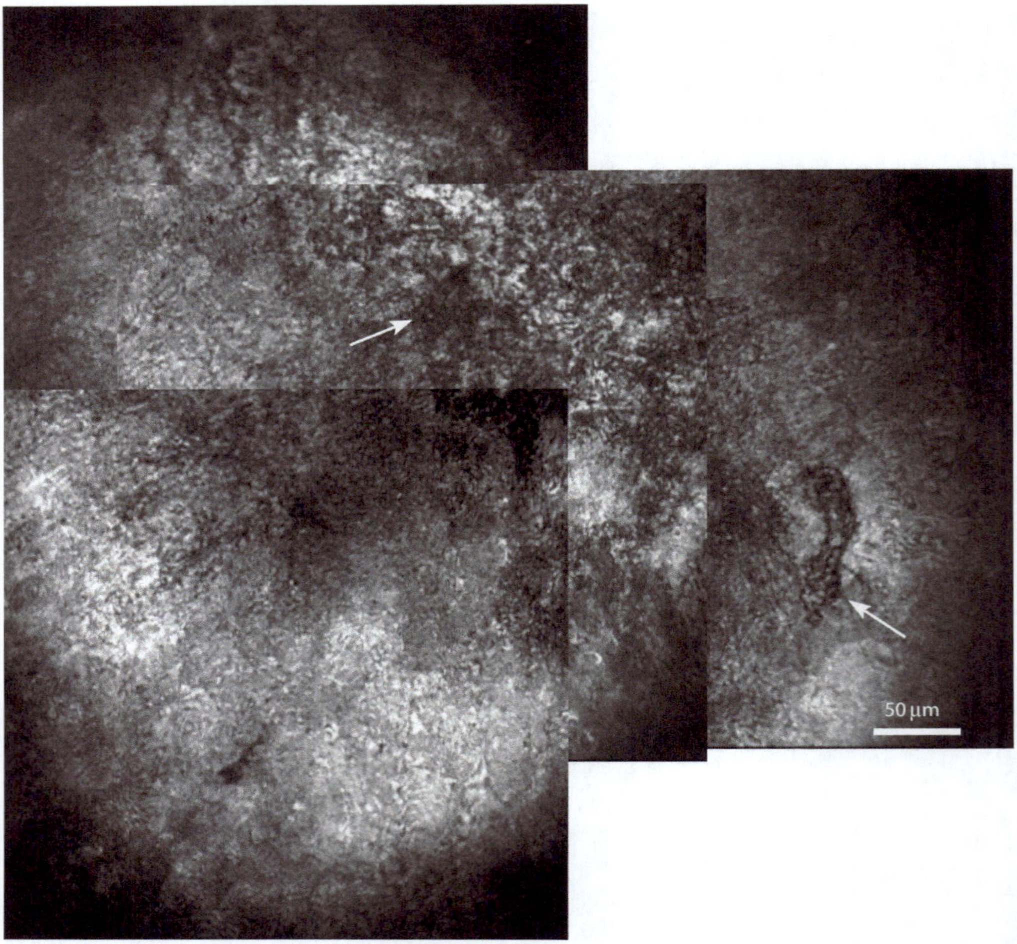

Fig. 5.113 Reverse side of the human lens (in vivo). Cell structures with defects (*arrows*)

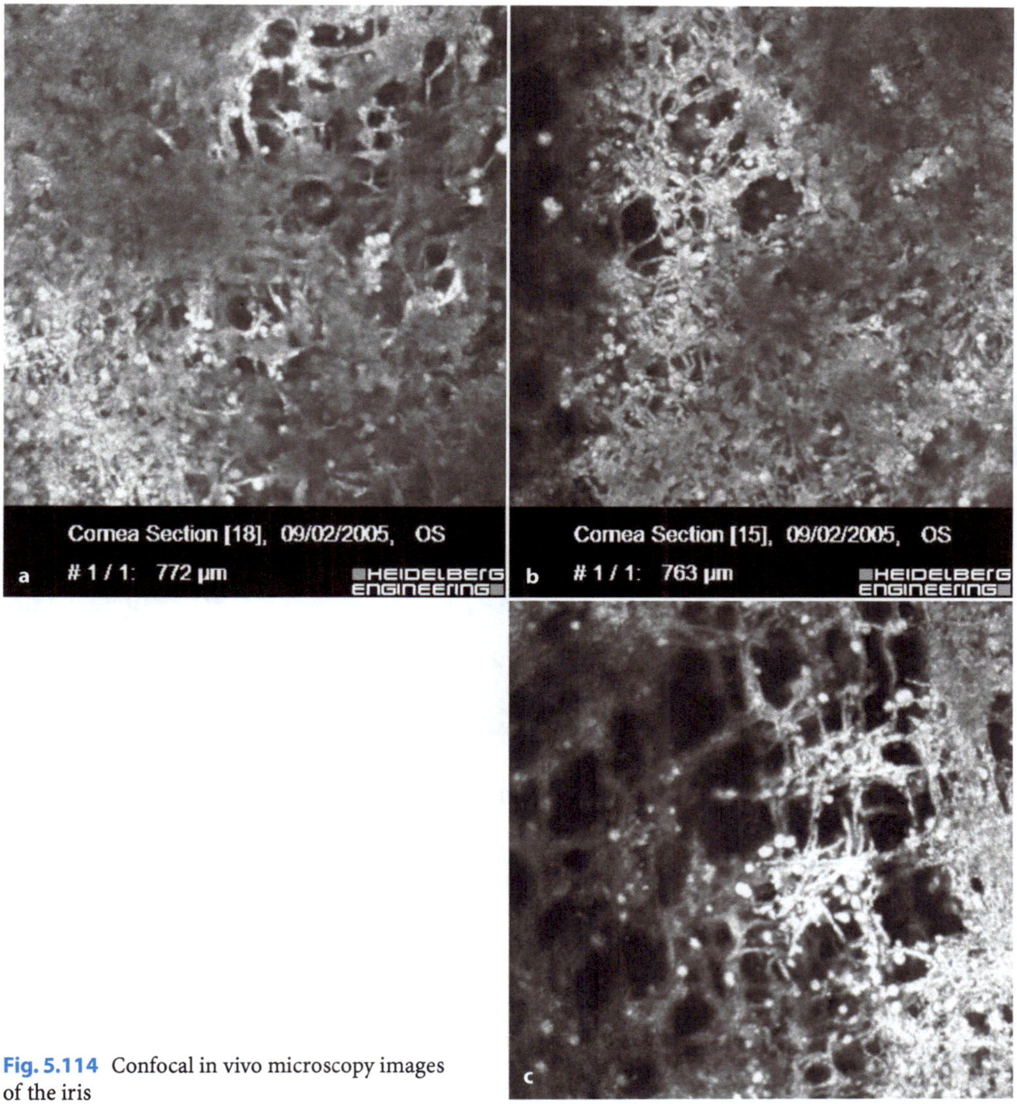

Fig. 5.114 Confocal in vivo microscopy images of the iris

5.8.2 Pathological Findings

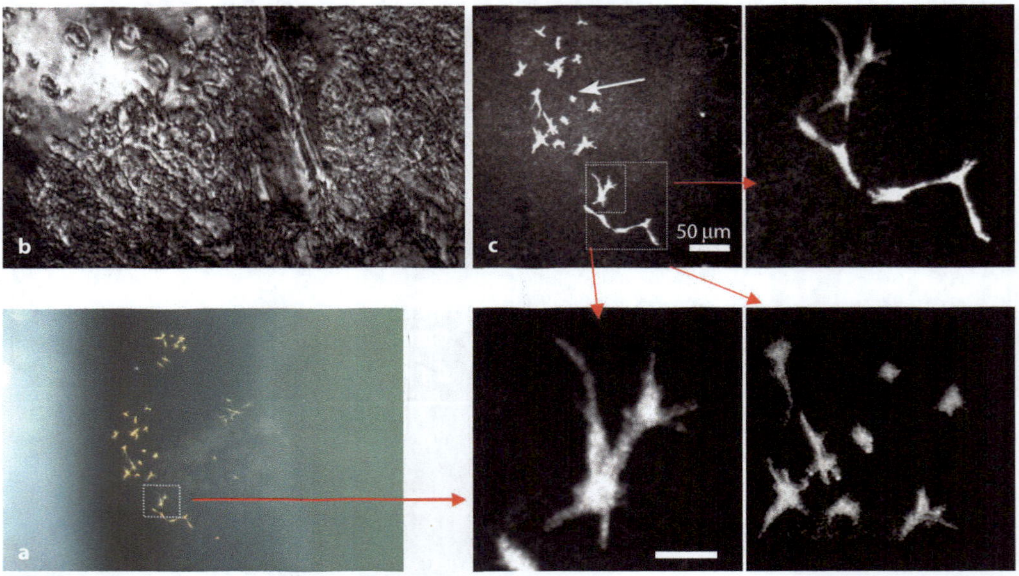

Fig. 5.115 Congenital ectopic iris. **a** Slit-lamp photograph. **b** Reverse side of the lens. **c** Pigment epithelial cells (*arrows*)

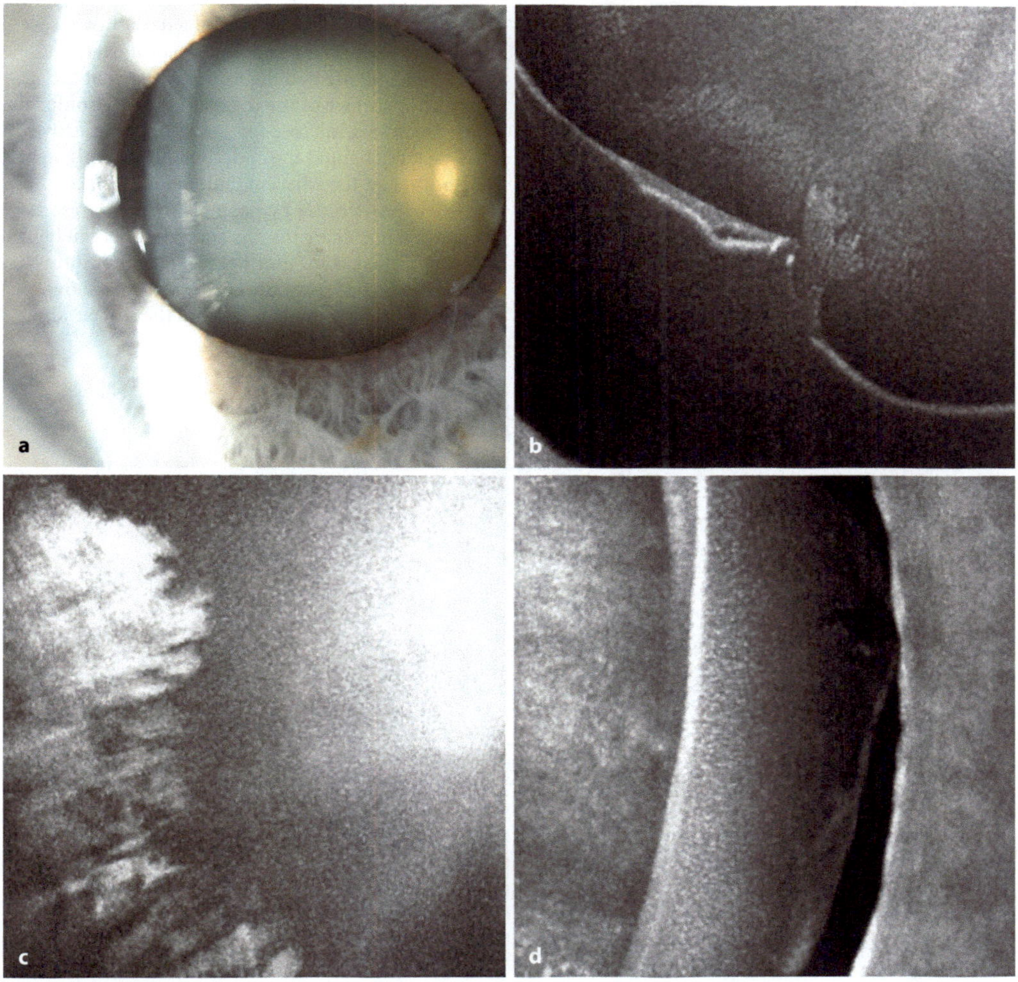

Fig. 5.116 Pseudoexfoliation glaucoma and cataract. **a** Slit-lamp photograph of an 80-year-old woman with pseudoexfoliation glaucoma and cataract, visual acuity 12/20 (right eye) and 16/20 (left eye); confocal images, noncontact examination. **b** Pseudoexfoliation line at the anterior lens surface. **c** Cataract formation of the peripheral lens. **d** Pseudoexfoliation line visible at the right side of the iris

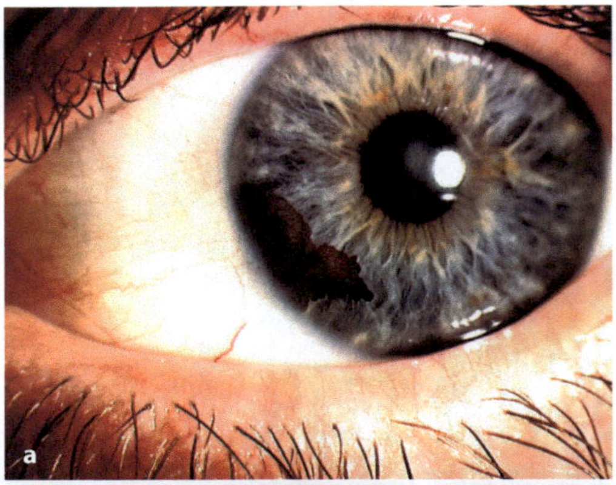

Fig. 5.117 Pigmented nevus of the iris. **a** A 43-year-old woman complained of a dark brown spot on the eye. Slit-lamp examination of the left eye revealed a pigmented flat tumor situated at the posterior surface of the cornea nasally. **b** Confocal laser scanning microscopy shows hyperreflective tissue lying behind the endothelium with pigment dispersed on the adjacent endothelial surface. **c** The hyporeflective cytosomal structures resembling cell nuclei are visible, and cell walls can be distinguished. **d, e** There are no signs of infiltration of corneal layers, the corneal stroma above the alterations is unchanged, and the surrounding endothelium is not distorted

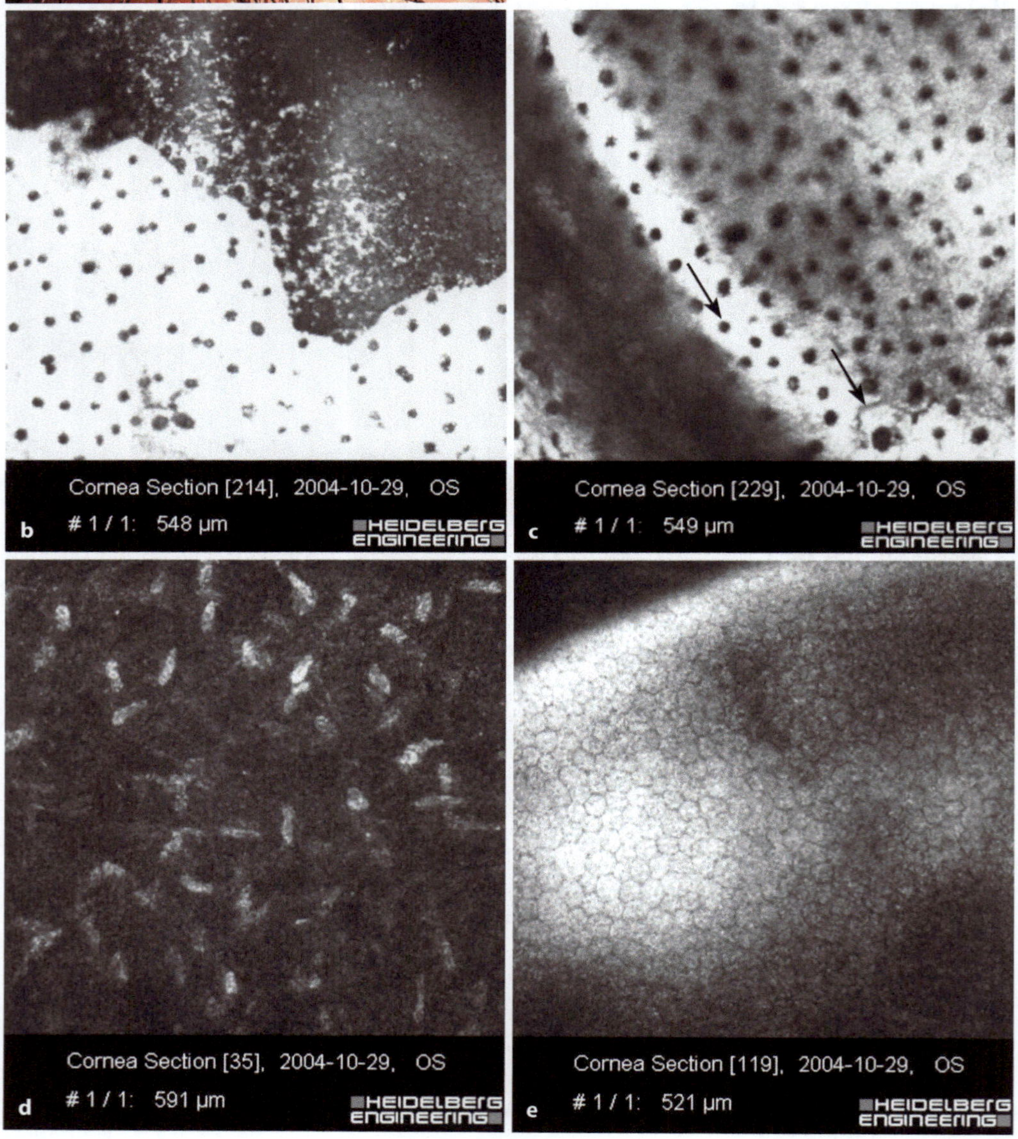

5.9 Limbal Region

5.9.1 Normal Anatomy

Because it forms a junctional zone with the conjunctiva, the limbal region is especially important. Inflammatory cells migrate across it into the cornea in immunological disease; it is the source of new inbudding corneal vessels; and, not least, it also plays an important role in corneal regeneration as the site of origin of corneal stem cells.

The limbal region is where the corneal epithelium forms a junction with the conjunctival epithelium, which comprises approximately 10–12 cell layers. This region also contains a radial arrangement of trabecular conjunctival processes (the limbal palisades of Vogt) that are considered to be the site of origin of corneal stem cells [16, 74]. Overall, the organization of the conjunctival epithelium is less uniform because different epithelial cell types (goblet cells, for example) occur here and the arrangement of the individual cell layers is also not so strictly parallel with the surface [17].

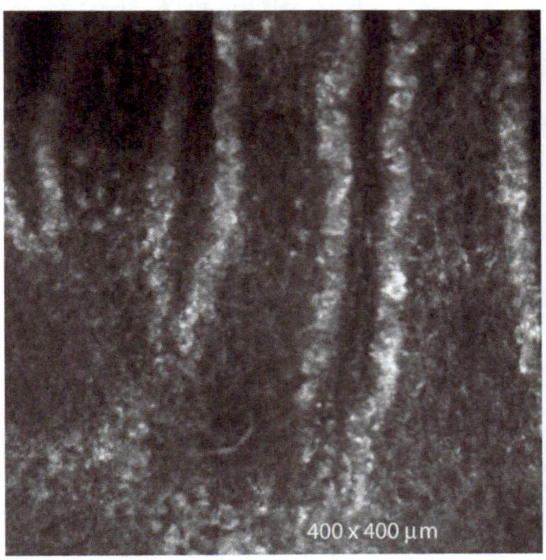

Fig. 5.118 Limbal palisades of Vogt (near limbus at 6 o'clock). Trabecular extensions of the conjunctiva are growing from outside (in this case, from below) in a radial pattern toward the cornea

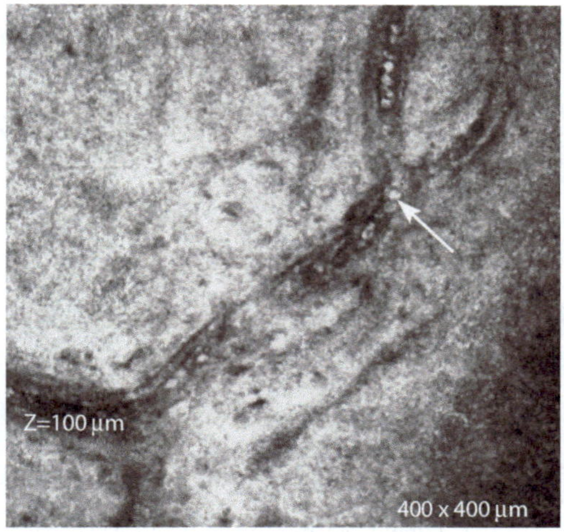

Fig. 5.119 Branched conjunctival vessel close to the limbus with erythrocytes visible (*arrow*), near limbus at 9 o'clock

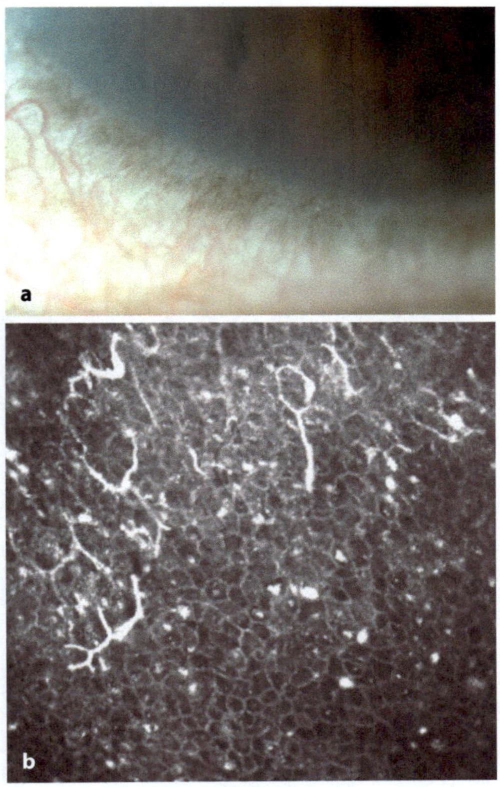

Fig. 5.120 Limbus. **a** Slit-lamp photograph of a pigmented limbus. **b** Peripheral corneal cells: numerous dendritic cells and hyperreflective structures corresponding to pigment deposits

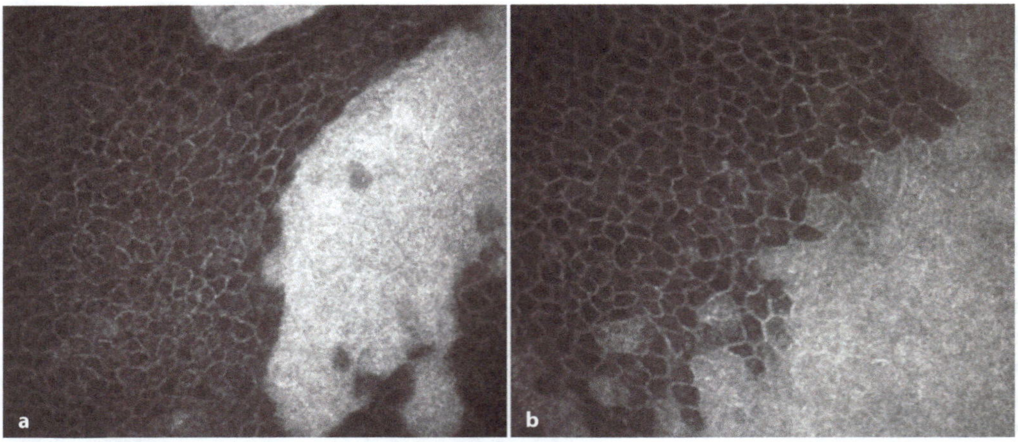

Fig. 5.121 Limbus. Normal junction between corneal cells (*black cells*) and conjunctival cells (*reflective cells*)

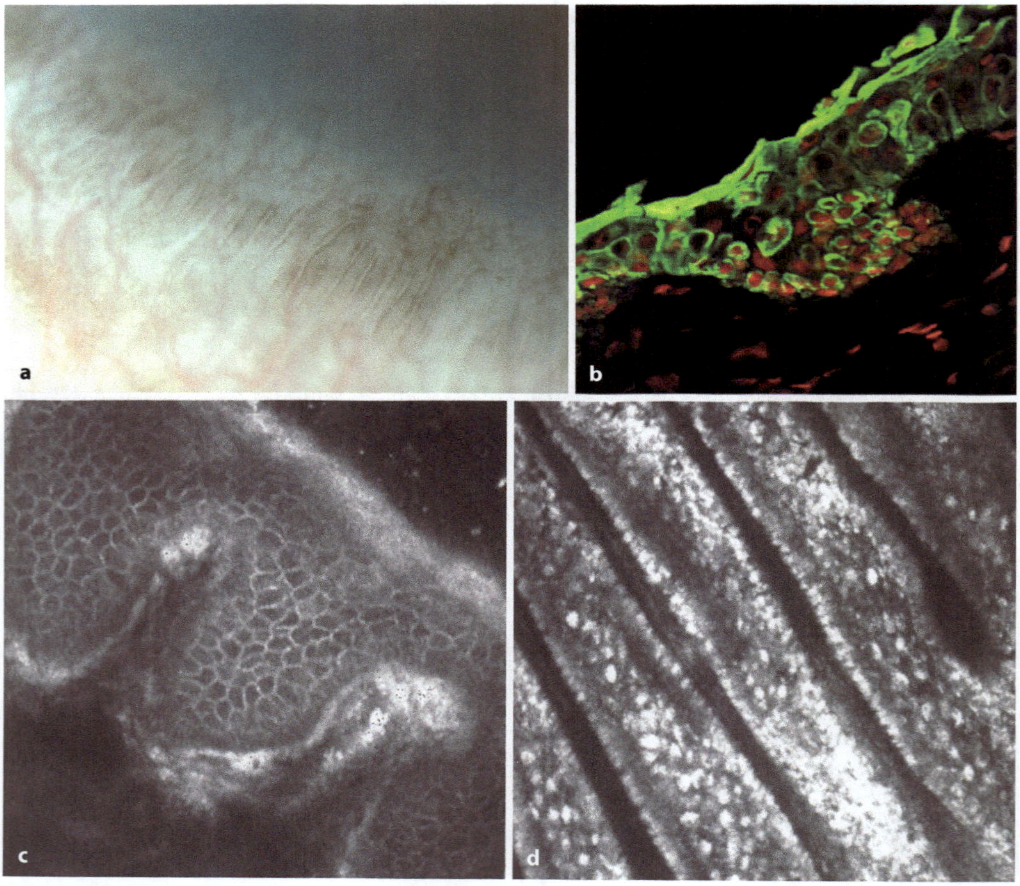

Fig. 5.122 Normal limbus. **a** Slit-lamp photography of normal limbus and palisades of Vogt. **b** Immunohistology of normal limbus: immunostaining with cytokeratin K19 (*green*) and propidium iodide (*red*). **c** Confocal in vivo microscopy image, oblique section: palisades of Vogt. **d** Confocal in vivo microscopy image: palisades of Vogt

On confocal microscopy, the epithelial cells of the conjunctiva, unlike those of the cornea, are more reflective, smaller, and less well demarcated. Their nucleus is relatively large and bright. The junctional zone is characterized by inhomogeneous reflectivity and marked variation in cell shape and size. The limbal palisades of Vogt can often be visualized as parallel trabecular extensions of the conjunctival epithelium (Fig. 5.118). In the immediate junctional zone, the conjunctival epithelium commonly exhibits tonguelike extensions that are mostly well demarcated, especially in the deeper layers, and at the end of which are located isolated cells or cell groups with very bright cell borders and a bright cytoplasm. These may be secretory cells. Subepithelially, in the region of the conjunctiva close to the limbus, are the blood vessels of the limbal vascular plexus, in the lumen of which flowing blood cells can be seen (Figs. 5.119–5.123).

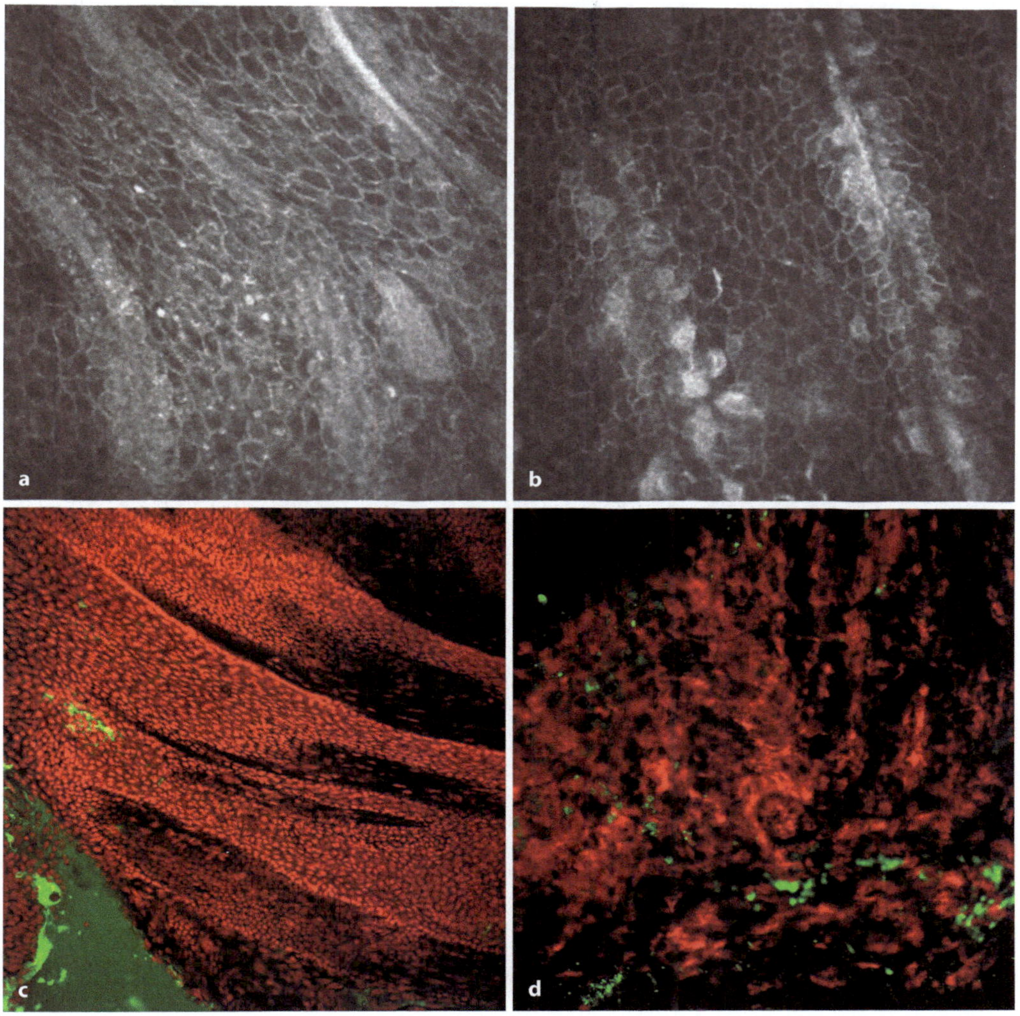

Fig. 5.123 Normal limbus and palisades of Vogt. **a, b** Palisades of Vogt at two different depths. Epithelial stem cells are thought to be located between the palisades of Vogt. **c, d** Immunohistological images of the palisades of Vogt: epithelial cells (*red*), stem cells stained with ABCG2 (*green*)

5.9.2
Pathological Findings

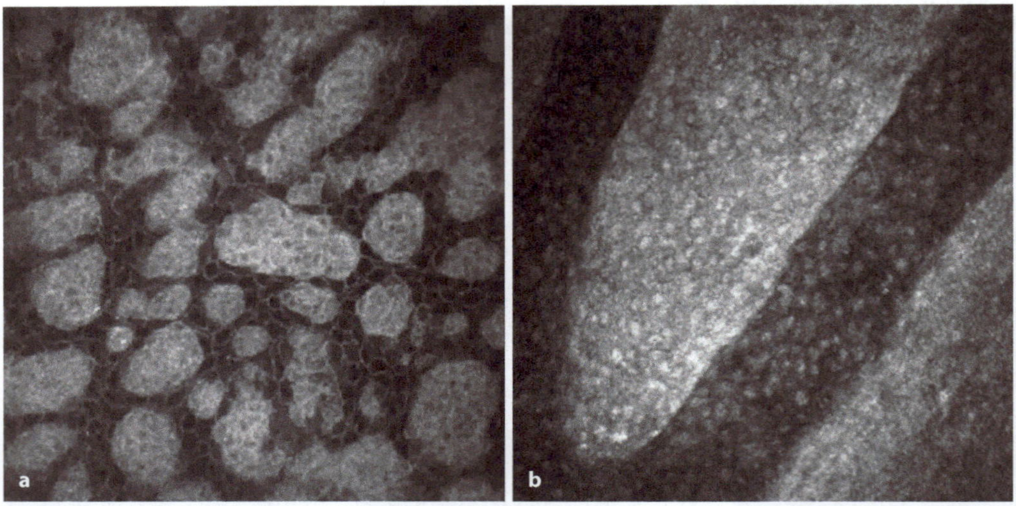

Fig. 5.124 Pathological limbus in inflammatory diseases: abnormal junction between corneal cells (*black cells*) and conjunctival cells (*reflective cells*)

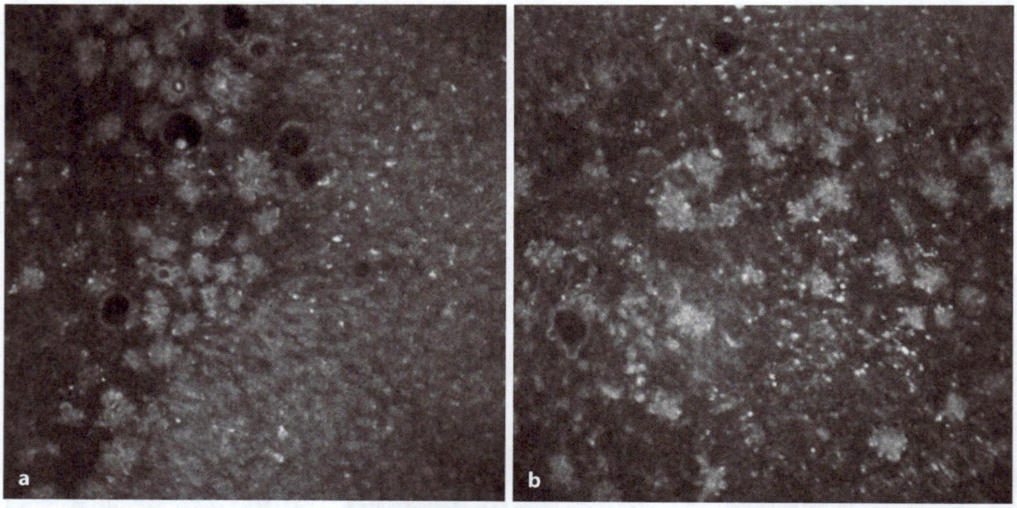

Fig. 5.125 Pathological limbus in a case of vernal keratoconjunctivitis. Islets of inflammatory cells and conjunctival cells within the corneal epithelium

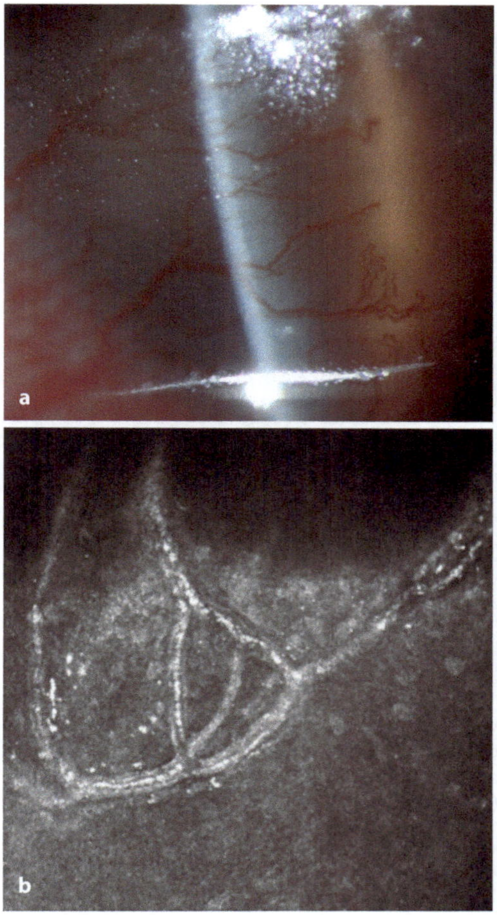

Fig. 5.126 Corneal neovascularization. a Slit-lamp photograph of corneal neovascularization. b Confocal in vivo microscopy image of corneal neovascularization: neovascular loops

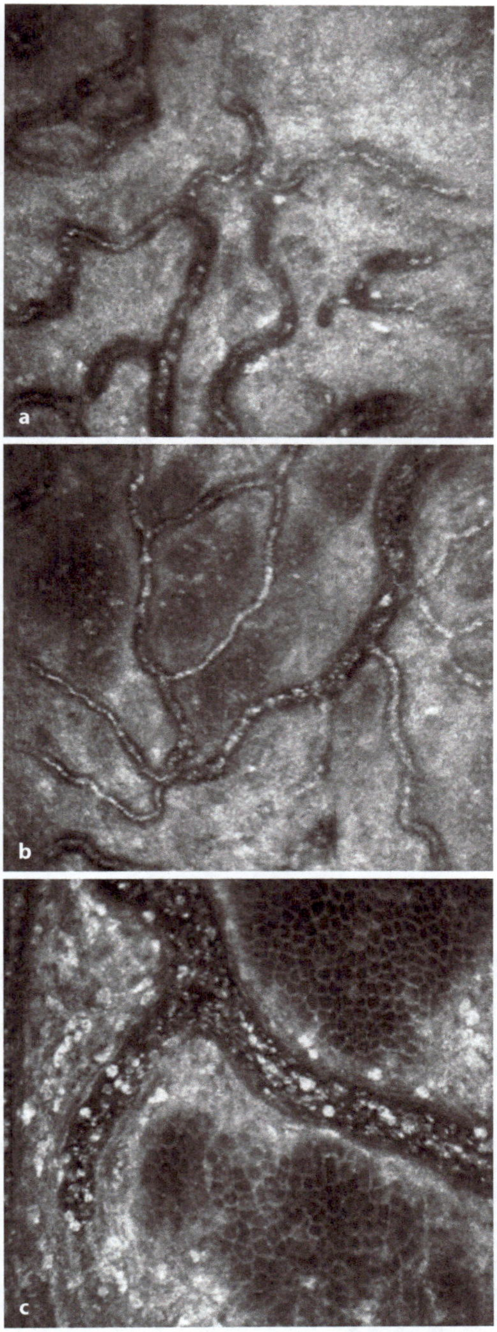

Fig. 5.127 Corneal neovascularization. a, b Corneal blood vessels. c Corneal neovascularization. Hyperreflective inflammatory cells surround the blood vessels and intravascular round reflective cells

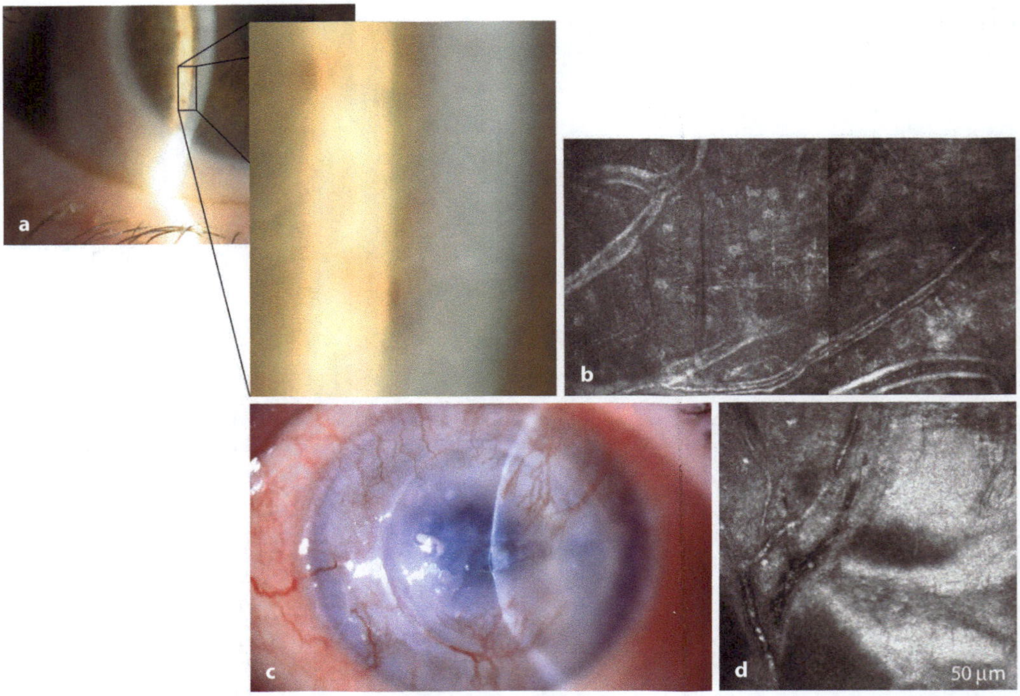

Fig. 5.128 Vascularization of the cornea. **a** Slit-lamp photograph of 53-year-old woman with ghost vessels. **b** Confocal microscopic images show empty vessels at the periphery of cornea at the level of the posterior stroma. **c** Slit-lamp photograph of 73-year-old woman with corneal vascularization of donor and recipient cornea after penetrating keratoplasty. **d** Confocal images show blood capillaries with bright cell bodies (erythrocytes) in the corneal epithelium

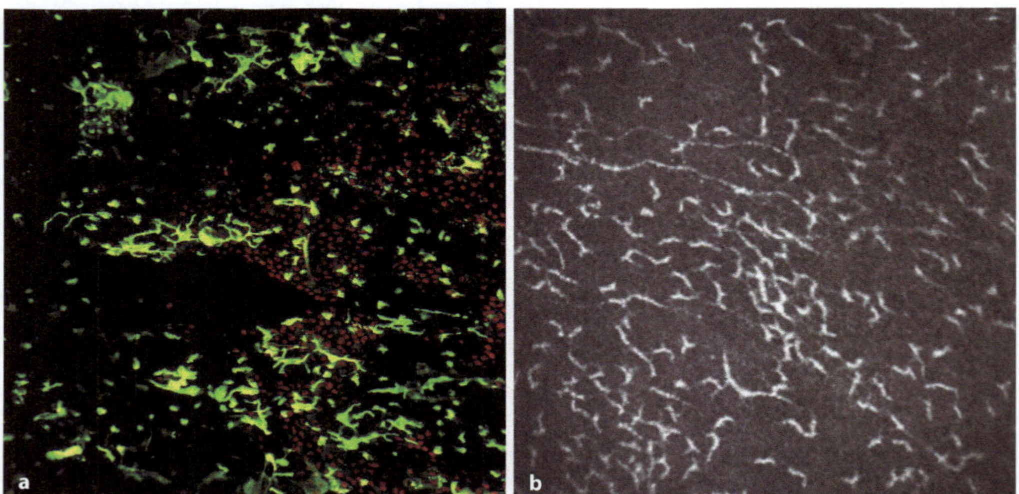

Fig. 5.129 Inflammation, dendritic cells. **a** Immunohistology: dendritic cells (*green*) in the peripheral cornea (*red*). **b** Confocal in vivo microscopy: hyperreflective dendritic cells

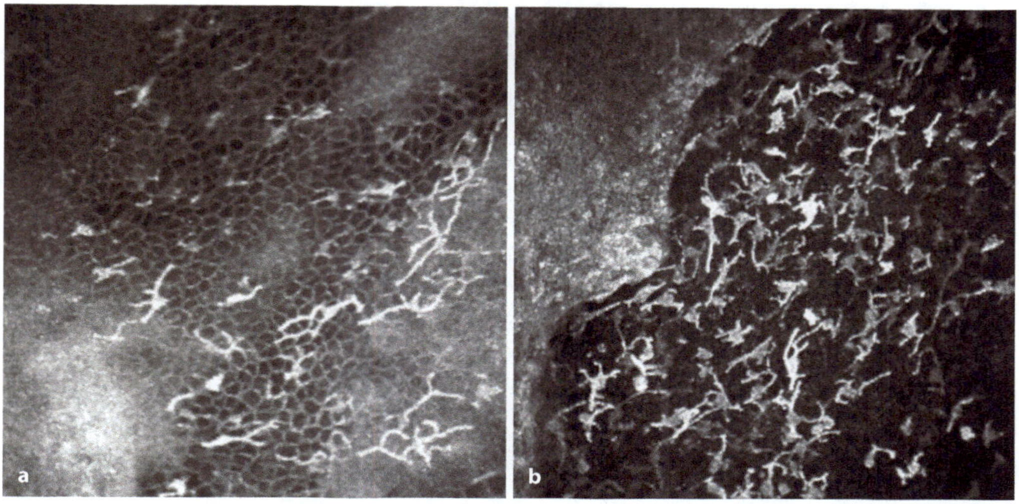

Fig. 5.130 Inflammation, dendritic cells. Numerous hyperreflective dendritic cells in the peripheral cornea of patients with ocular surface inflammatory diseases

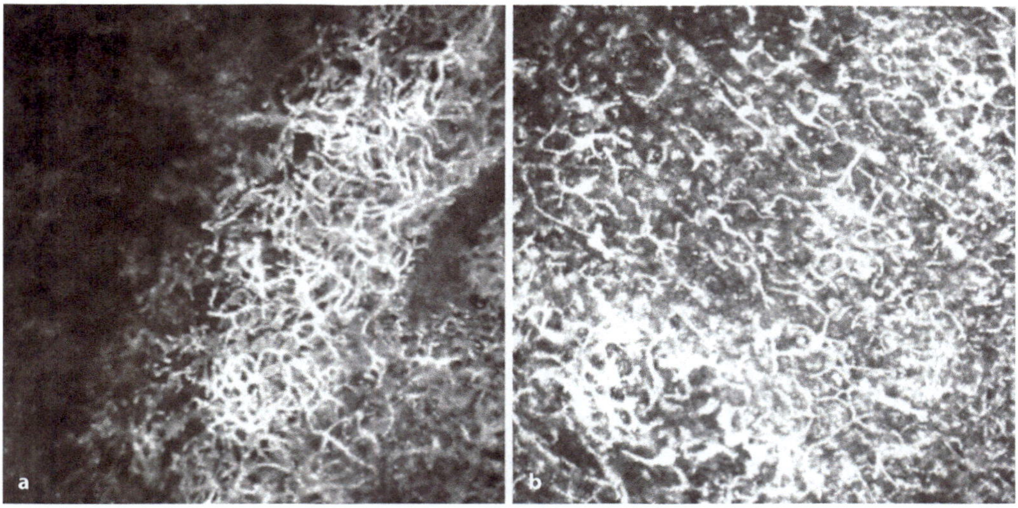

Fig. 5.131 Inflammation, dendritic cells. Major corneal infiltration of dendritic cells in a severe case of vernal keratoconjunctivitis

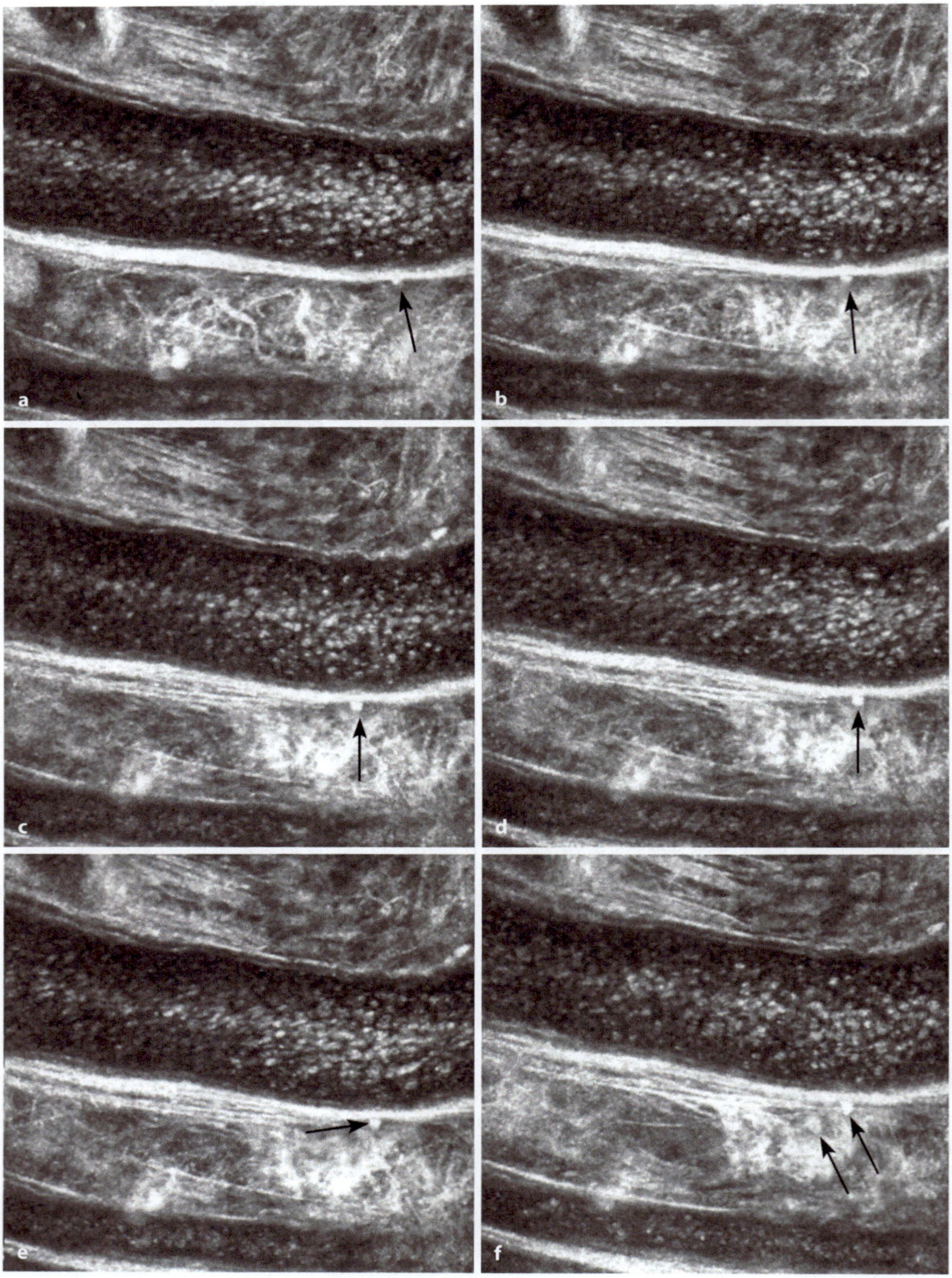

Fig. 5.132 Inflammation, rolling diapedesis. **a–e** Confocal in vivo microscopy sequence showing diapedesis. An inflammatory cell (*arrows*) crosses the vascular endothelium and then acquires the mature phenotype of a dendritic cell (**e, f**)

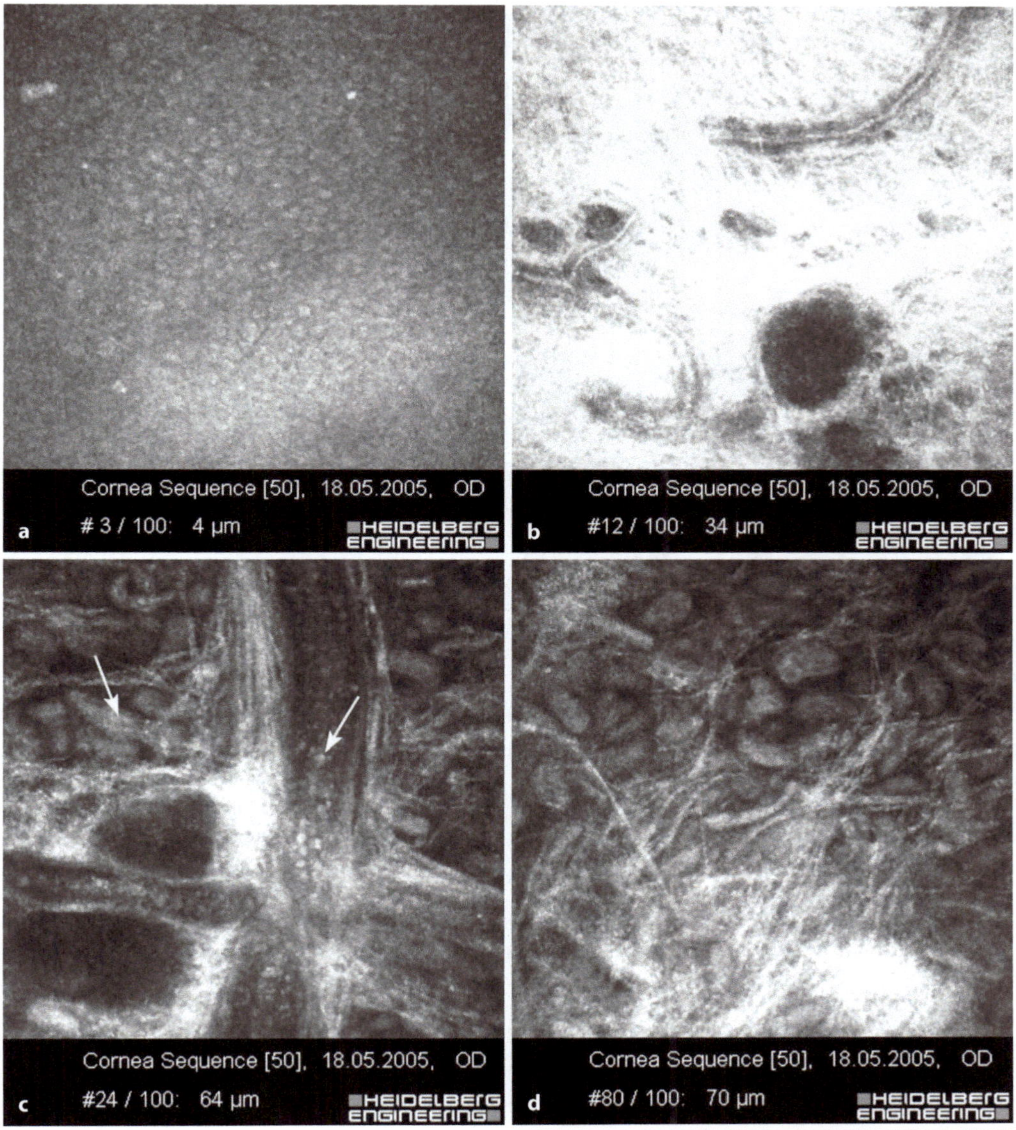

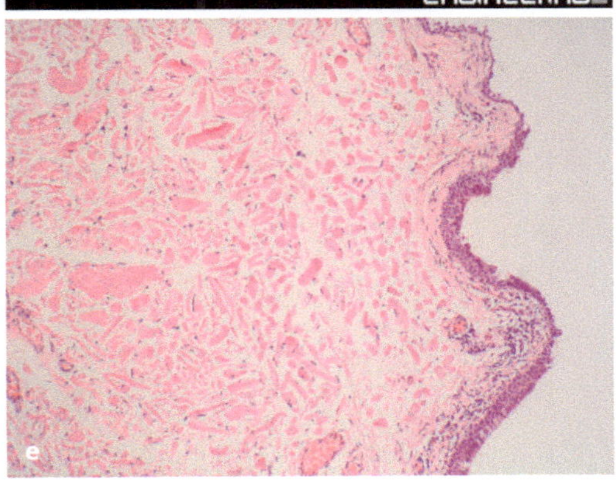

Fig. 5.133 Amyloidosis. **a–e** Under normally structured conjunctiva (**a**) and in the subconjunctival space (**b**) between blood vessels (*right arrow*), a lobular tissue pattern (*left arrow*) is visible (**c, d**), which shows a high degree of consistency with histology after invasive biopsy

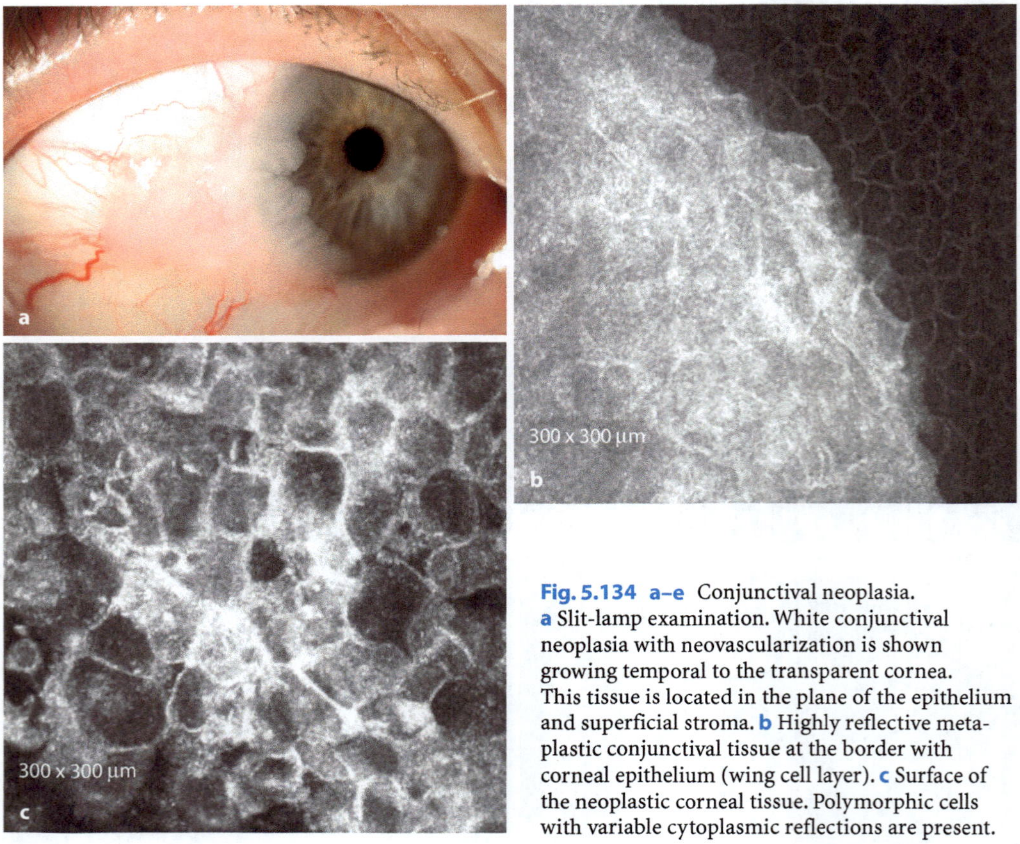

Fig. 5.134 a–e Conjunctival neoplasia. **a** Slit-lamp examination. White conjunctival neoplasia with neovascularization is shown growing temporal to the transparent cornea. This tissue is located in the plane of the epithelium and superficial stroma. **b** Highly reflective metaplastic conjunctival tissue at the border with corneal epithelium (wing cell layer). **c** Surface of the neoplastic corneal tissue. Polymorphic cells with variable cytoplasmic reflections are present.

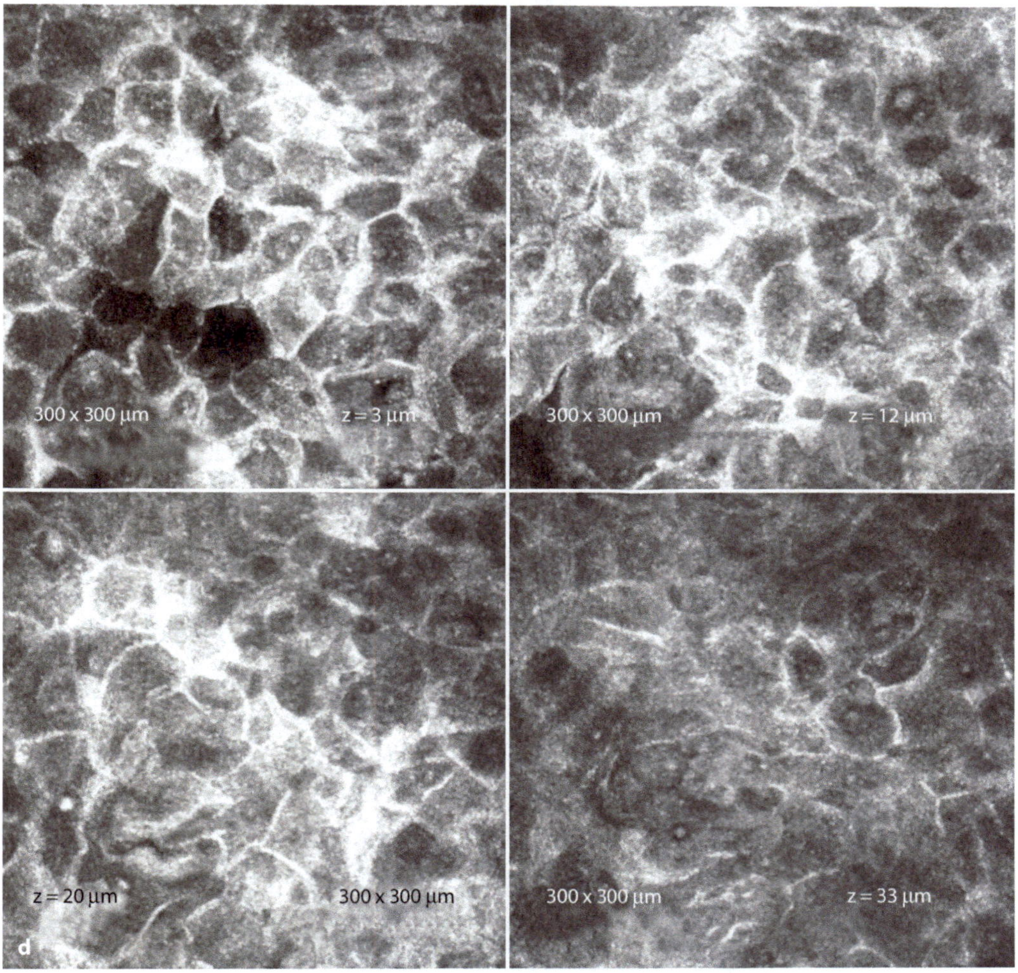

Fig. 5.134 (*continued*) **d, e** Image series with increasing focal depth from the loose structured epithelial surface to deeper low-contrast neoplastic tissue

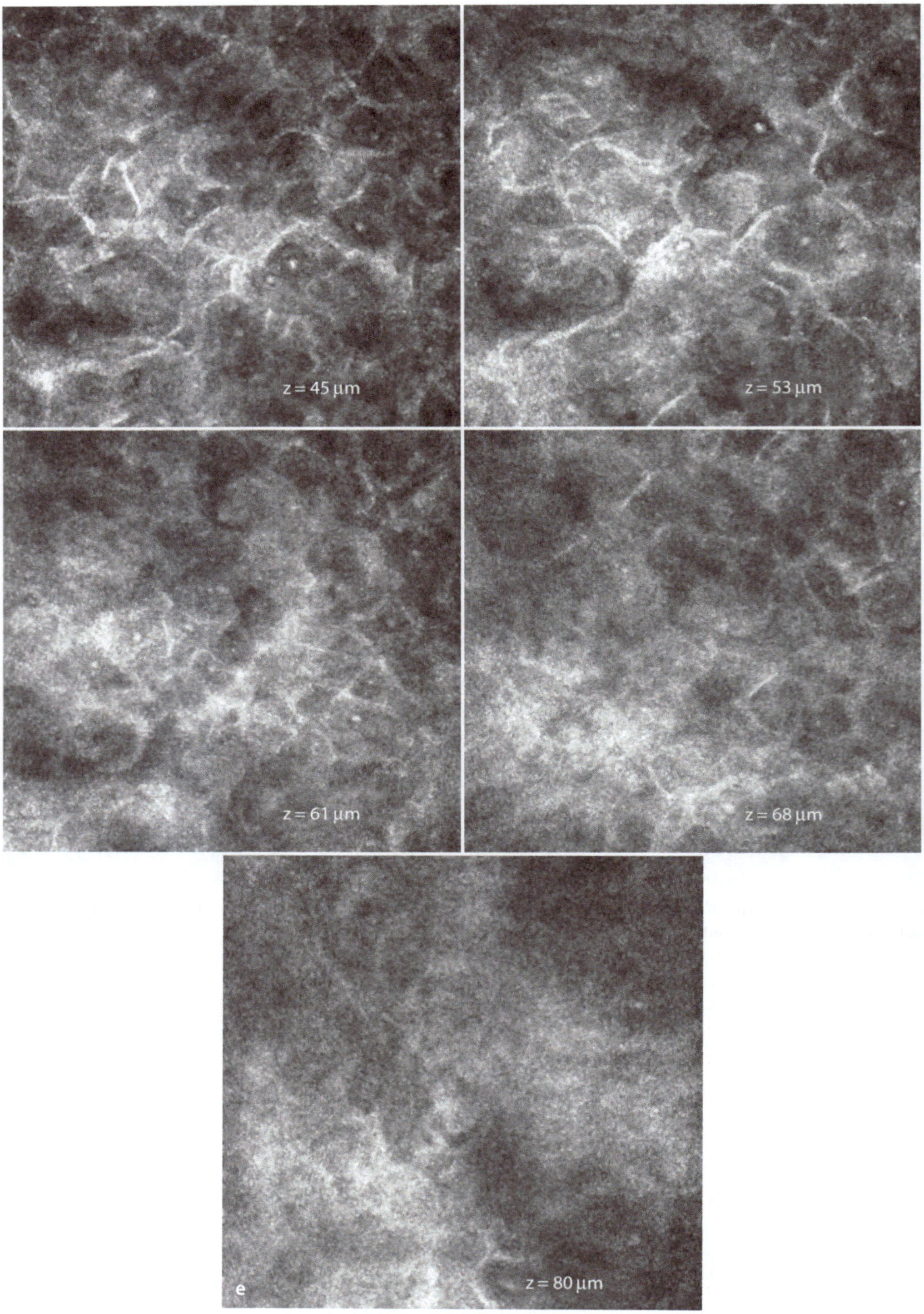

Fig. 5.134 (*continued*) **d, e** Image series with increasing focal depth from the loose structured epithelial surface to deeper low-contrast neoplastic tissue

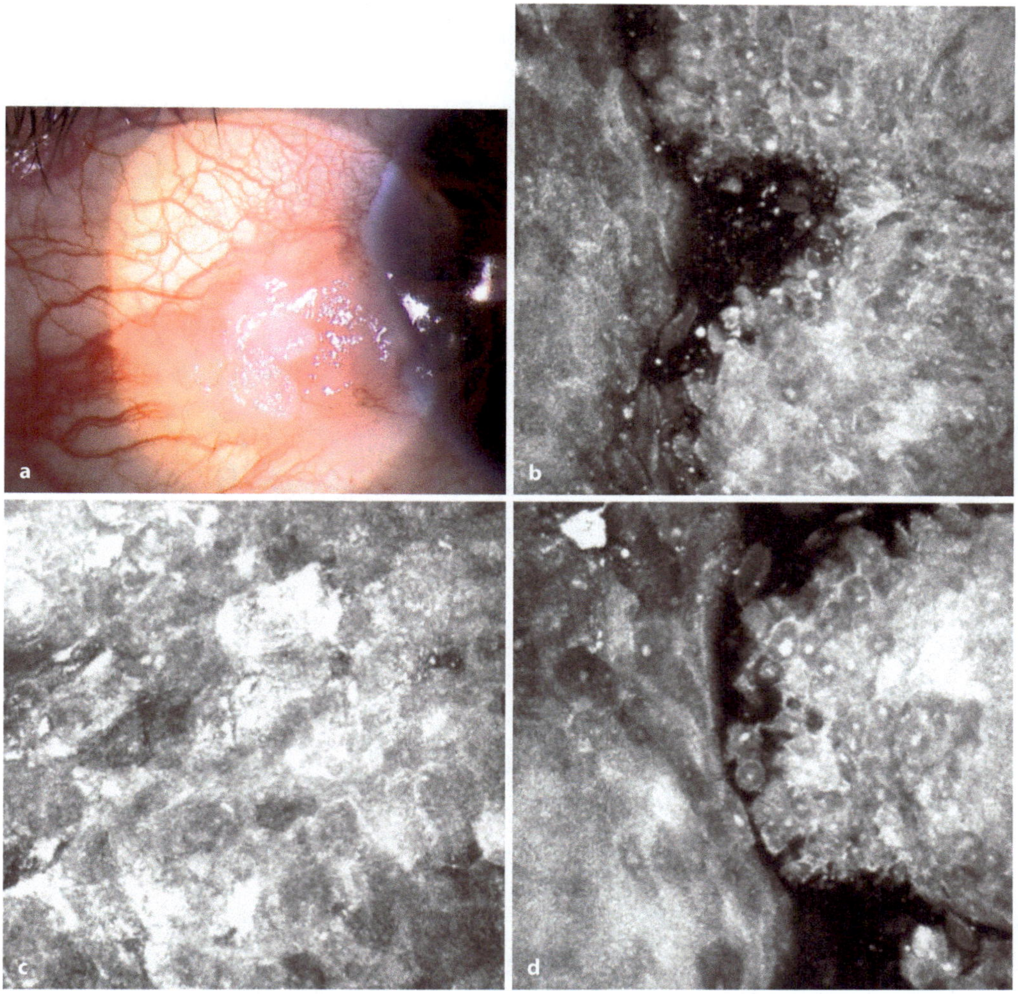

Fig. 5.135 Conjunctival and limbal tumors. **a** Slit-lamp photograph of a 55-year-old woman with a limbal tumor. **b–d** Numerous cellular abnormalities: nuclear abnormalities, polymorphism

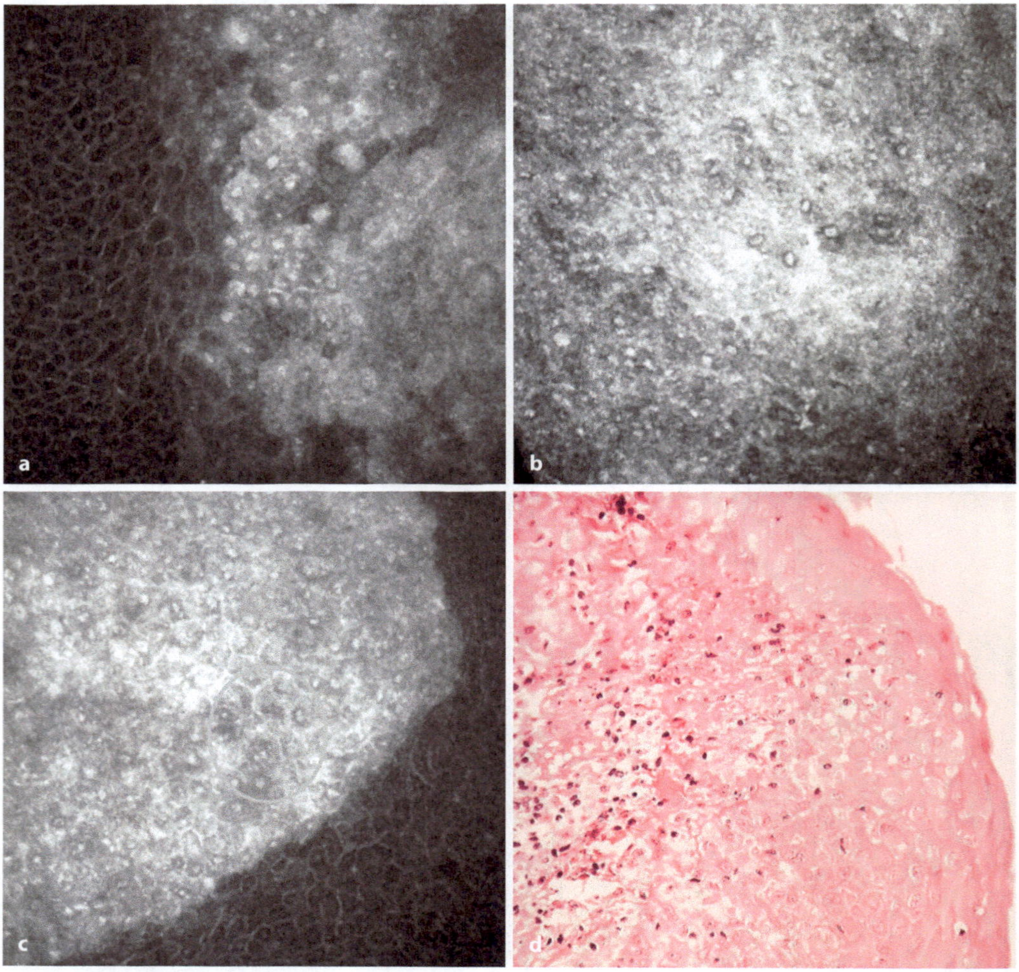

Fig. 5.136 Conjunctival and limbal tumors, epithelioma. **a–c** Numerous cellular abnormalities. **d** Histology

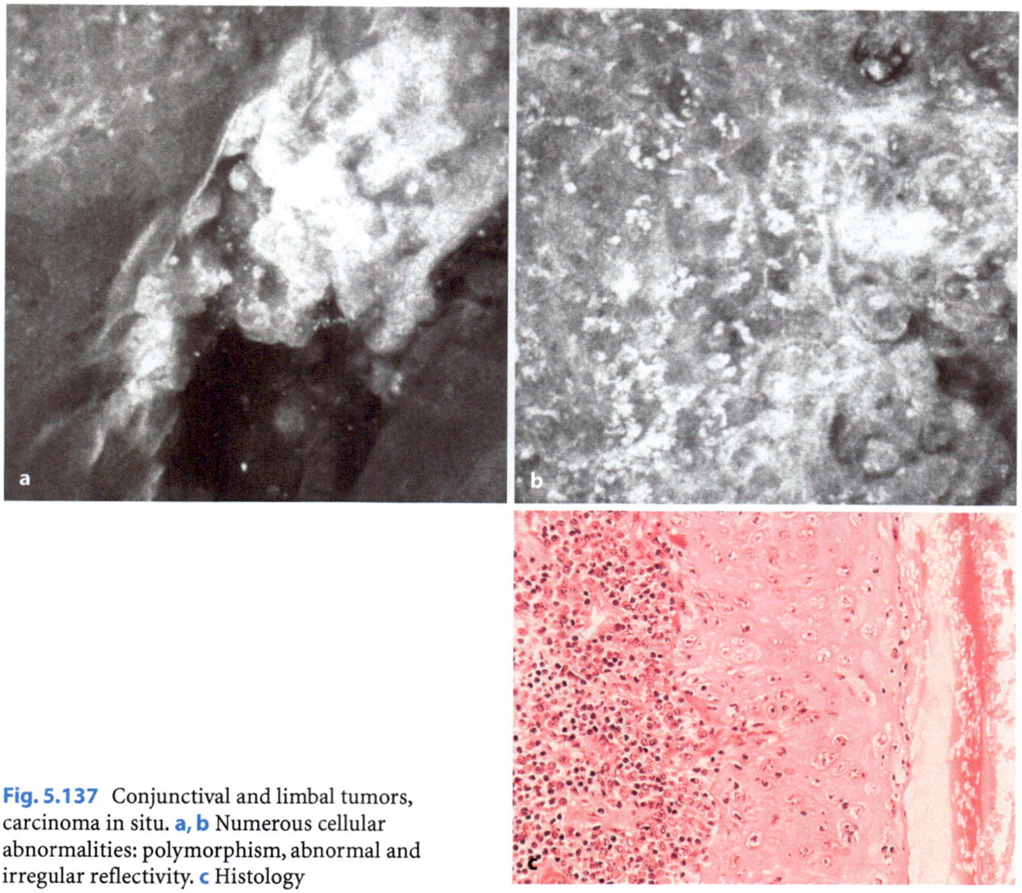

Fig. 5.137 Conjunctival and limbal tumors, carcinoma in situ. **a, b** Numerous cellular abnormalities: polymorphism, abnormal and irregular reflectivity. **c** Histology

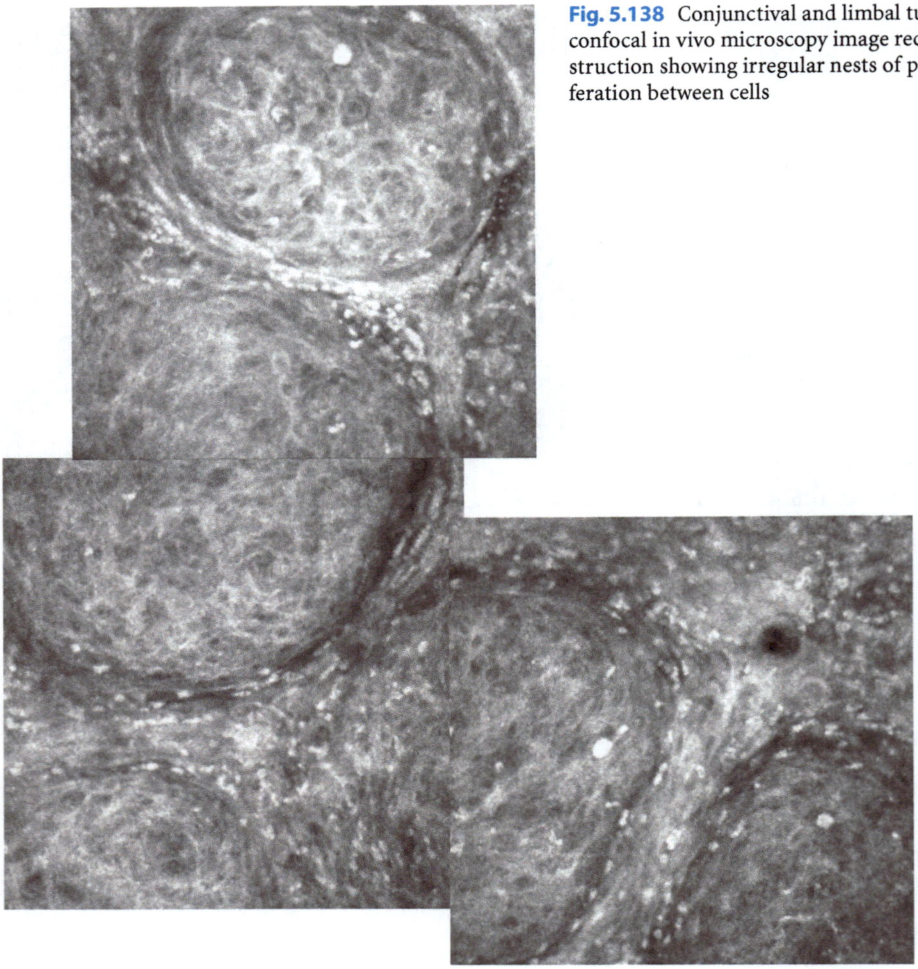

Fig. 5.138 Conjunctival and limbal tumors: confocal in vivo microscopy image reconstruction showing irregular nests of proliferation between cells

5.10 Conjunctiva

The different areas of the conjunctival epithelium display varying morphology.

The tarsal conjunctiva consists of two or three layers of cylindrical epithelium and contains individual goblet cells and lymph vessels. Deep to the epithelium is located the loosely meshed, highly vascular fiber lattice of the lamina propria with its papillary structure; this contains free cells (lymphocytes, plasma cells) and occasional lymph follicles, which are present in vastly increased numbers in inflammatory conditions, as well as smaller glandular packages. The lamina propria is continuous with the deeper-lying subconjunctival lamina, which is denser and is tightly fused with the tarsal plate.

The bulbar conjunctiva, in contrast, is characterized by several layers of squamous epithelium with numerous goblet cells. Deep to the epithelium is a very loosely meshed, highly mobile connective tissue with a pronounced capillary network.

5.10.1 Normal Anatomy

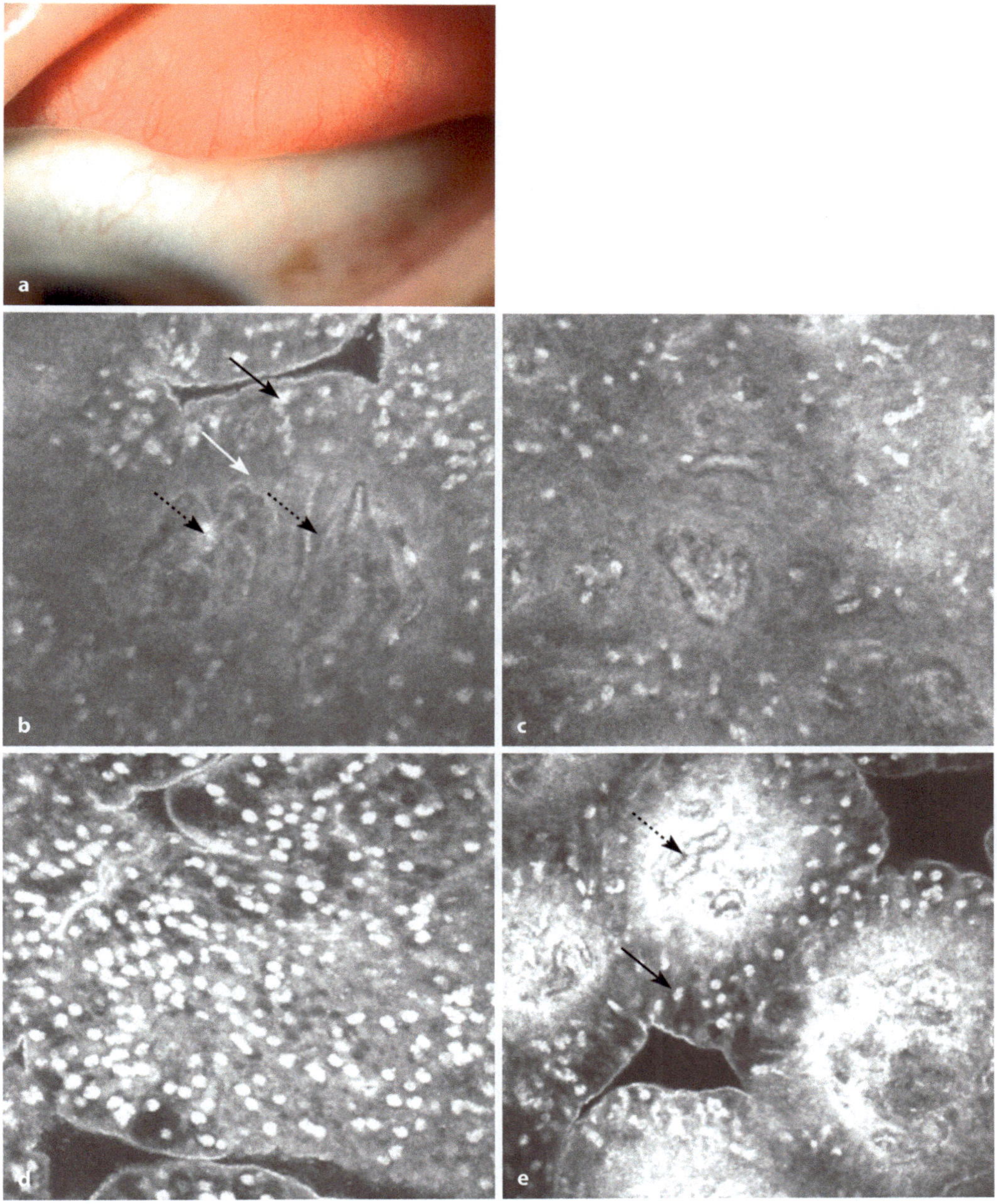

Fig. 5.139 Tarsal conjunctiva. **a** Slit-lamp photograph of the healthy tarsal conjunctiva of the upper lid. **b** Conjunctival epithelium with some highly reflective cells, possibly goblet cells (*black arrow*); under the basal membrane (*white arrow*) the lamina propria of the underlying connective tissue is seen with numerous small blood vessels and some highly reflective round cells, possibly lymphocytes (*dashed-line arrows*). **c** Subconjunctival connective tissue with blood vessels and highly reflective cells (lymphocytes, plasma cells). **d** Lymphatic follicle of the conjunctiva in a healthy person. **e** Epithelium (*black arrow*), lamina propria with blood vessels (*dashed-line arrow*)

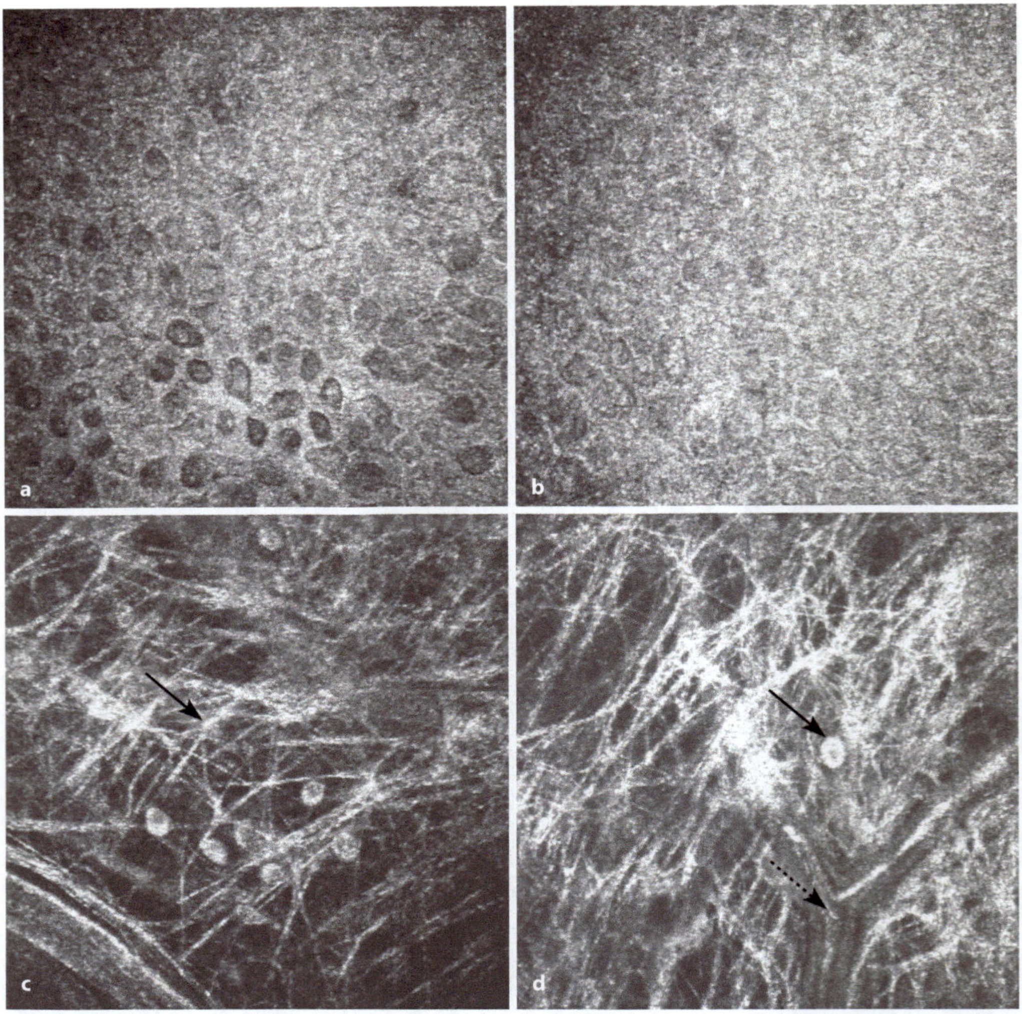

Fig. 5.140 Bulbar conjunctiva. **a** Superficial conjunctival epithelium (z=3 µm). Tthe epithelial structure is not as well organized as in the cornea; the cell borders are brighter than the cytoplasm, and in some cells a bright nucleus is visible. **b** Deeper conjunctival epithelium (z=10 µm). **c–e** Lamina propria with highly reflective latticelike connective tissue and some highly reflective cells (*black arrows*) and blood vessels (*dashed-line arrows*). **f, g** Oblique sections: epithelium (*black arrow*), underlying connective tissue (*dashed-line arrow*), blood vessels (*white arrow*)

5.10.2
Pathological Findings

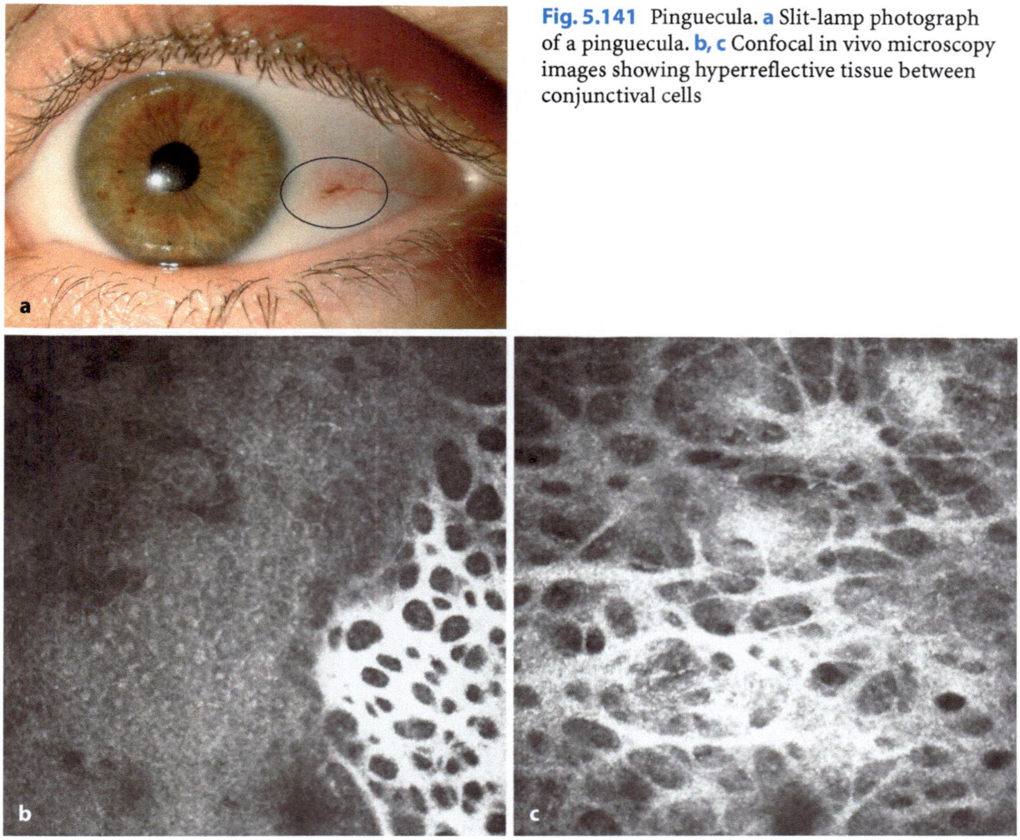

Fig. 5.141 Pinguecula. a Slit-lamp photograph of a pinguecula. b, c Confocal in vivo microscopy images showing hyperreflective tissue between conjunctival cells

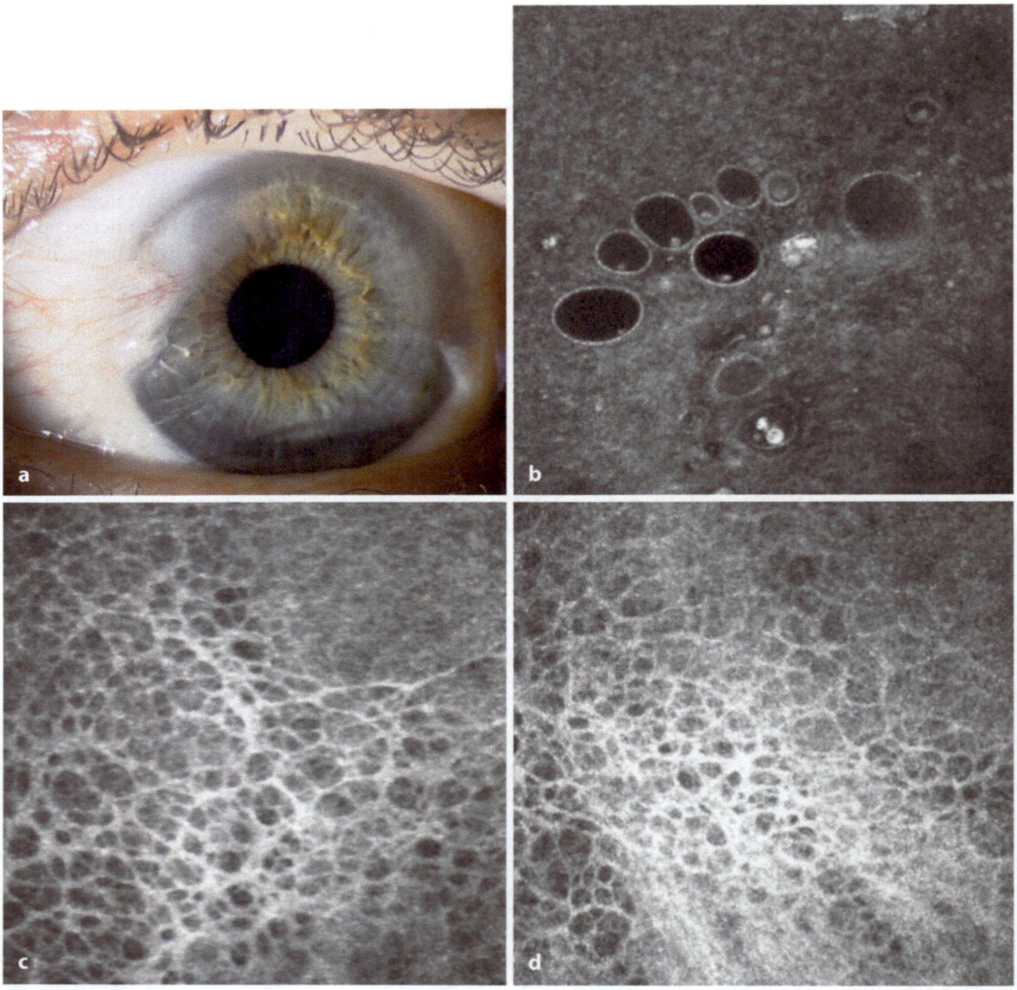

Fig. 5.142 Pterygium. **a** Slit-lamp photograph of a 33-year-old man with pterygium. **b** Superficial confocal in vivo microscopy image of pterygium: microcysts between conjunctival epithelial cells. **c, d** Reflective stroma of pterygium

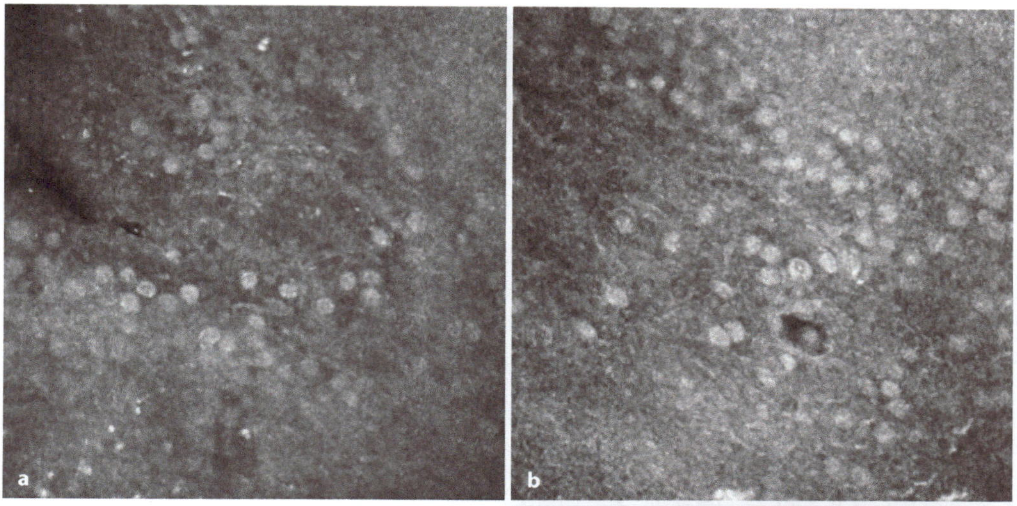

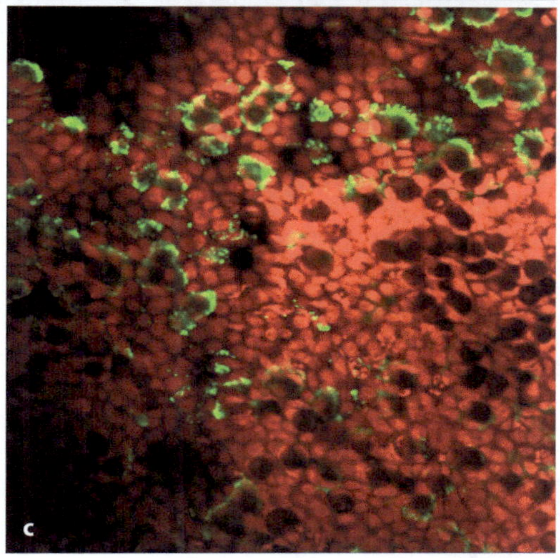

Fig. 5.143 Pterygium. **a, b** Superficial images of pterygium: numerous goblet cells (*reflective round cells*). **c** Impression cytology: confirmation of the high number of goblet cells (*green*) within the epithelium (*red*)

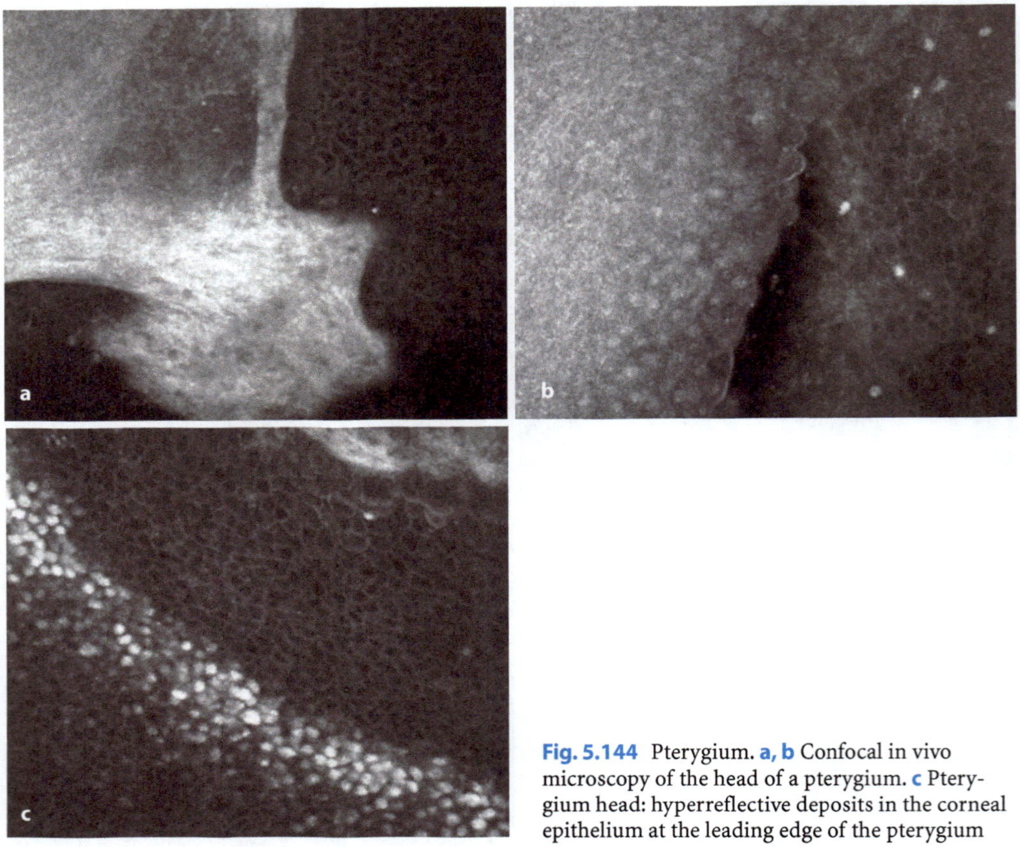

Fig. 5.144 Pterygium. **a, b** Confocal in vivo microscopy of the head of a pterygium. **c** Pterygium head: hyperreflective deposits in the corneal epithelium at the leading edge of the pterygium

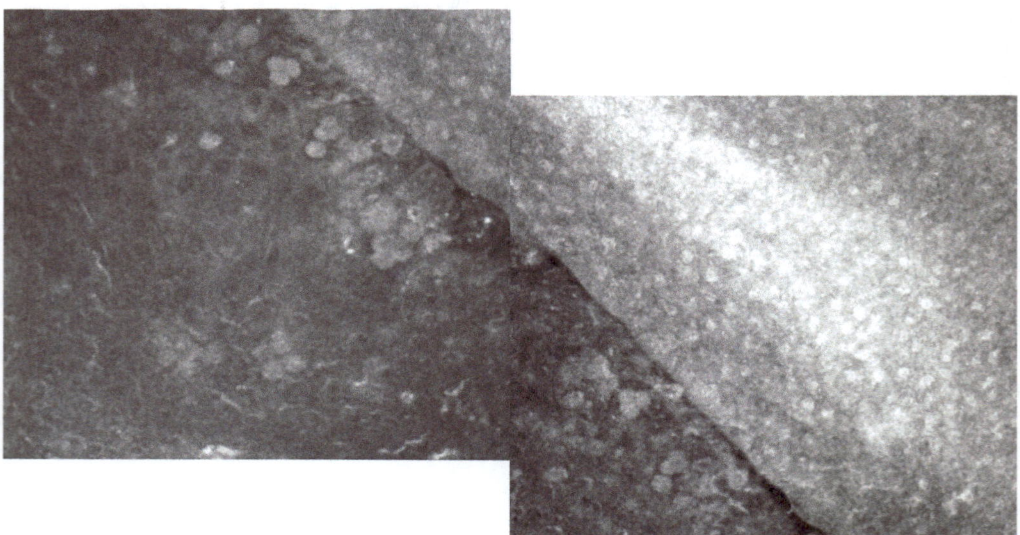

Fig. 5.145 Pterygium. Head of pterygium, infiltration of conjunctival cells (*reflective cells*) within the corneal epithelium at the edge of the pterygium

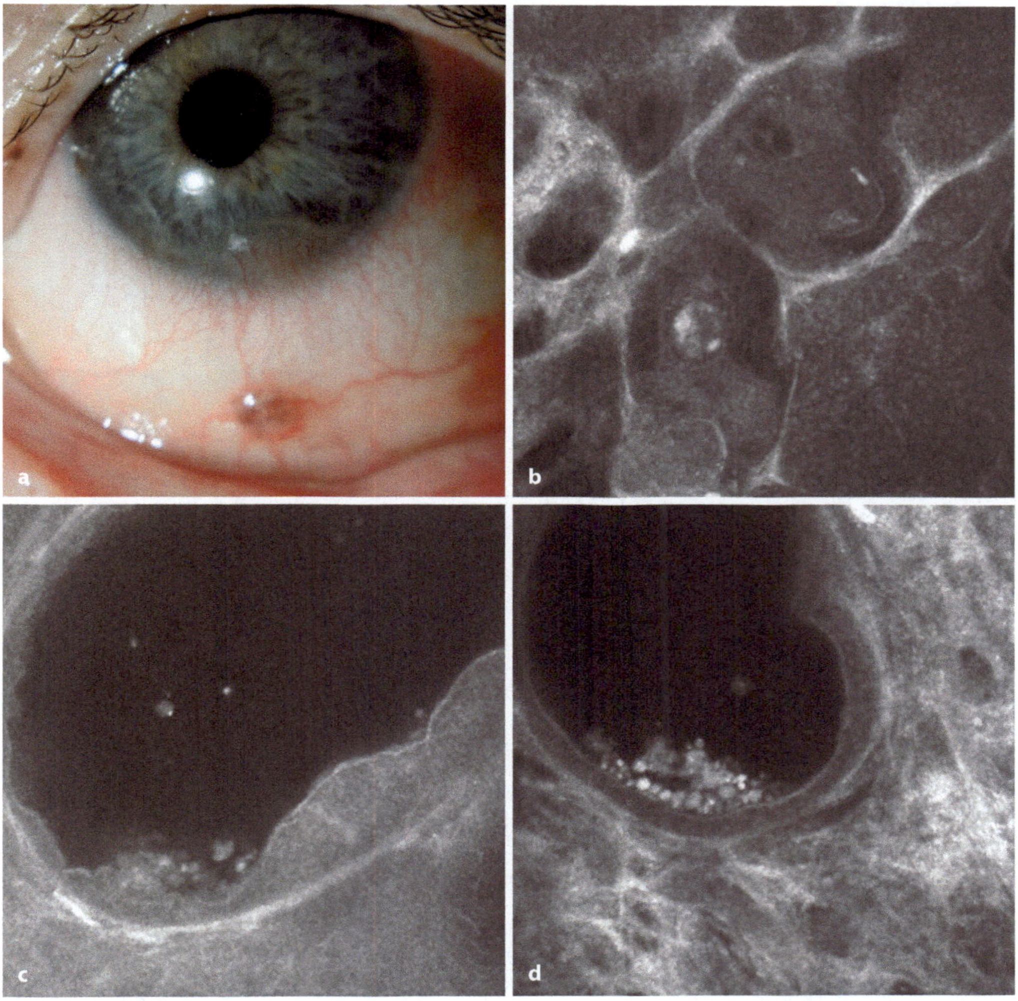

Fig. 5.146 Nevus and conjunctival melanosis. **a** Slit-lamp photograph of a 50-year-old man with an inferior conjunctival nevus. **b–d** Confocal in vivo microscopy images showing cysts within the nevus

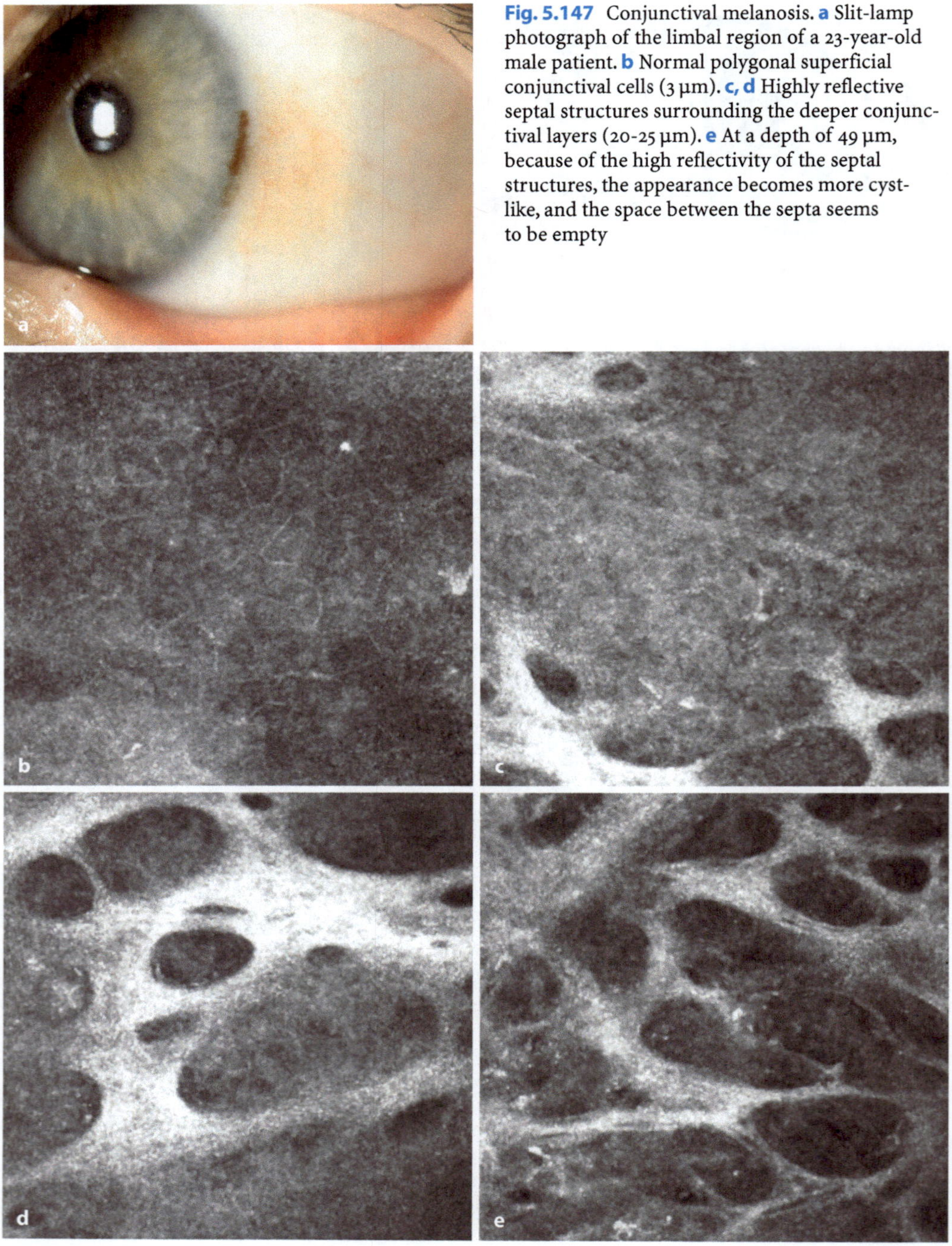

Fig. 5.147 Conjunctival melanosis. **a** Slit-lamp photograph of the limbal region of a 23-year-old male patient. **b** Normal polygonal superficial conjunctival cells (3 µm). **c, d** Highly reflective septal structures surrounding the deeper conjunctival layers (20-25 µm). **e** At a depth of 49 µm, because of the high reflectivity of the septal structures, the appearance becomes more cyst-like, and the space between the septa seems to be empty

5.11 Eyelid

5.11.1 Normal Anatomy

The human eyelid consists of four layers [67]. The innermost layer is the conjunctiva (refer to Sect. 5.10), which is separated from the eyeball by the tear film. The conjunctival tissue comprises a thin layer of nonkeratinized epithelium scattered with goblet cells and a stromal portion containing many blood vessels as well as immunocompetent cells.

Applied closely to the conjunctiva is the tarsal plate of the eyelid. This is a layer of dense fibrous connective tissue that stabilizes and confers shape on the human eyelid [56]. The tarsal plate is one of the main insertion points for the levator palpebrae muscle [2].

A third, muscular layer consists of the orbicularis oculi muscle and parts of the levator palpebrae muscle. The orbicularis oculi muscle is notionally divided into three zones: orbital, preseptal, and pretarsal.

The outermost layer of the eyelid is the epidermis. Compared with the surrounding epidermal structures, the eyelid epidermis is extremely thin, comprising only three to four cell layers. The eyelid margin is covered with cutaneous epithelium [96]. A mucocutaneous junction with the conjunctiva, known as Marx's line or the gray line, is located at the inner edge of the margin [15]. At its anterior edge the eyelashes emerge, and the inner edge contains the excretory ducts of the meibomian glands. A small elevation known as the lacrimal point is located about one-sixth of the distance along the eyelid margin from the medial canthus. The eyelashes are thick, short, curved hairs, characteristically arranged in two or three rows [21].

The human eyelid contains different types of glands, the largest being the meibomian glands. Arranged in 20–25 noncommunicating tubes in each eyelid, these are found in the tarsal plates. Each gland tube has its own orifice through which lipid is secreted. Images of the meibomian glands obtained by confocal in vivo microscopy have been presented in the literature [36, 55]. However, confocal visualization appears to be problematic because of the deep location of these glands in the tarsal plate.

Smaller glands found in the eyelid are the accessory glands of Wolfring, located at the upper border of the tarsal plate, and the accessory glands of Krause, located near the fornix. The glands of Zeis, which are attached directly to the follicles of the eyelashes, produce a sebaceous secretion. The glands of Moll, essentially a type of specialized sweat gland, are situated close to the eyelid margin. Sweat glands are found scattered throughout the dermal layer of the eyelid [67].

The function of the eyelid is to protect the eye from trauma and excessive light. With each blink of the eyelids, the tear film is spread over the eyeball to protect it from drying (see Sect. 5.1).

The superficial structures of the eyelids can be examined by confocal microscopy, and pathologic changes, such as tumors or inflammation, display changes in microscopic morphology, as illustrated in the following section. Confocal microscopy facilitates objective assessment of the clinical situation in blepharitis, not merely in terms of diagnosis and disease severity but also for the purposes of classification and as an aid in deciding whether additional anti-inflammatory therapy is necessary.

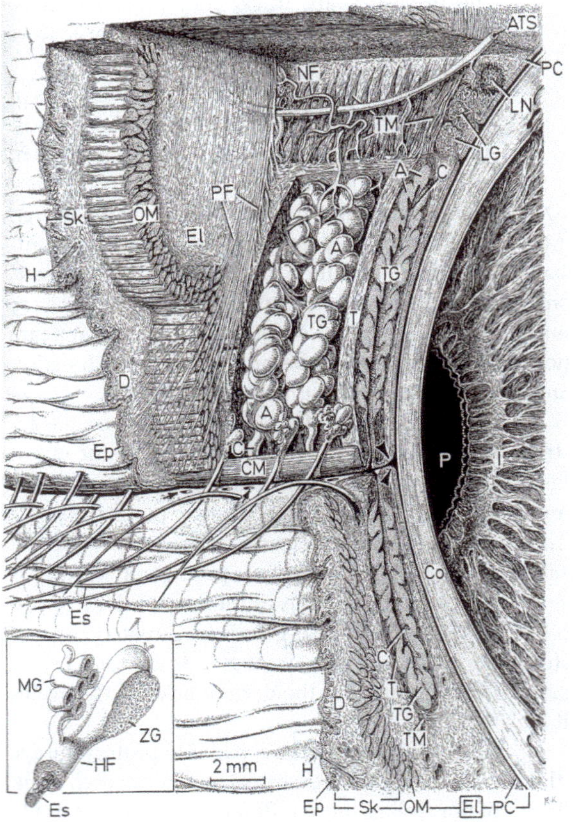

Fig. 5.148 Schematic illustration of the eyelid structures (adapted from Krstić RV. *Human Microscopic Anatomy: An Atlas for Students of Medicine and Biology*. Berlin, Heidelberg, New York: Springer-Verlag, 1991 [39])

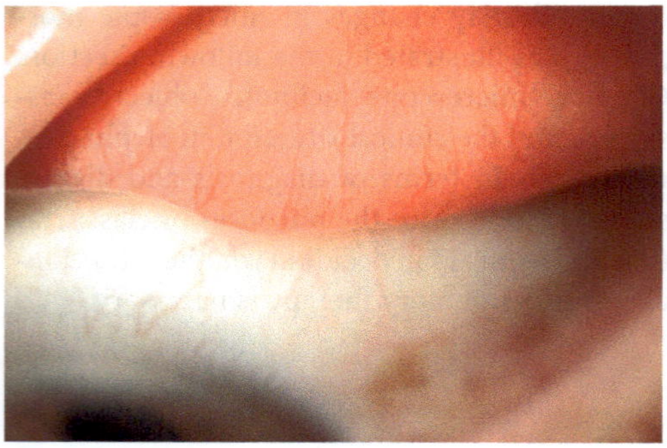

Fig. 5.149 Slit-lamp photograph of the normal eyelid

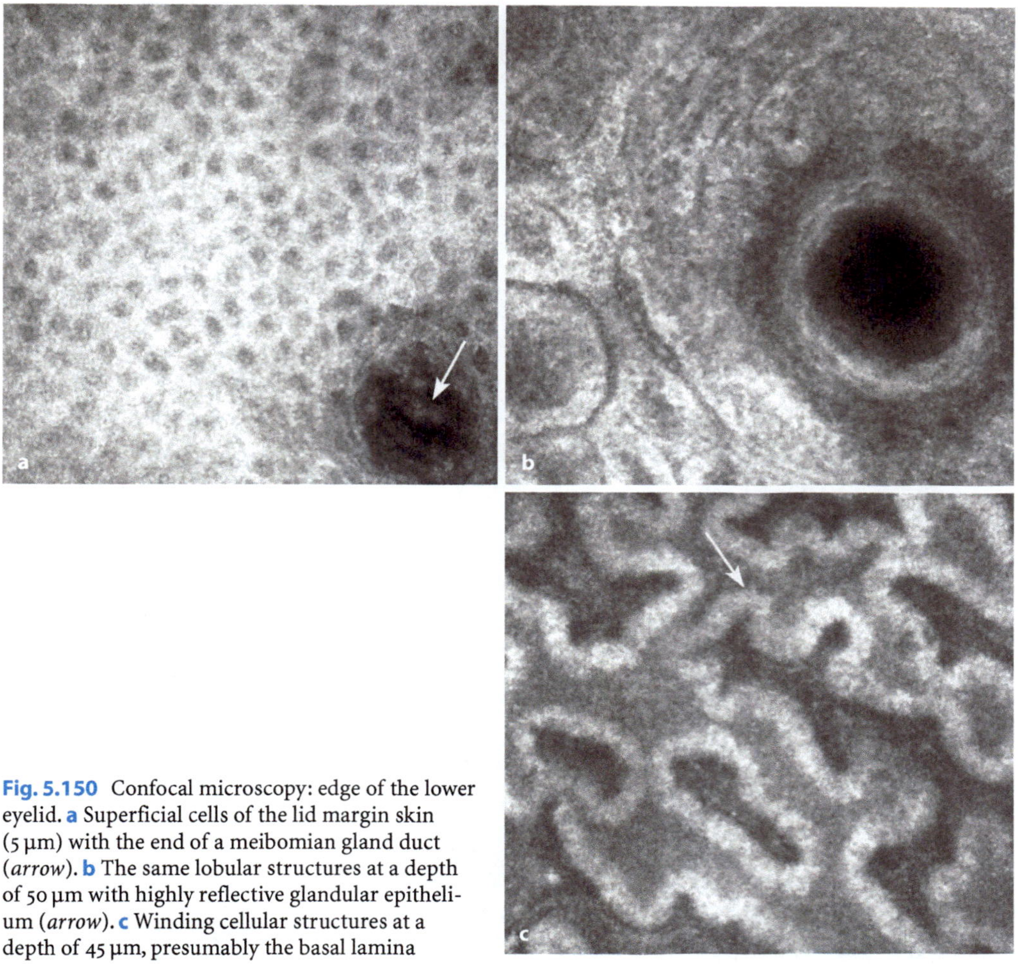

Fig. 5.150 Confocal microscopy: edge of the lower eyelid. **a** Superficial cells of the lid margin skin (5 µm) with the end of a meibomian gland duct (*arrow*). **b** The same lobular structures at a depth of 50 µm with highly reflective glandular epithelium (*arrow*). **c** Winding cellular structures at a depth of 45 µm, presumably the basal lamina

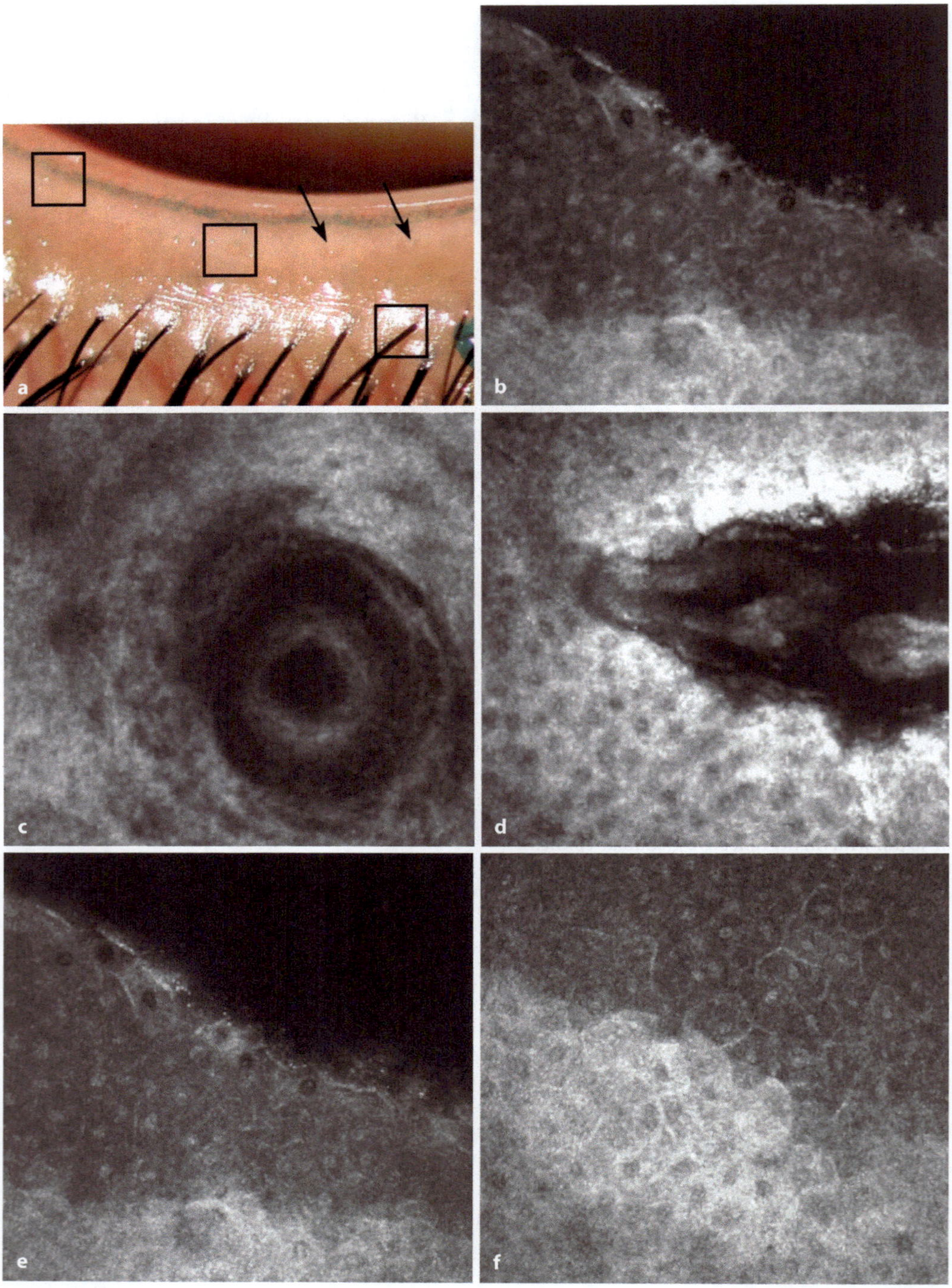

Fig. 5.151 **a** Slit-lamp photograph of the lower eyelid. Marx's line is stained with lissamine green. *Arrows* indicate the location of the orifices of the meibomian glands. **b** Confocal image of the inner edge of the eyelid and the mucocutaneous junction. **c** Excretory duct of a meibomian gland. **d** Hair follicle of an eyelash. **e, f** Confocal image of Marx's line, the junction between the cutaneous eyelid epithelium (*lighter cells*) and the tarsal conjunctiva (*darker cells*)

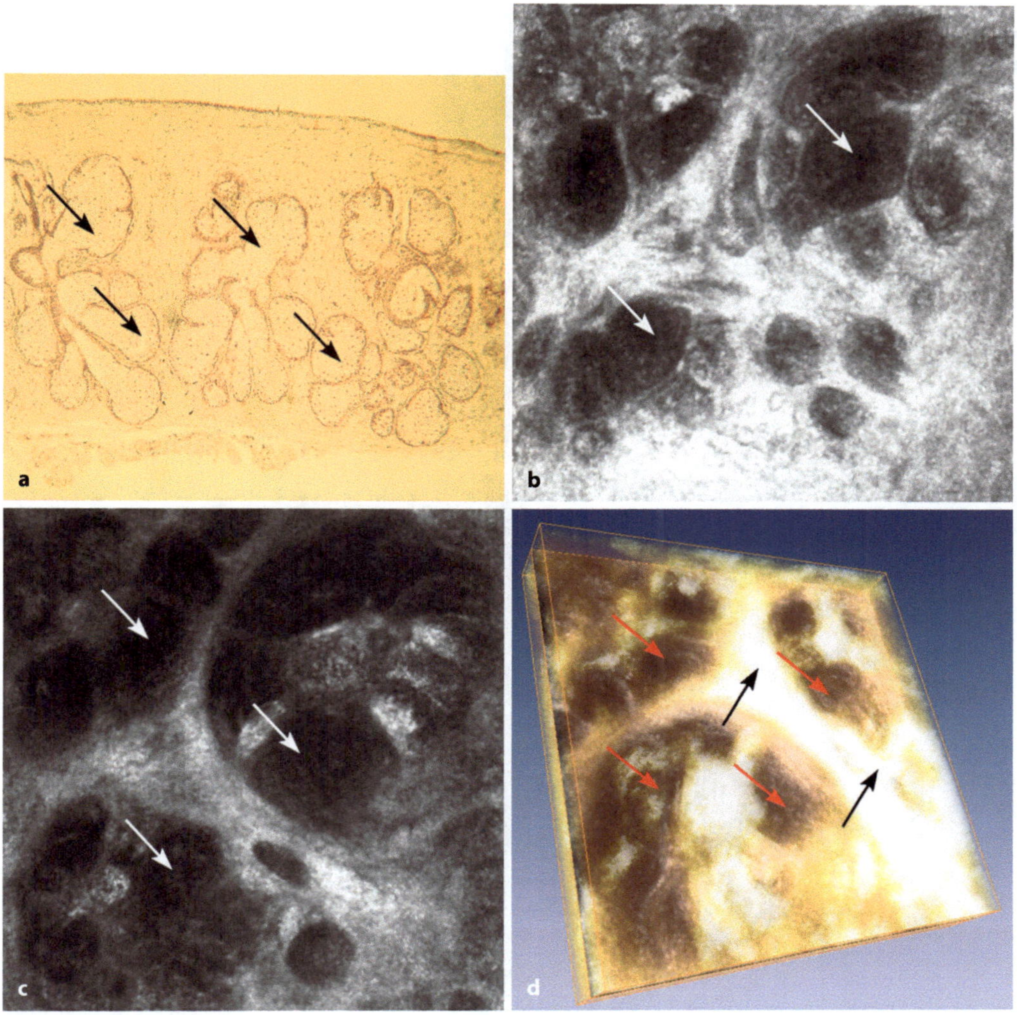

Fig. 5.152 Meibomian glands. **a** Histological sagittal section of the human eyelid. Meibomian glands are marked with *arrows*. **b, c** Confocal images obtained from a sagittal eyelid section ex vivo immediately after excision. Depth difference between the pictures is 20 µm. The meibomian glands are marked with *arrows*. **d** Three-dimensional reconstruction of the image series, showing the meibomian glands (*red arrow*) within the fibrous tissue of the tarsal plate (*black arrows*)

5.11.2 Pathological Findings

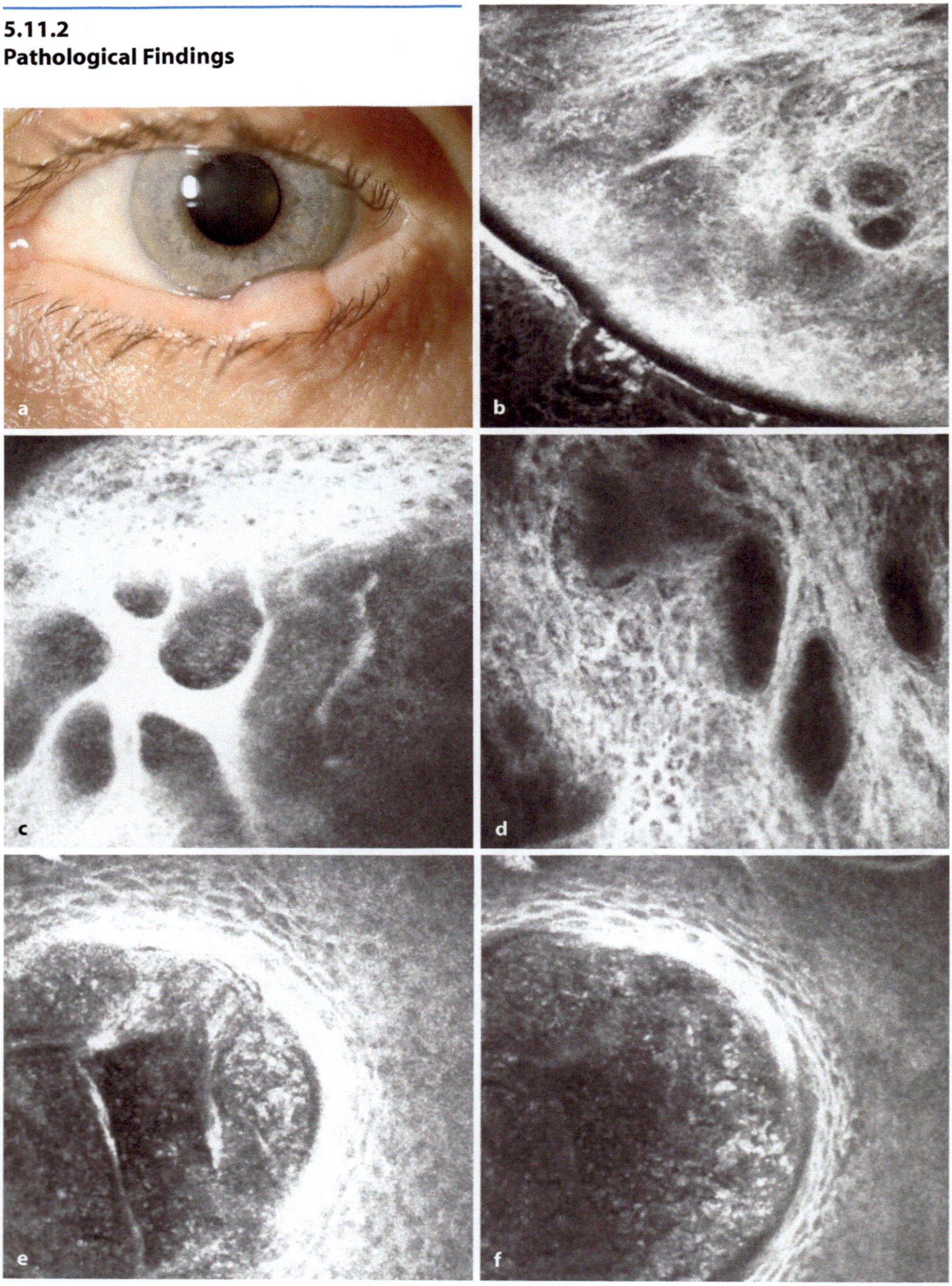

Fig. 5.153 Papilloma. **a** Slit-lamp photograph of the lower lid of a 64-year-old woman. **b** Confocal microscopy (45 µm) of the edge of the lesion (oblique section): the tear film is visible in the lower left corner. **c** Confocal microscopy (51 µm): highly reflective septa due to the lobular structure of the lesion. **d** In the deeper layers the lesion appears show cystic organization. **e, f** Center of the lesion, superficial (7–12 µm): cystic structure surrounded by epithelium. This may be an enlarged meibomian or other gland

Fig. 5.154 Basal cell carcinoma. **a** Slit-lamp photograph of the lower lid of the right eye of a 69-year-old man. **b–e** Confocal microscopy: irregular epithelium with varying cell size and shape, especially in the deeper layers

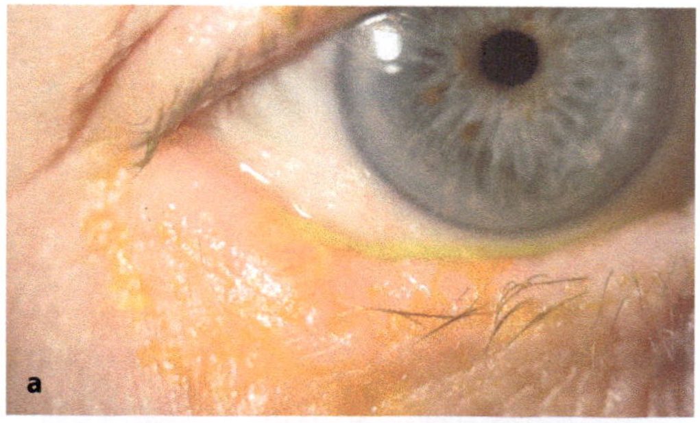

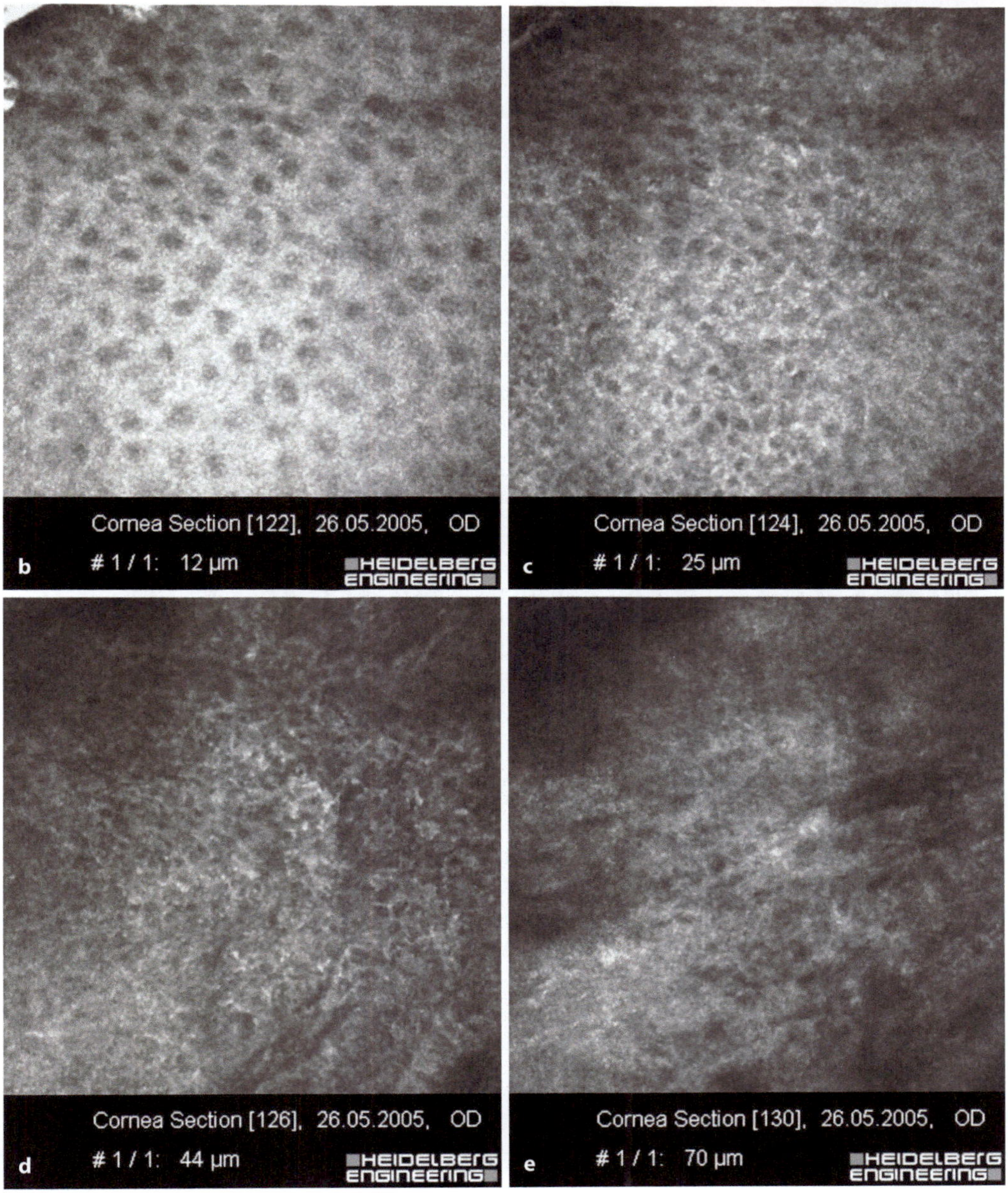

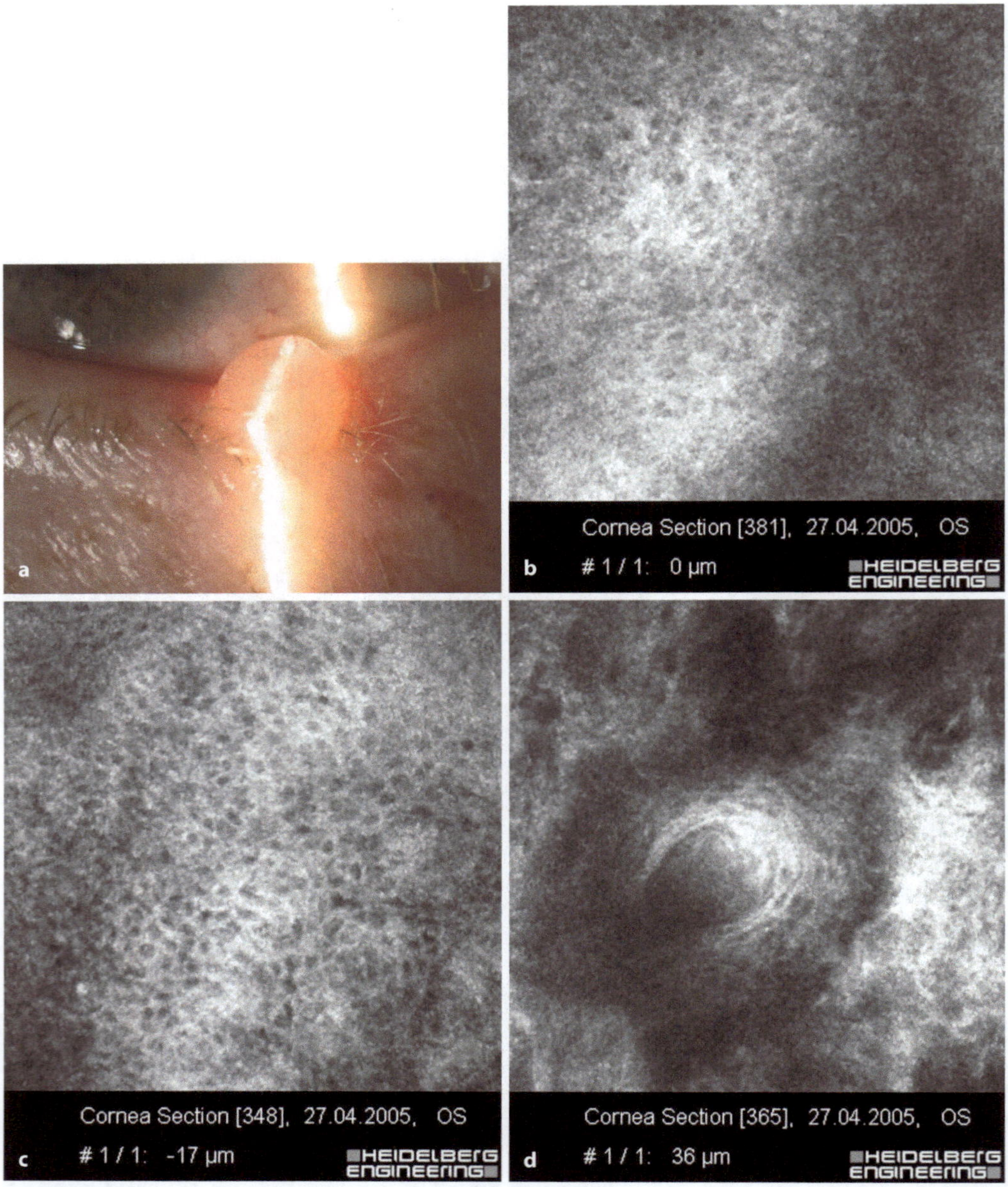

Fig. 5.155 a–f Basal cell carcinoma. **a** Nodular basal cell carcinoma of the lower lid of the right eye in a 76-year-old male patient. **b, c** Confocal microscopy: irregular epithelium with varying cell size and shape. **d–f** Completely irregular and polymorphic histology in the deeper layers with glandlike and cystlike structures

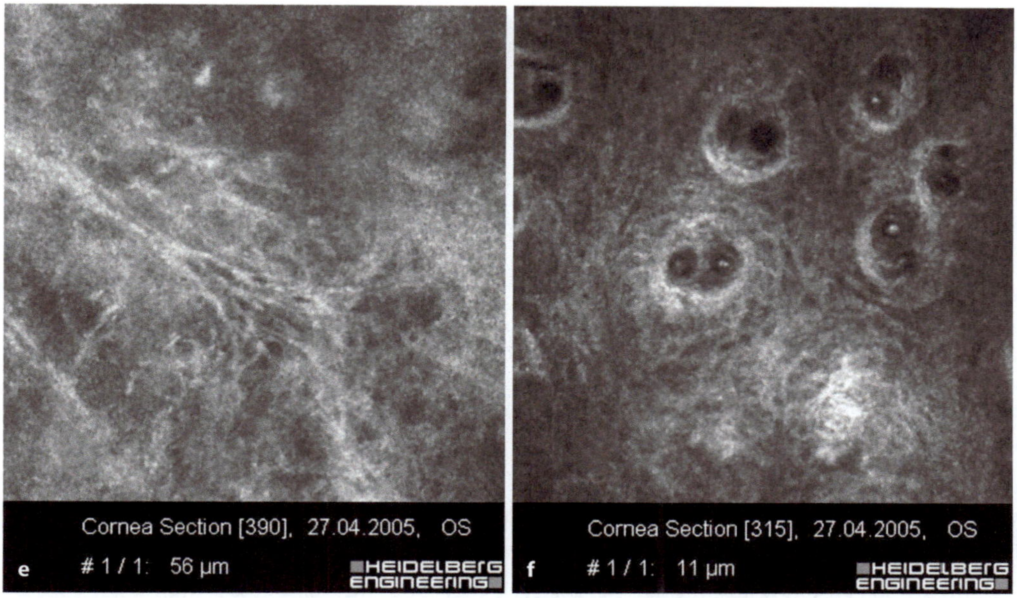

Fig. 5.155 (*continued*) **d–f** Completely irregular and polymorphic histology in the deeper layers with glandlike and cystlike structures

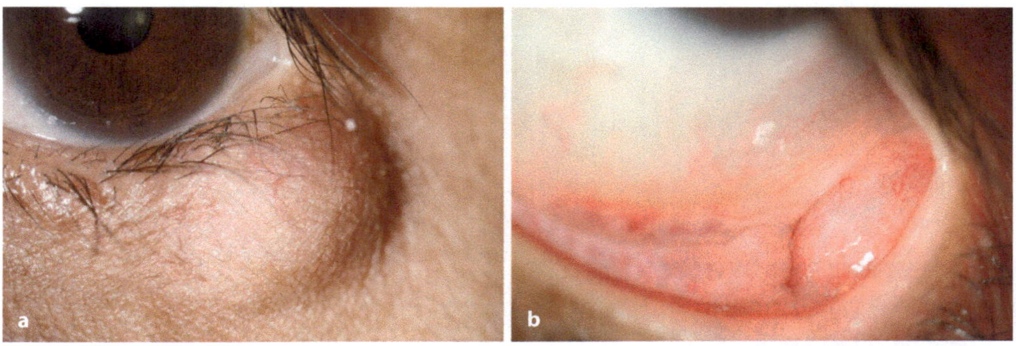

Fig. 5.156 a–e Chalazion. **a, b** Slit-lamp photographs of a 45-year-old woman with a chalazion

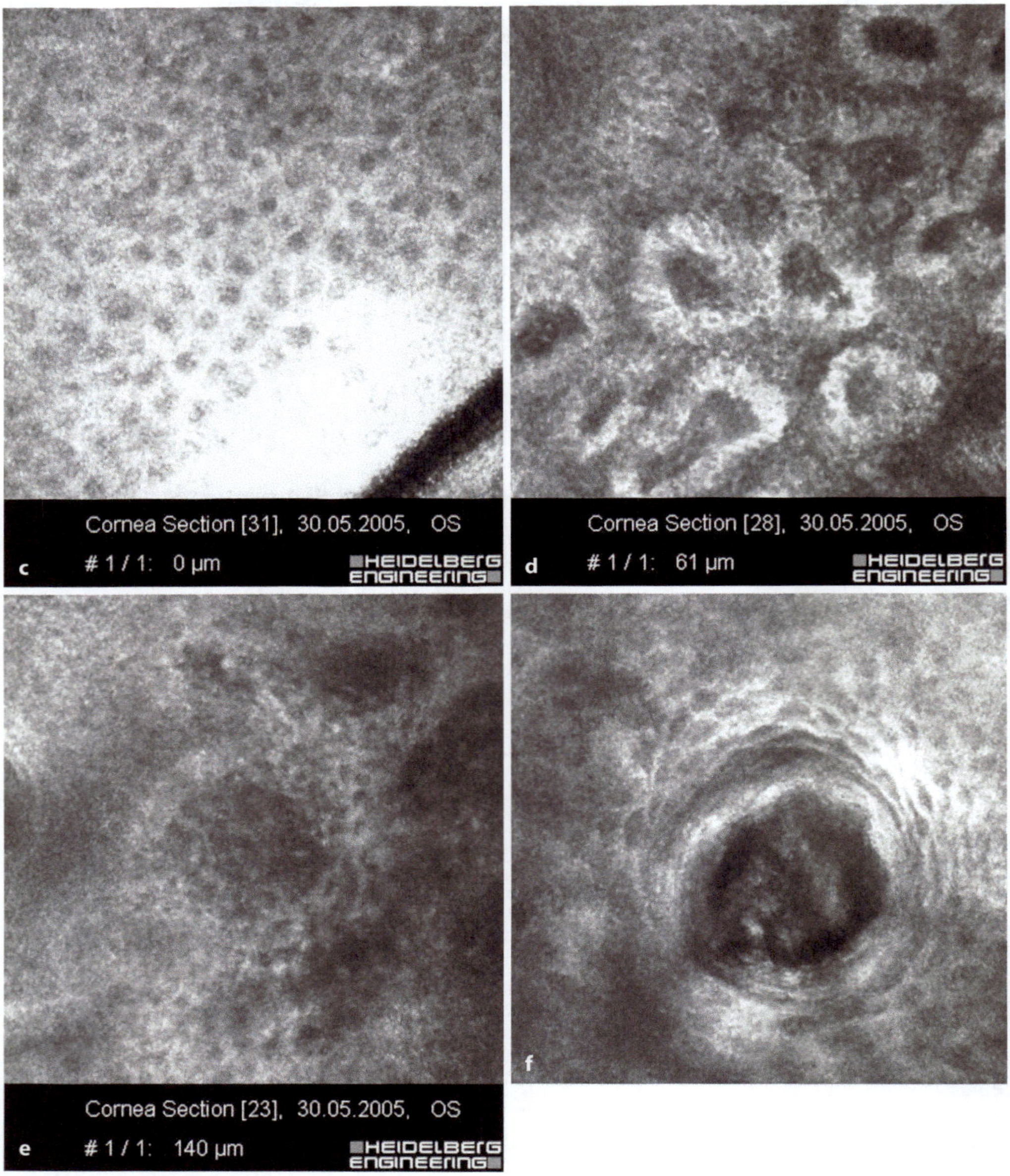

Fig. 5.156 (*continued*) **c–e** Confocal microscopy through the conjunctiva: conjunctival epithelium (**c**), deeper lobular structures, presumably horizontal cone sections of the epithelial layers (**d, e**). **f** Obstructed excretory duct of a meibomian gland

Promising Applications

Because slit-lamp biomicroscopy based on focal illumination and binocular observation is the standard diagnostic tool in ophthalmology, it is now possible to translate these moderately magnified optical sections to the cellular level.

Cells of local origin as well as those that have migrated into the area of interest can be identified, differentiated, and placed in the context on the basis of cellular interactions. New populations that were previously invisible with in vivo techniques, such as antigen-presenting LCs, can now be localized and quantified. Because of the high time resolution with 30 frames per second, cell migration through vessel walls can be observed, and dynamic processes can be included in the description of in vivo physiological and pathophysiological phenomena.

6.1 Glaucoma Surgery

The development of a filtering bleb, determined by the postoperative wound healing process, is a major factor in the efficiency and long-term success of surgical procedures in glaucoma. Many authors have investigated the clinical [10, 68] and, in a few cases, the histological appearance [1, 69] of these blebs. In vivo confocal microscopy allows clinicians and researchers to visualize functioning or nonfunctioning blebs at the cellular level [40].

Functioning blebs have numerous optically clear spaces corresponding to microcysts between superficial conjunctival cells (Fig. 6.1a–c). Subepithelial connective tissue images of these blebs reveal loosely arranged tissue (Fig. 6.1d, e).

Very few microcysts are observed between superficial conjunctival cells in nonfunctioning blebs (Fig. 6.2a–c). The subepithelial tissue of these blebs shows a dense connective tissue (Fig. 6.2d, e). The encapsulation of nonfunctioning blebs has also been clearly observed (Fig. 6.2g, h).

Functioning blebs with mitomycin C have numerous large microcysts in the superficial epithelial layer (Fig. 6.3a, b), and some of these microcysts contain hyperreflective microdots (Fig. 6.3c, d). The subepithelial tissue of these blebs shows a loose arrangement (Fig. 6.3e).

Confocal in vivo microscopy images of filtering blebs reveal good consistency with the findings from previous ex vivo studies. By permitting the in vivo study of wound healing mechanisms after filtering surgery, this technique enhances understanding of the histological processes responsible for filtration or failure.

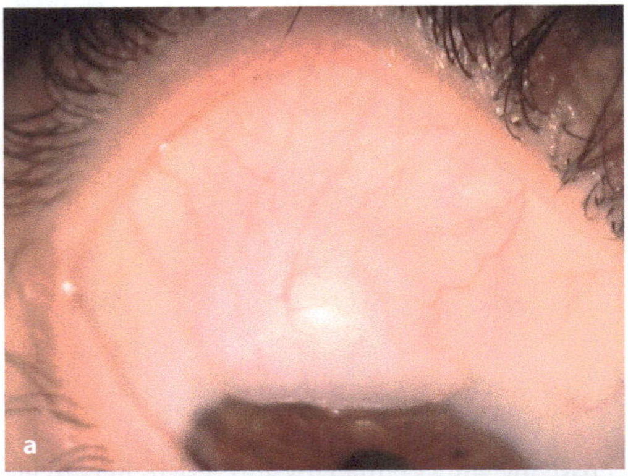

Fig. 6.1 Glaucoma surgery: functioning blebs. **a** Slit-lamp photograph of a functioning bleb. **b, c** Numerous clear spaces, corresponding to microcysts in the conjunctival epithelium. **d,e** Widely spaced subepithelial connective tissue

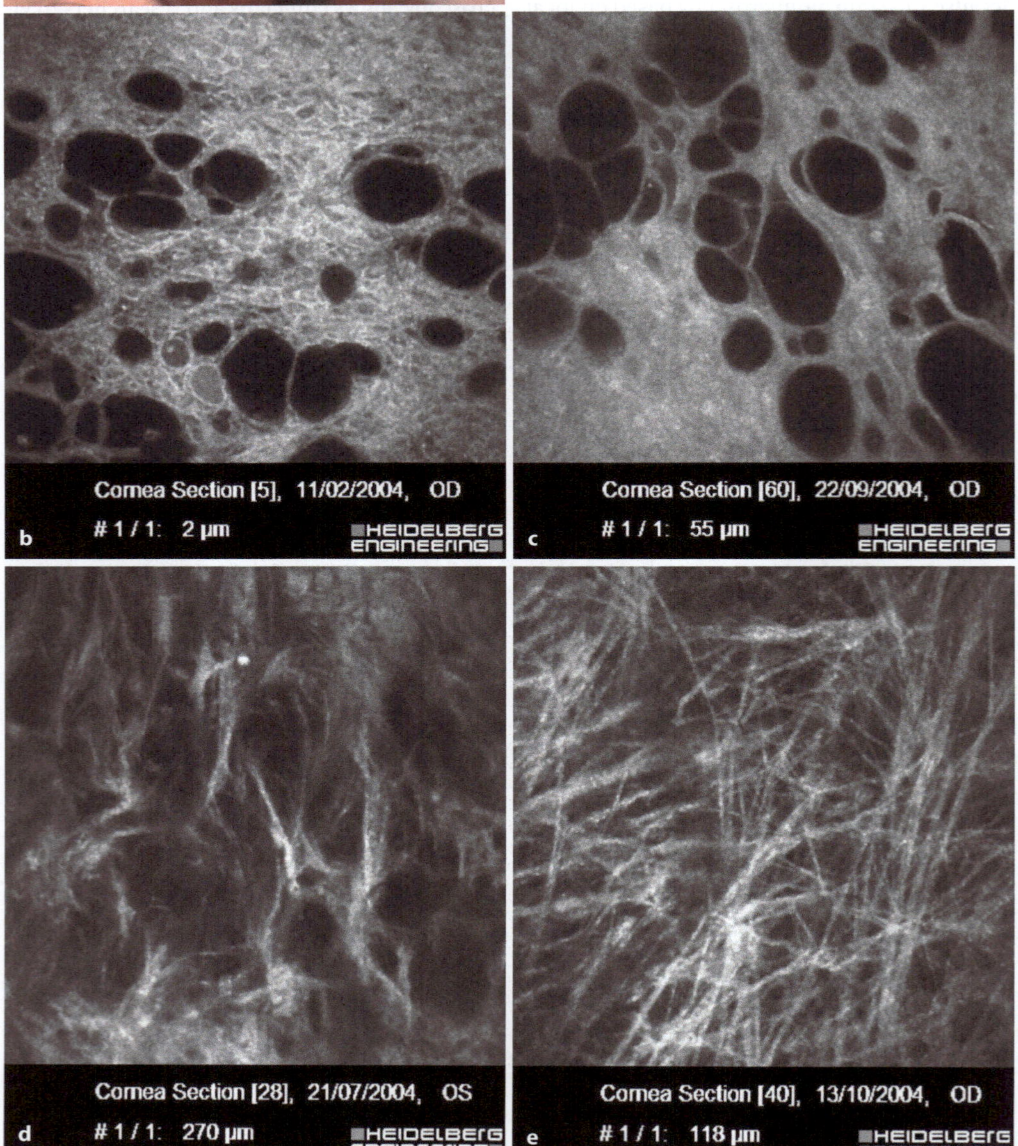

Fig. 6.2 a–h Nonfunctioning blebs. **a** Slit-lamp photograph of a nonfunctioning bleb. **b, c** Few or no microcysts in the conjunctival epithelium. **d, e** Dense subepithelial connective tissue

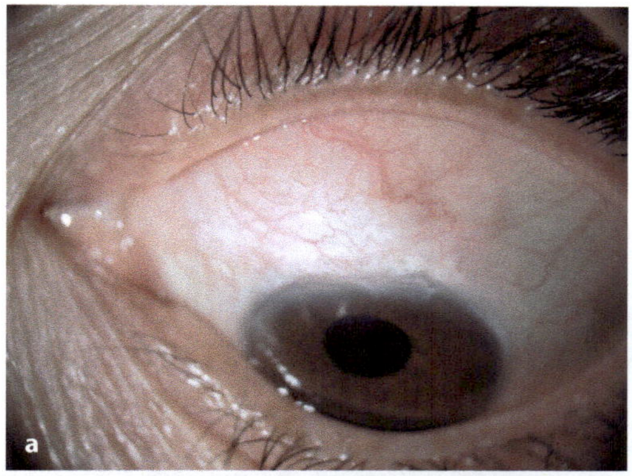

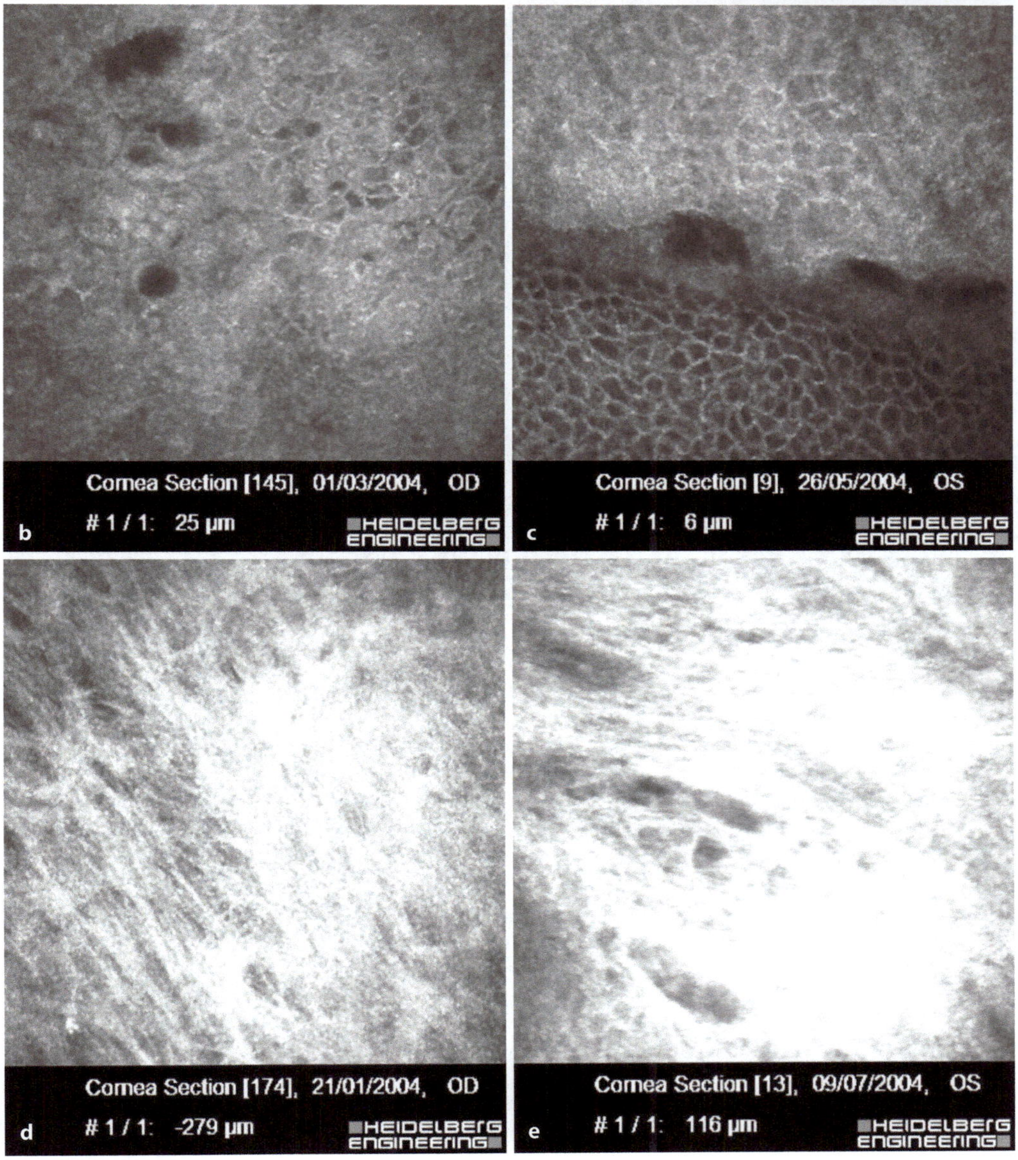

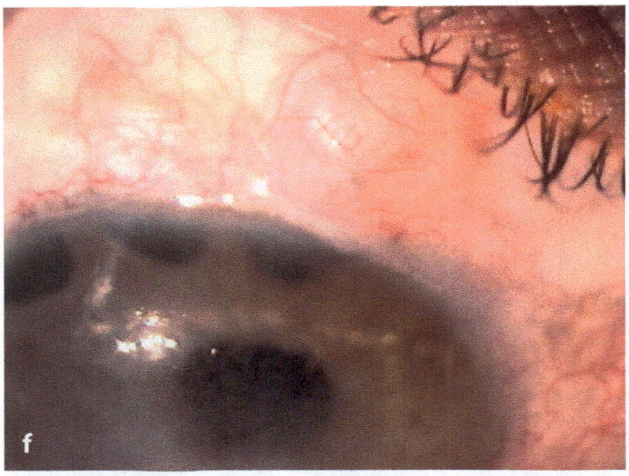

Fig. 6.2 (*continued*) **f** Slit-lamp photograph of an encapsulated nonfunctioning bleb. **g** Few microcysts at the periphery of the bleb. **h** Encapsulation visible at the periphery of the bleb

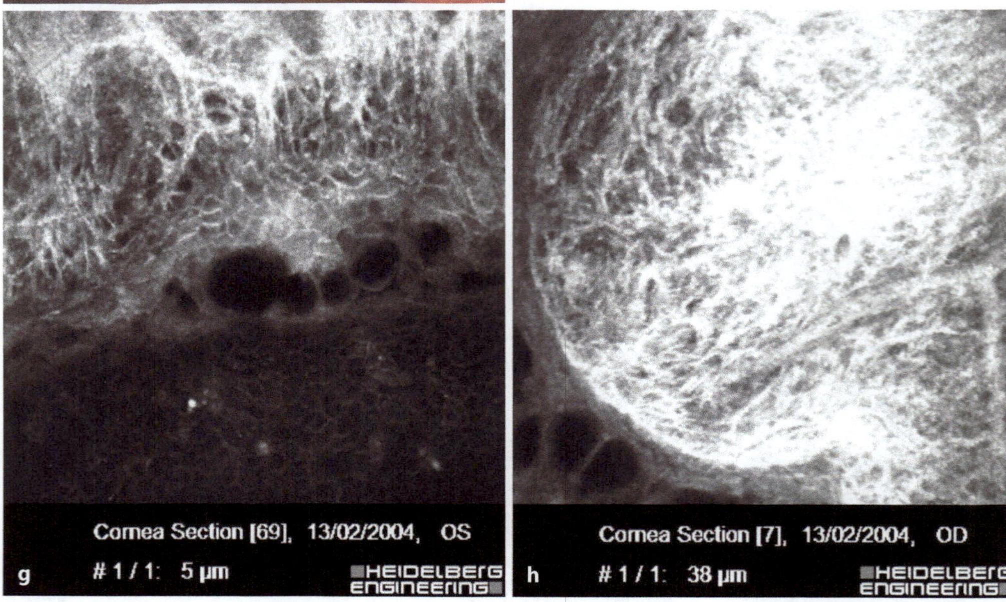

Fig. 6.3 Functioning blebs with mitomycin C. a Slit-lamp photograph of a functioning bleb with mitomycin C. b, c Numerous microcysts in the conjunctival epithelium. d Microcysts containing hyperreflective microdots. e Loosely arranged connective tissue

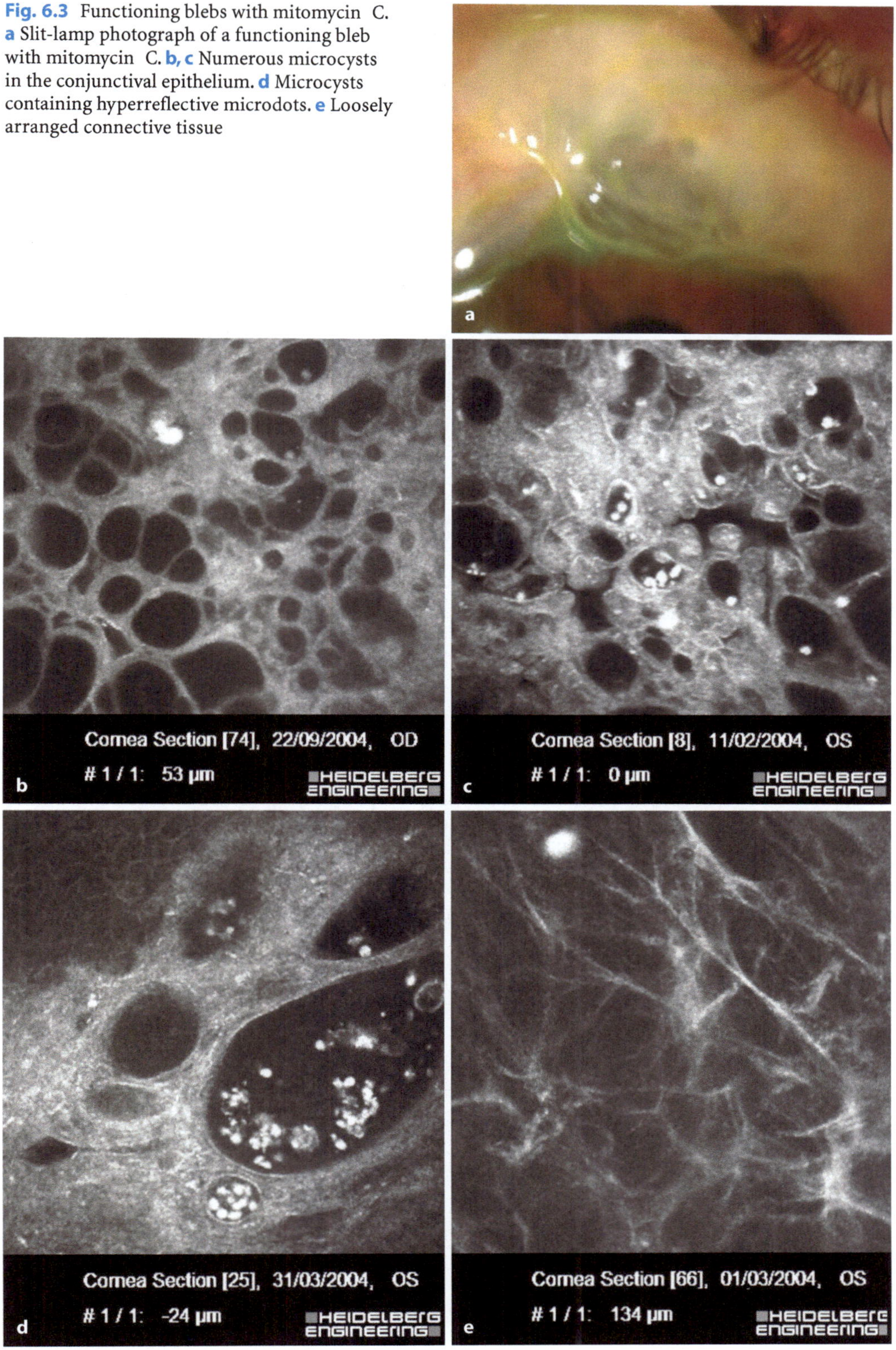

6.2 Alterations of the Cornea in Contact Lens Wearers

Distinct changes in corneal morphology, pachymetry, and structure in contact lens wearers can be demonstrated by confocal laser scanning in vivo microscopy. These findings are best interpreted as resulting from mechanical or metabolic disturbances of the cornea (Fig. 6.4a.)

All cell layers (superficial, intermediate, and basal cells) are present and characterized by bright cell borders and uniformly dark cytoplasm. The cell count increases with layer depth because of a decrease in cell diameter. Bowman's membrane and the subepithelial plexus display boundary structures between the epithelium and stroma (Fig. 6.4b)

Superficial cells are characterized by a dark nucleus, and the cytoplasm is generally darker than in the normal cornea. The polygonal structure is retained, but cell bodies are generally smaller (30 µm in contact lens wearers and up to 50 µm in the normal cornea; Fig. 6.4c and Fig. 6.5a). Our data (unpublished results) show a significant increase in superficial cell density ($p<0.05$) both centrally and peripherally.

The intermediate cells show no morphological changes compared with findings in normal subjects. Pale cell borders, invisible nuclei, and dark cytoplasm were detected in both the lower and upper wing cells (Fig. 6.4d and Fig. 6.5b). A significant reduction in the cell count was noted only in the periphery ($p<0.05$).

Basal cell structure is characterized by an inhomogeneous cytoplasm and invisible nucleus; cell diameters are approximately 8–10 µm (Fig. 6.4e and Fig. 6.5c). A significant reduction in the cell count was also detected in the peripheral cornea ($p<0.05$).

Analysis of the pachymetry data revealed reduced corneal thickness in the periphery compared with that in normal volunteers, especially in patients who had worn contact lenses for longer than 10 years. There were no age-related changes in cell count or epithelial thickness, but stromal thickness was reduced.

The type of contact lens (hard versus soft) has no influence on corneal morphology; the duration of contact lens wear is the factor with the greatest impact. Corneal microdeposits in stroma (Fig. 6.5d), signs of polymegathism, pleomorphism (Fig. 6.5e, f), and endothelium precipitates (Fig. 6.5e) are the most common findings. Alterations in LCs also occur because of contact lens wearing.

In light of this, when investigating the cornea in contact lens wearers, attention must be focused on the cell density of each layer and on the thickness of the corneal epithelium, and results between the center and periphery must always be compared.

In addition to these indications, which are already well defined, all patients with ocular surface diseases, atypical infections of the cornea and conjunctiva, and metabolic diseases possibly associated with corneal deposits, as well as patients requiring follow-up after corneal surgery, may be considered for confocal in vivo microscopy.

Ocular surface neoplasms, such as acquired and congenital naevi and possibly malignant transformations of such tumors, may be an interesting field of research, and further evaluation is needed before useful clinical indications can be defined.

Table 6.1 summarizes the authors' experience and may stimulate users of this atlas to continue clinical work and research involving confocal imaging.

Table 6.1 Summary of applications of confocal imaging

Useful application	Structures of interest	Quantifiable parameters	Possible consequences
Glaucoma filtration surgery	Microcysts, inflammatory cells, red/white blood cells, unidentified white highly reflective spherical structures 2–3 μm in diameter	Microcyst density	Administration of wound-healing modulators
Contact lens wearers	Conjunctivalization of the peripheral cornea, goblet cells in atypical location, increased numbers of dendritic cells including the corneal center, increased numbers of superficial corneal epithelial cells	Langerhans cell density correlated with location (center/periphery) Superficial cells per mm^2 (double density compared with noncontact lens wearers)	Change of contact ens type or change lof lens care products
	Microdots in corneal stroma, keratocyte nuclei with reduced contrast		
Ocular surface disease	Langerhans cells, inflammatory cells in epithelial layers, inflammatory cells surrounding meibomian glands, conjunctivalization of the cornea Nerve fibers present?	Langerhans cell density and location, goblet cell and inflammatory cell density Nerve fiber density	Change of artificial tear medication; consider additional anti-inflammatory therapy Consider neuroparalytic component
Ocular infections	Acanthamoeba cysts present? Integrity of Bowman's membrane? Cellular infiltrates? Fungal elements present?		Acanthamoeba treatment to be included in therapeutic regimen Specific antifungal treatment
Ocular allergic reactions	Langerhans cells and lymphocytic infiltrates in cornea, tarsal plate and conjunctival structures	Number and type of dendritic cells/mm^2	Reduction of antigens, antiallergy treatment
Metabolic diseases with microdeposits in cornea and conjunctiva: mucolipidosis, cystinosis, Fabry's disease, tyrosinemia, gangliosidosis	Deposits detectable prior to slit-lamp biomicroscopy in cornea and conjunctiva		Possible change in treatment strategy
Systemic therapy possibly associated with corneal and conjunctival deposits			

Table 6.1 (*continued*) Summary of applications of confocal imaging

Useful application	Structures of interest	Quantifiable parameters	Possible consequences
Corneal transplantation	Corneal thickness, endothelial cell density, endothelial deposits, corneal infiltrates preferably close to suture material Interface structures in lamellar grafts	Cell densities Characterization of corneal infiltrates: leukocytes/ lymphocytes	Antibiotic versus anti-inflammatory versus immuno-suppressive treatment
Corneal refractive surgery	Epithelial thickness, flap thickness Keratocytes activated? Interface micromor-phology, corneal nerves Quality of LASIK cut, depending on cutting devices	Mapping of flap thickness Nerve regeneration	Improvement of microsurgery quality; analysis of wound healing reactions and treatment monitoring

Fig. 6.4 Epithelium in contact lens wearers. **a** Slit-lamp photograph of hard contact lens on a left eye. **b** Oblique corneal section: superficial, intermediate and basal cells, Bowman's membrane and anterior stroma. **c** Superficial cells. **d** Intermediate cells. **e** Basal cells

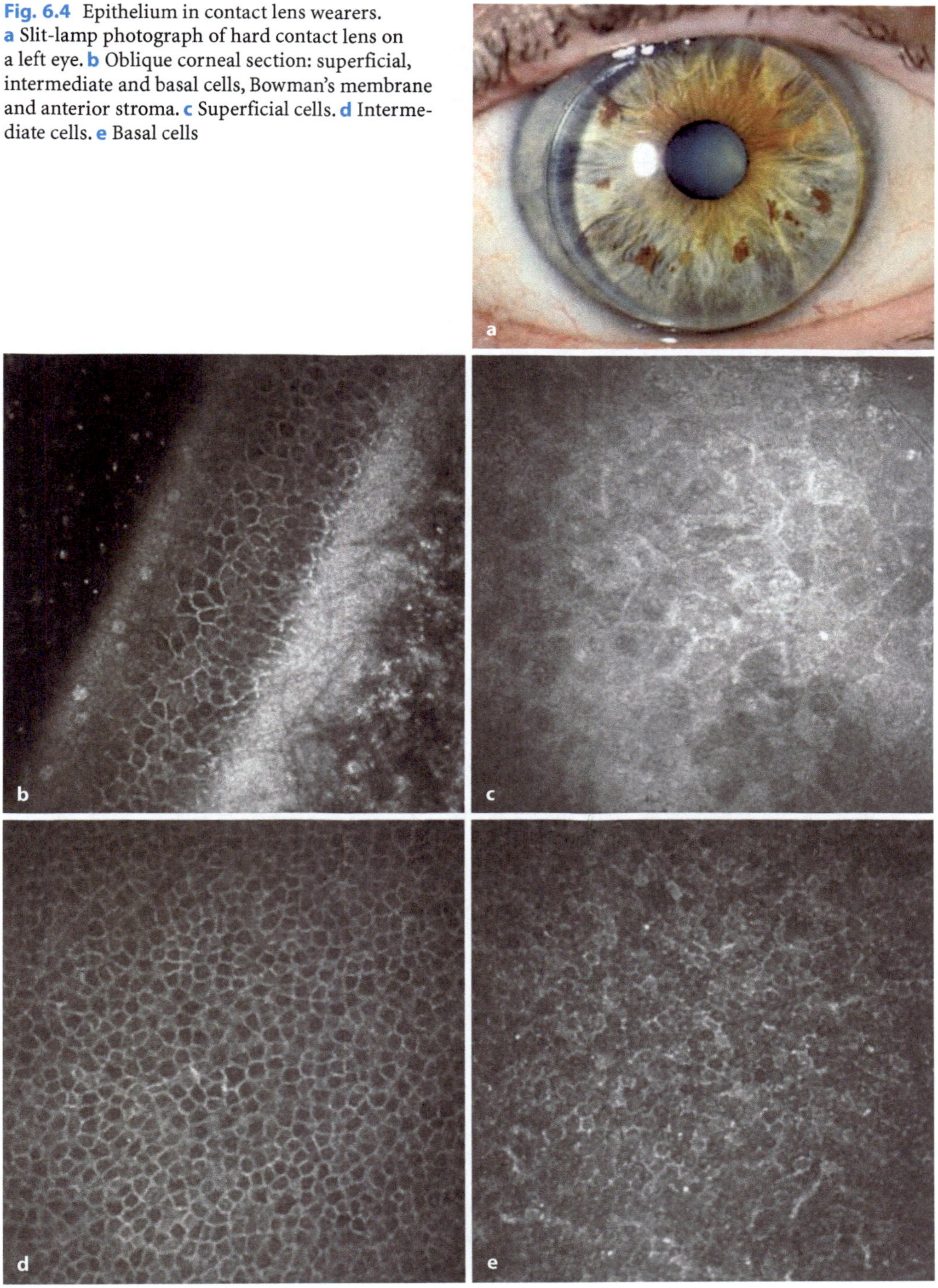

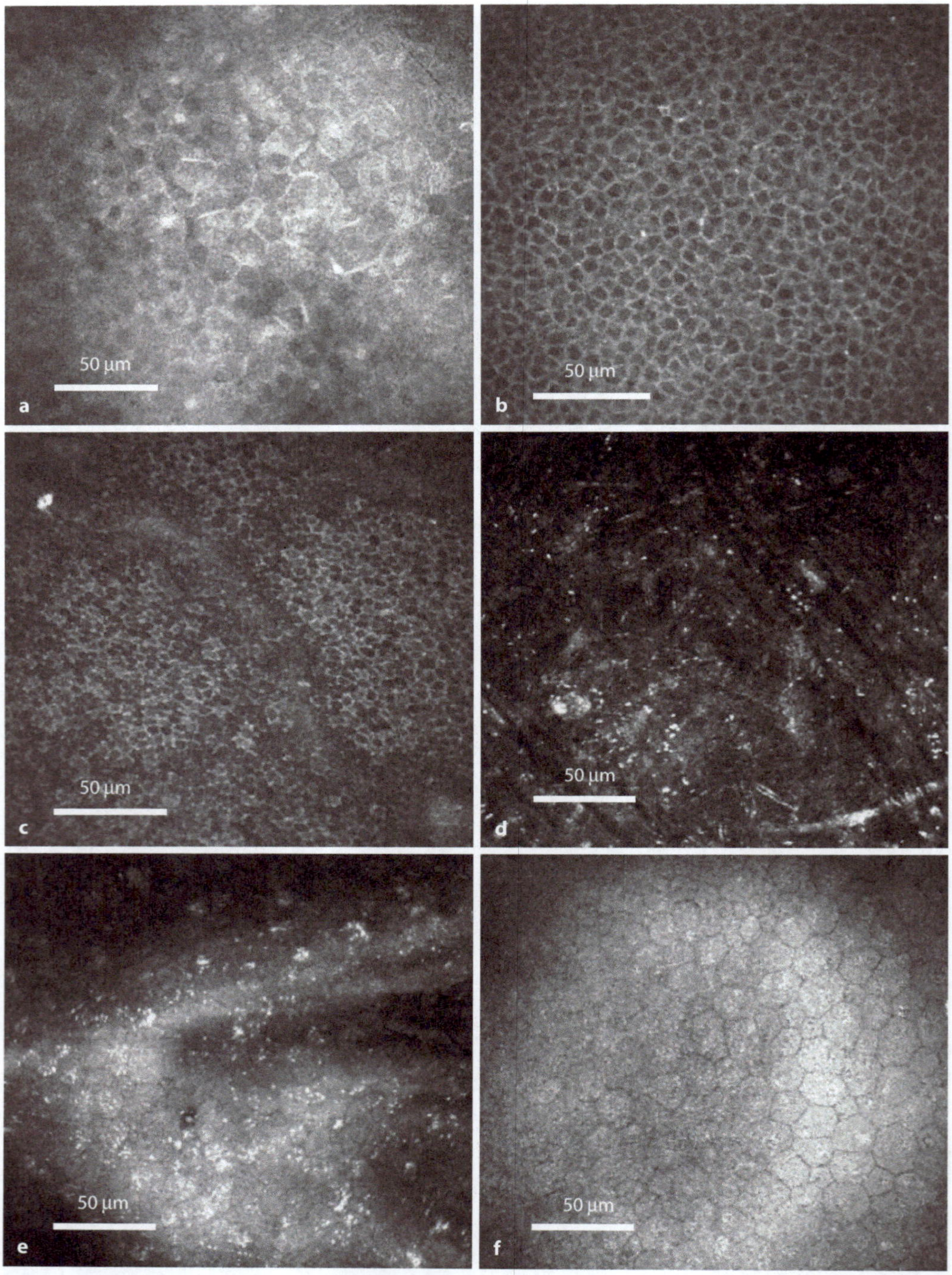

Fig. 6.5 Hard contact lens wearer. **a** Superficial cells, corneal epithelium, depth 0 µm. **b** Intermediate cells, corneal epithelium, depth 30 µm. **c** Basal cells, corneal epithelium, depth 52 µm. **d** Stroma, hyperreflective microdots and destruction of normal structure, depth 350 µm. **e** Deposits just about endothelium, depth 480 µm. **f** Endothelium, polymegathism and pleomorphism, depth 520 µm

Nonophthalmological Applications

Although the RCM is designed for ocular and ocular adnexal surface structures, it is well suited for analyzing any surface of the human body that can be reached by the instrument. It has proved its clinical relevance in the evaluation of oral and pharyngeal mucosal membranes [30] and shows promising results in dentistry applications. Other fields of application may emerge in due course.

7.1 Skin

Human skin consists of the epidermis and dermis, together with its appendages such as glands (Fig. 7.1). The epidermis, which is accesible to confocal laser scanning microscopy to a depth of approximately 200 µm, consists of an avascular layer of squamous epithelium.

The skin, especially that of the fingers, is innervated by myelinated, sensory, and nonmyelinated branches of the peripheral nerves, which form a subpapillary plexus and penetrate to the surface as free nerve endings (cutaneous sensation). These nerve endings are terminal branches of a neuron. A distinction is drawn here between axonal presynaptic endings and receptor endings. These may terminate in the epidermis as free endings, or telodendra. These are present in fairly high densities, especially in the skin of the fingers and the soles of the feet.

Figure 7.2 shows the squamous epithelium with these nerve endings as telodendra (*arrow*) in high density on the inner surface of the fingertips at different depths.

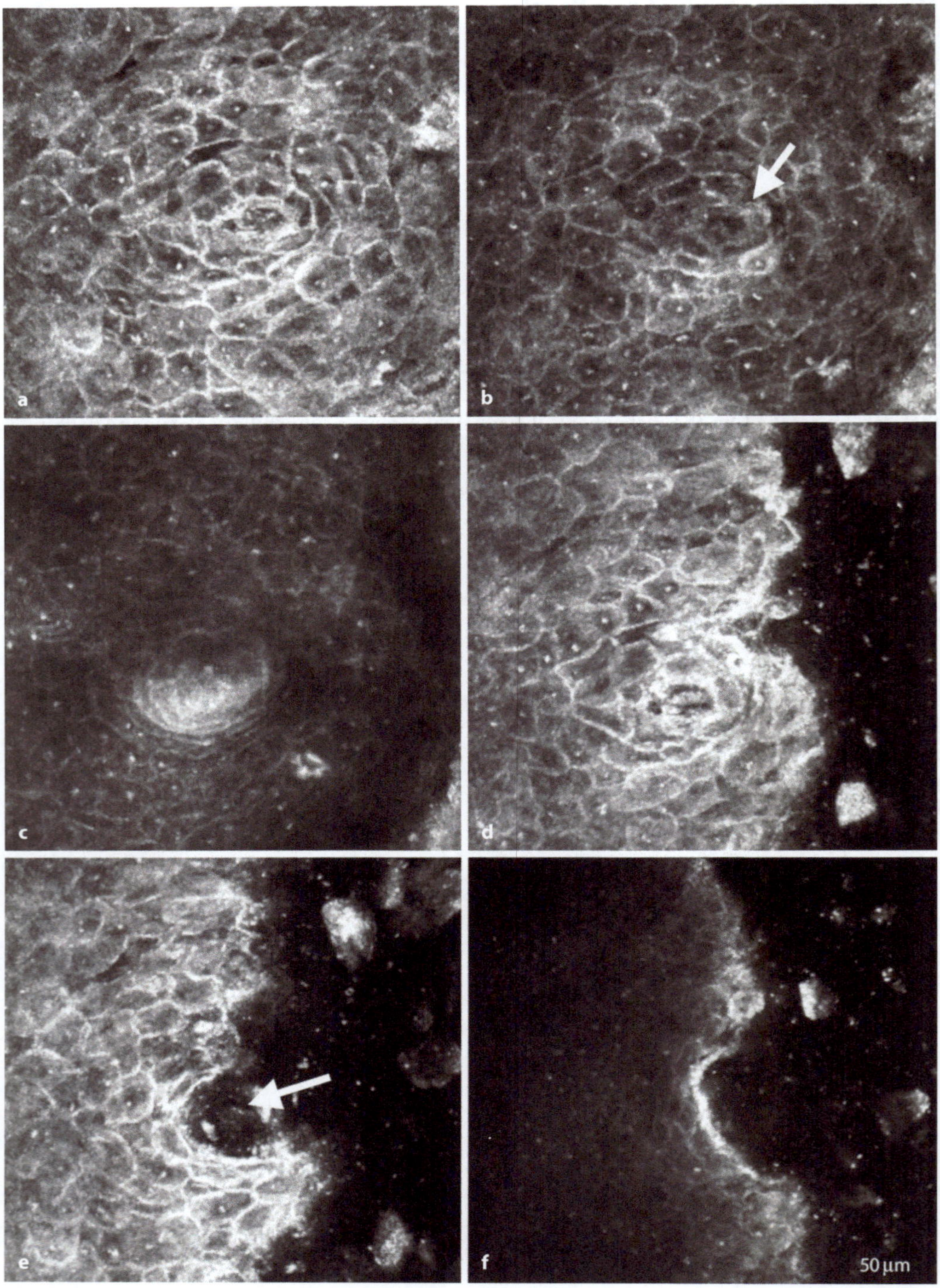

Fig. 7.1 Structures in human epidermis (fingertip). **a–c** Skin with sweat gland opening on to finger pad (*arrow*) at various depths to approximately 200 µm. **d–f** Sweat gland opening; oblique section (*arrow*)

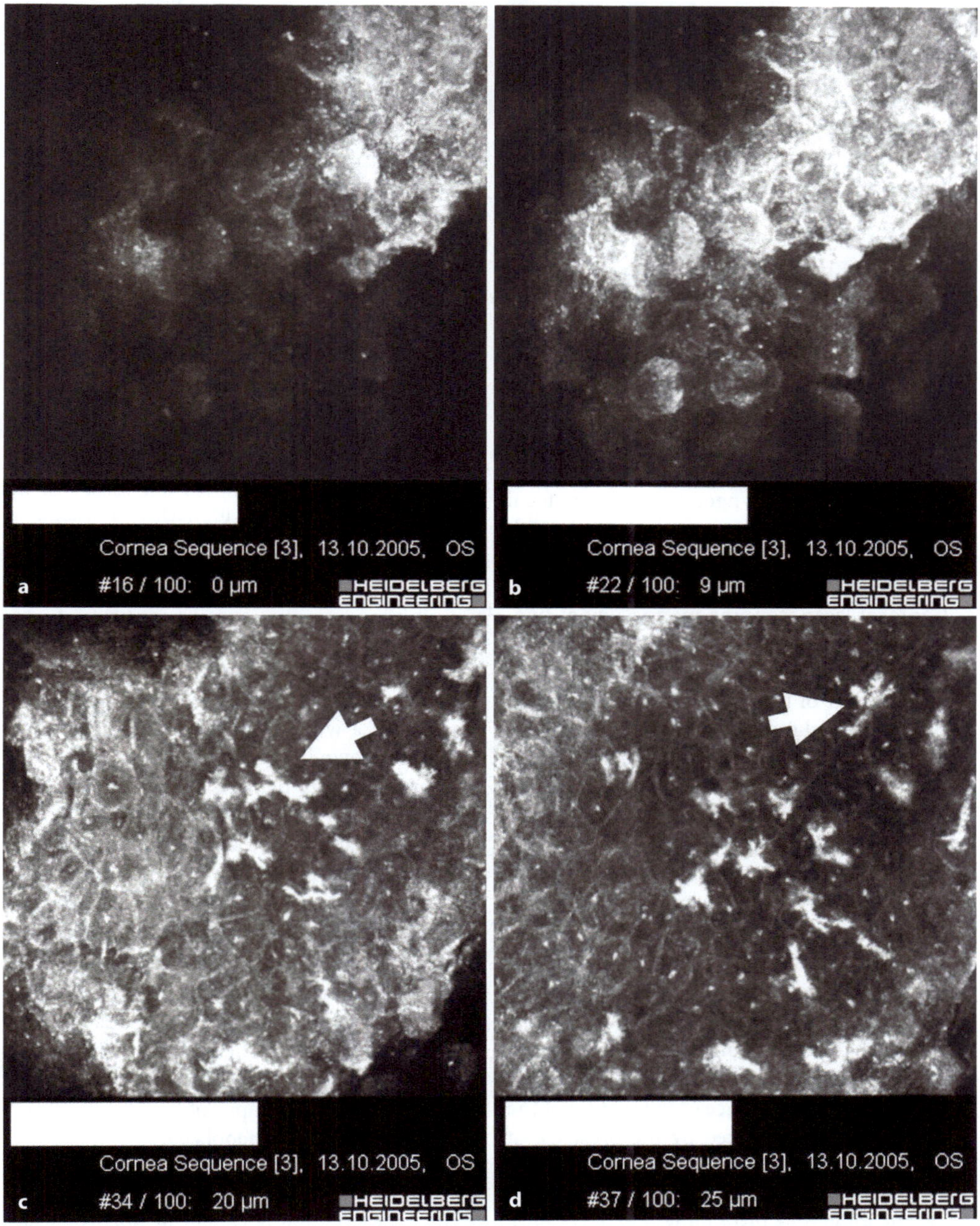

Fig. 7.2 Image of epidermal layer with nerve endings (*arrow*): Heidelberg Retina Tomograph II/RCM; 63× (Zeiss), *z*-scan internal, 5-µm steps from *z*-series

7.2 Oral Mucosa and Tongue

The main focus of research in this area in the department of otorhinolaryngology in Rostock is to assess the suitability of confocal microscopy for the noninvasive diagnosis of precancerous lesions.

A series of in vivo studies was initially conducted to examine the oral epithelia of the cheek, tongue, and lips in healthy subjects [30, 31]. The findings obtained using confocal microscopy were compared with histological findings. In a second series, ex vivo studies were started to examine the epithelia of the larynx [32]. Biopsies were obtained during microlaryngoscopy and after laryngectomy. All confocal microscopic findings were compared with histopathological findings (horizontal sections) and discussed with an experienced pathologist.

Confocal laser scanning microscopy of buccal epithelia was performed at standardized locations (in the vestibule of the mouth 1 cm lateral to the corner of the mouth, buccal mucosa in the premolar and molar region).

Figure 7.6a–c shows buccal epithelium from the vestibule of the mouth at a depth of 5 µm. Epithelial cells without nuclei (keratinocytes) are visible. The differing appearance of buccal epithelia from different regions – the premolar region (Fig. 7.6b) and the molar region (Fig. 7.6c) – should be noted.

The buccal mucosa contains an inconstant number of epithelial papillae (Fig. 7.7), which are similar to the filiform papillae of the tongue. Neither structure is involved in the peripheral taste system.

The external appearance of the surface of the tongue tip is characterized by the presence of filiform and fungiform papillae (Fig. 7.8). These latter structures contain the peripheral gustatory organs known as taste buds. The taste pore can be identified clearly at the surface of the fungiform papilla (Fig. 7.9a). Figure 7.9b–d shows taste pore and epithelial cells around the longitudinal axis. The subepithelial vessels supply the taste bud structure (Fig. 7.9e–f). The corresponding histopathological section (toluidine blue stain) of three taste buds (Fig. 7.108a) is shown in Fig. 7.10b.

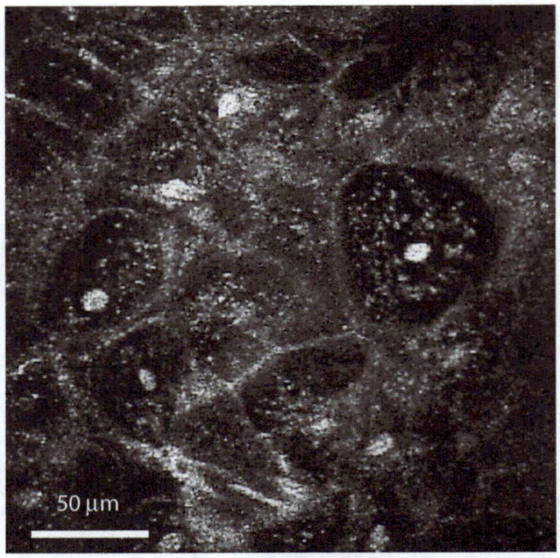

Fig. 7.3 Confocal laser in vivo microscopy of lip mucous membrane: cells with nuclei, mitochondria, and desmosomes

Epithelial changes in the labial and buccal mucosa of a woman after breast cancer treatment are shown in Figs. 7.11a–c [33]. Contact endoscopy findings demonstrate filiform papillary structures (red arrow in Fig. 7.11d, e) that are recognizable by the keratinized tip and cells without nuclei (inset). The subepithelial vessels typical of the papillary structures are also visible. Confocal microscopy reveals epithelial papillae (Fig. 7.11f) and cells without nuclei to a depth of 80 µm. Histopathologic investigation of papillomatosis and keratosis correlated with the in vivo findings. The extensive changes in oral and oropharyngeal epithelia occurred several weeks prior to relapse of breast cancer and were reported as paraneoplasia.

An endoscopic image of the buccal mucosa in a patient with severe mucositis undergoing radiochemotherapy is shown in Fig. 7.12. Cells with two nuclei indicate a higher rate of mitosis and proliferation [34].

The epithelium of the true vocal cord (5-µm depth) is shown in Fig. 7.13a–c. Epithelial cells with a centrally positioned nucleus are visible in the superficial layers of the larynx (Fig. 7.13a).

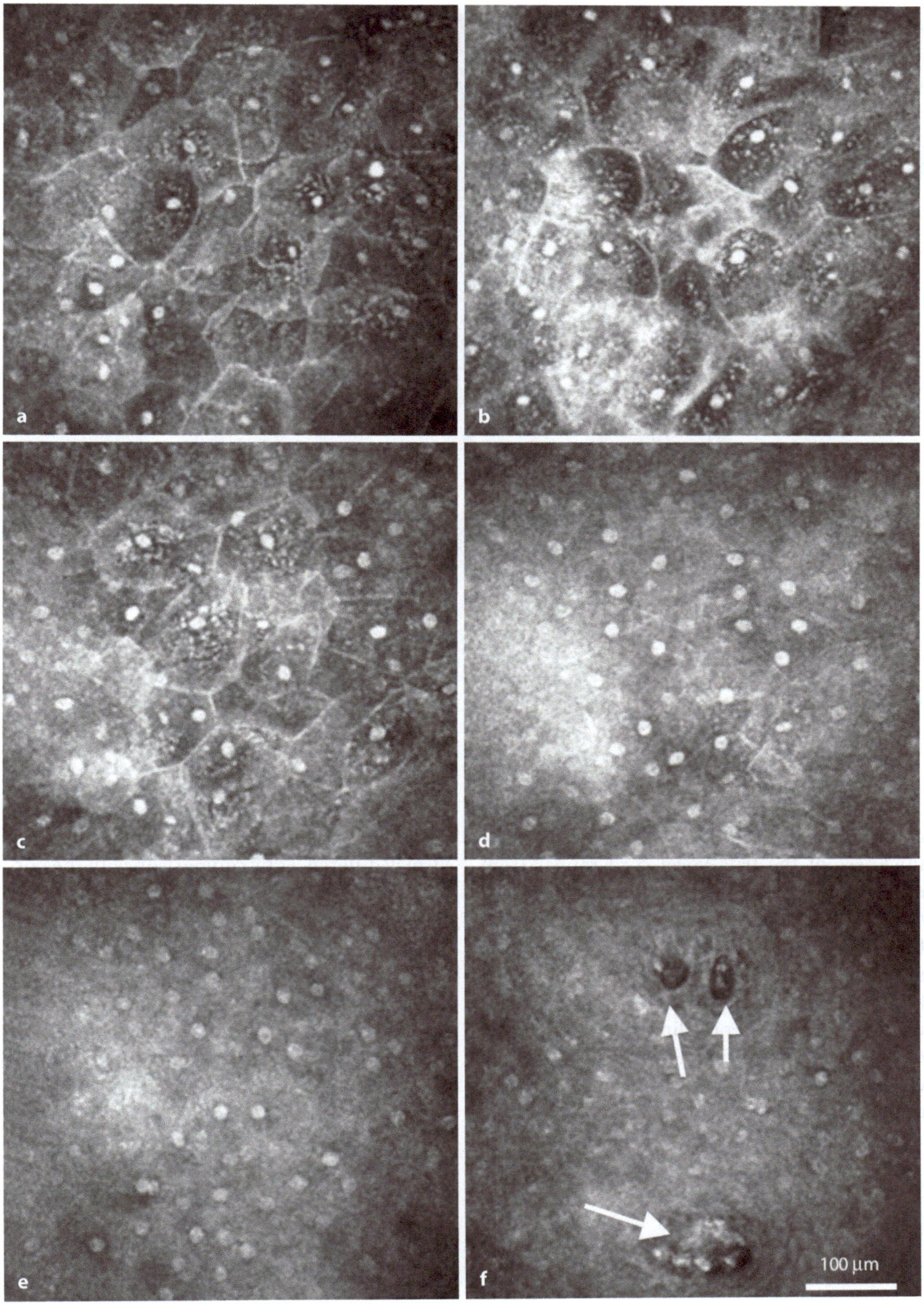

Fig. 7.4 Confocal laser in vivo microscopy of lip mucous membrane. **a–f** Cells with nuclei, *z*-scan series; blood vessels with blood flow (*arrow*; **f**)

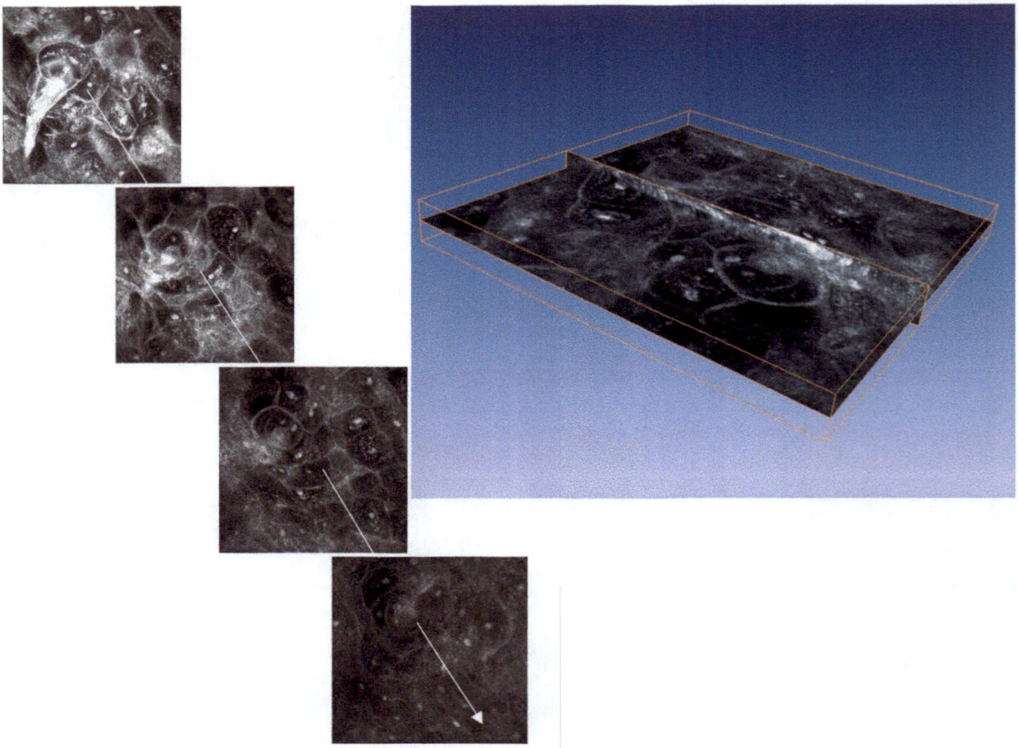

Fig. 7.5 Three-dimensional reconstruction of mucous membrane, z-scan =50 µm

The cells of the basement cell layer (not shown) appear smaller than the superficial cells. The thickness of the true vocal cord ranges from 65 µm to 130 µm between individuals. The transition to the subepithelial space can be identified in Fig. 7.13b. The parallel elastic fibers of the subepithelial space can be easily distinguished (Fig. 7.13c). By contrast with squamous epithelium, the respiratory epithelium of the supraglottal region bears kinocilia (Fig. 7.13d). The beating of the cilia and the directed mucous transport can be observed ex vivo for a few hours.

The confocal microscopic images of cancer of the tongue (Fig. 7.14b) and larynx (Fig. 7.15b) display differences by comparison with the normal epithelial in corresponding regions. On confocal microscopy, all malignancy criteria can be detected, including enlarged nuclei, enlarged cells with variable cell shapes, cell clustering, increased nucleus/cytoplasm ratio, irregular cell architecture, and loss of cellular junctions characterized by nonvisualization of the cell membrane.

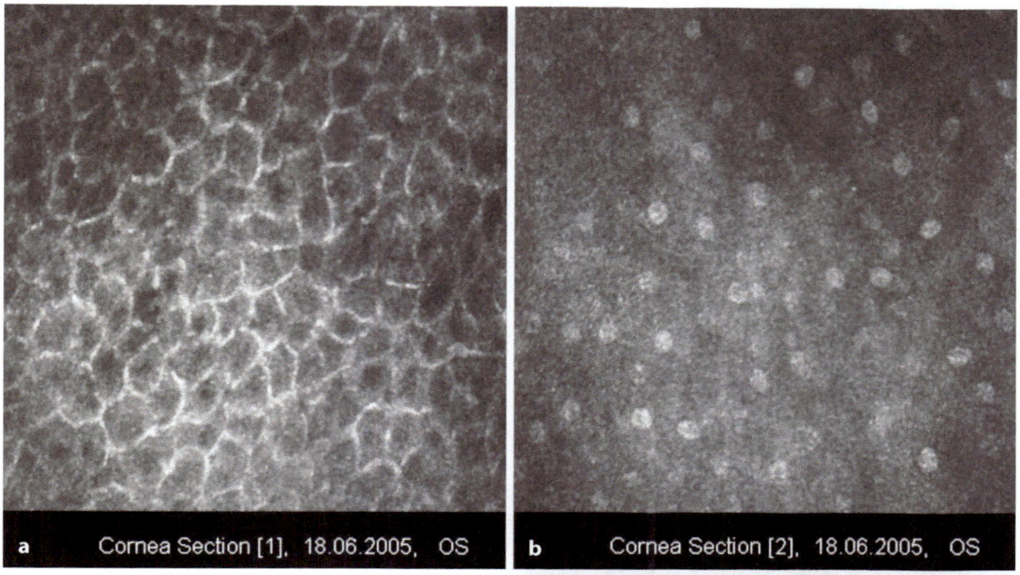

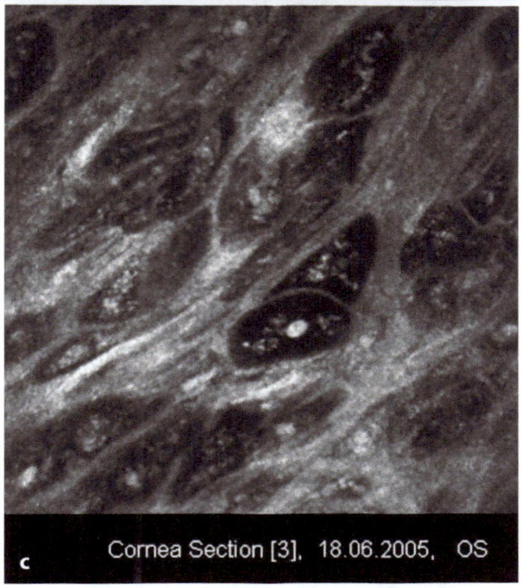

Fig. 7.6 Normal buccal mucosa at a depth of 5 μm in three different regions. **a** Oral vestibule, 1 cm lateral to the corner of the mouth: the epithelium is characterized by cells without nuclei (keratinocytes). The cell membrane is clearly visible. **b** Premolar region: nonkeratinized cells, recognizable by their nuclei. The cell membrane cannot be distinguished. **c** Molar region: cell membrane, nuclei and intracellular organelles are clearly visible

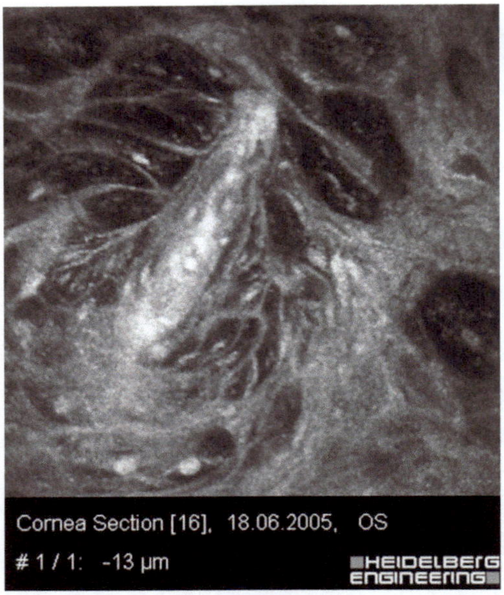

Fig. 7.7 Normal buccal mucosa at a depth of 5 µm. The connective tissue papillae shown are comparable with the filiform papillae of the tongue

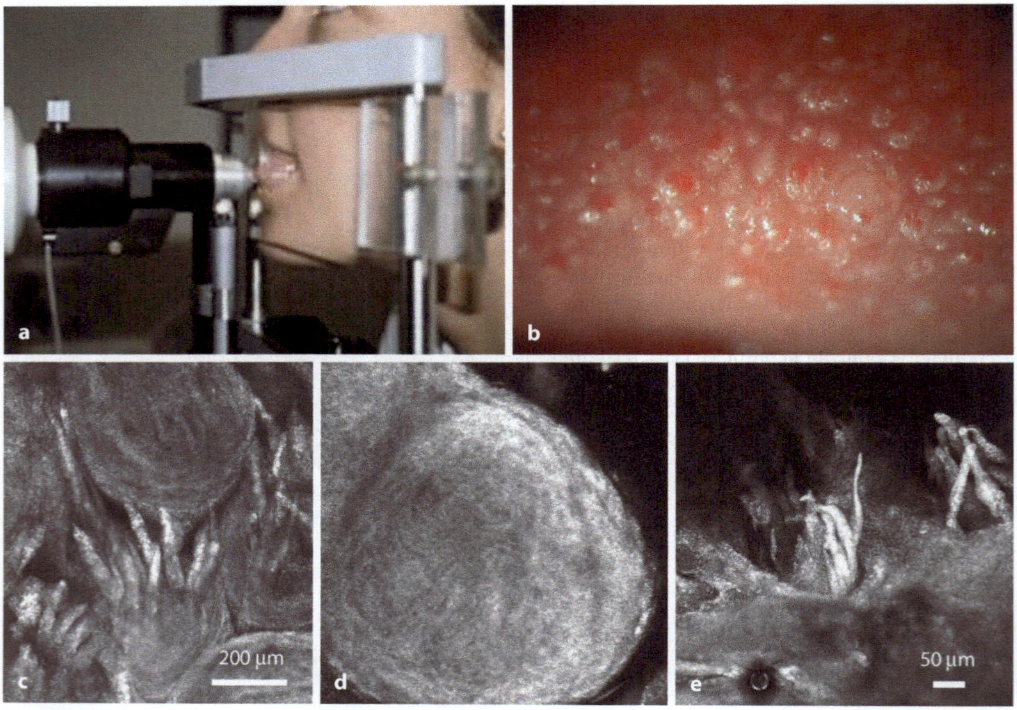

Fig. 7.8 Papillae of the human tongue. **a** Clinical setting. For stabilization, the tongue is pressed through the hole in a Plexiglas disk. **b** Photography of the tongue tip. The red-colored surface of the fungiform papillae can be easily identified. **c** The external appearance of the tongue is characterized by filiform (fingerlike, bending) and fungiform (rounded) papillae. **d** A single fungiform papilla is shown. **e** Two filiform papillae are visible

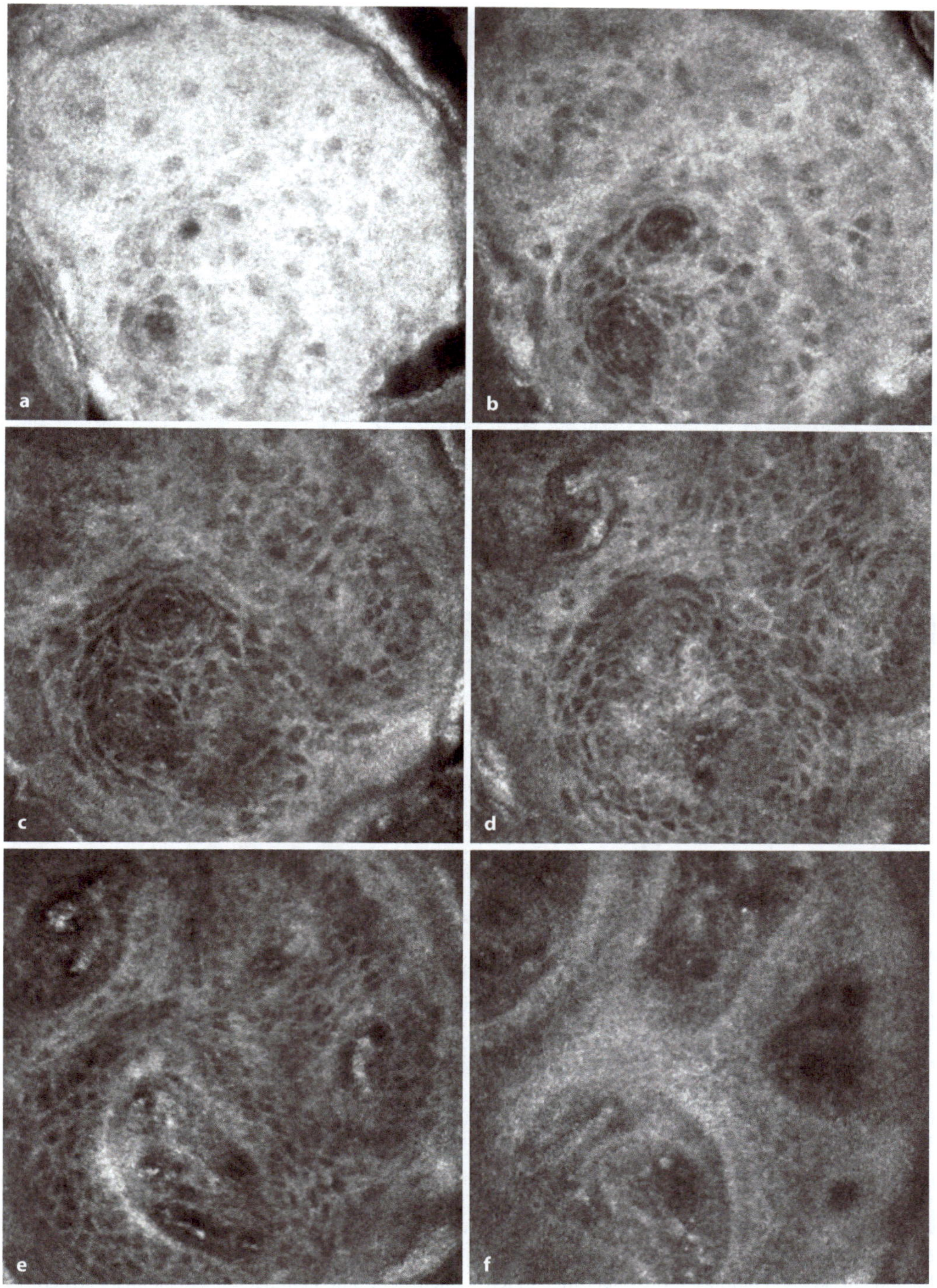

Fig. 7.9 Single fungiform papilla of the tongue at different depths. **a** 5 µm: Two rounded taste pores are visible. **b, c, d** The corresponding taste bud structure can be identified at 15 µm, 35 µm, and 50 µm, respectively. **e, f** The subepithelial vessels supplying the peripheral taste organ are shown (at 75 and 120 µm, respectively)

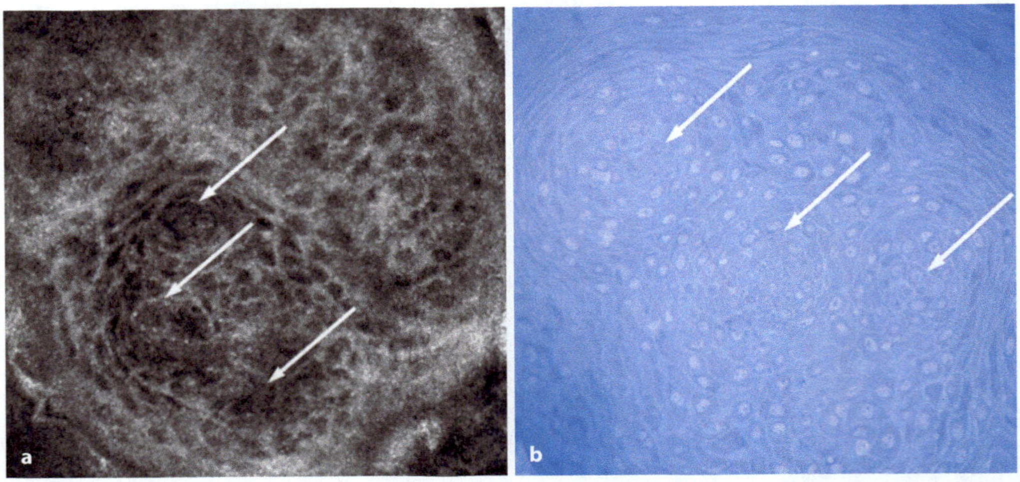

Fig. 7.10 Peripheral taste organs within a fungiform papilla. **a** Three taste bud structures. **b** Corresponding histopathological section (toluidine blue staining) with taste bud structures

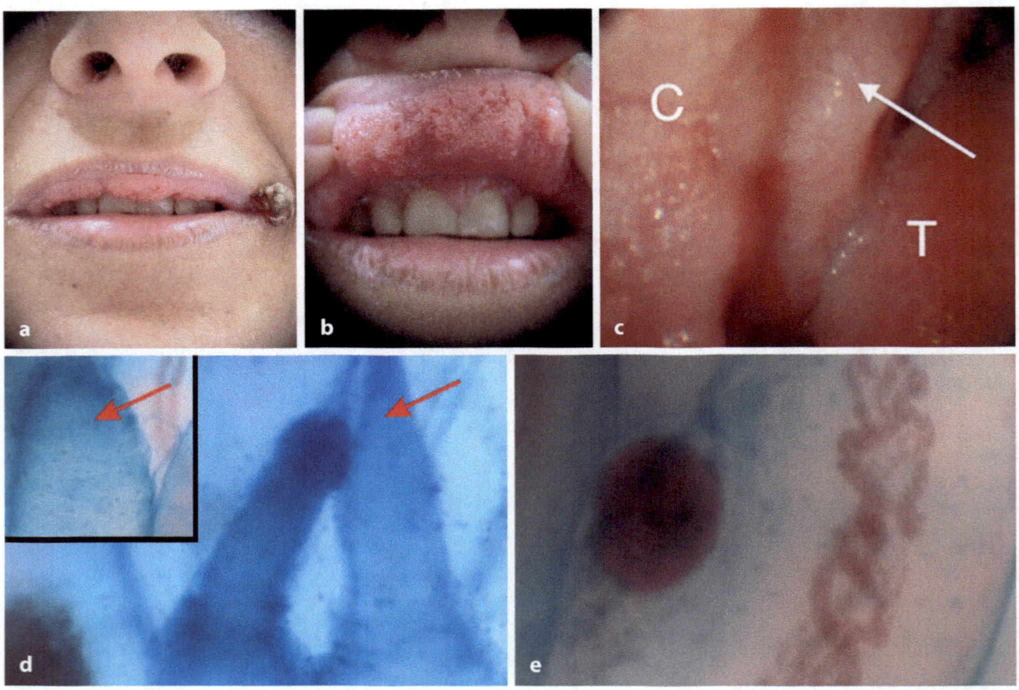

Fig. 7.11 a–f Paraneoplastic epithelial changes of lip and cheek. **a, b** Findings at the upper lip and left labial angle. **c** Thickened and hyperemic mucosa of the cheek (*arrow*). (*T* tongue, *C* cheek). **d, e** Contact endoscopy findings of the buccal mucosa following methylene blue staining: filiform papillary structures (*arrows*) are indicated. The subepithelial vessels are visible.

Fig. 7.11 (*continued*) **f** Corresponding images of the oral vestibule. By contrast with healthy buccal epithelium, fingerlike bending structures with keratotic epithelium are apparent

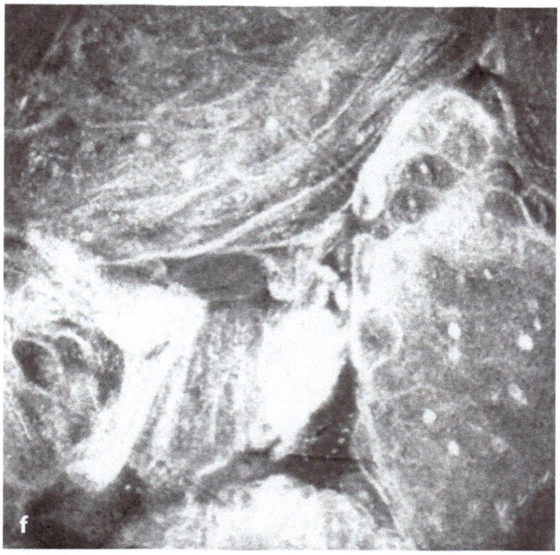

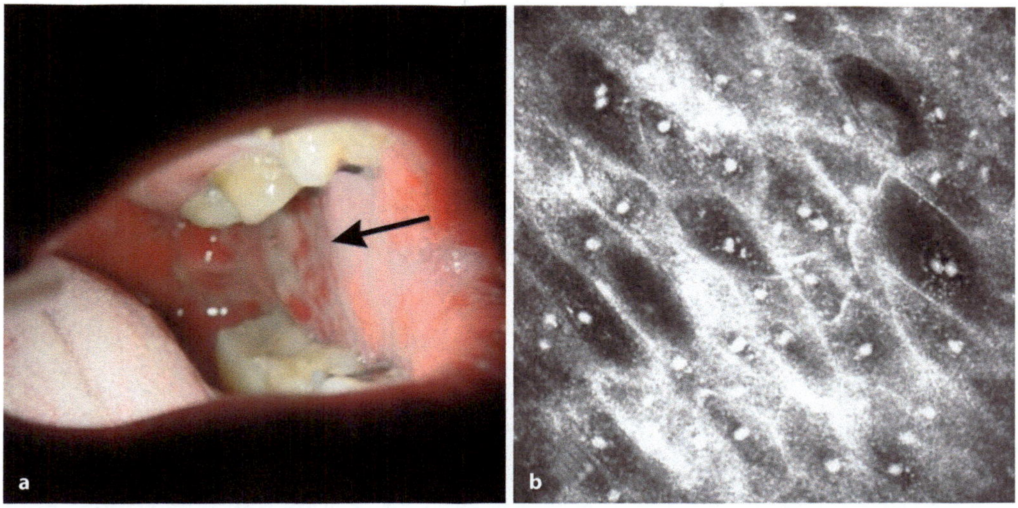

Fig. 7.12 Mucositis following radiochemotherapy. **a** Photograph of oropharyngeal mucosa of a patient with severe mucositis (*arrow*). **b** Buccal epithelium: cells with more than one nucleus indicate a higher proliferation rate

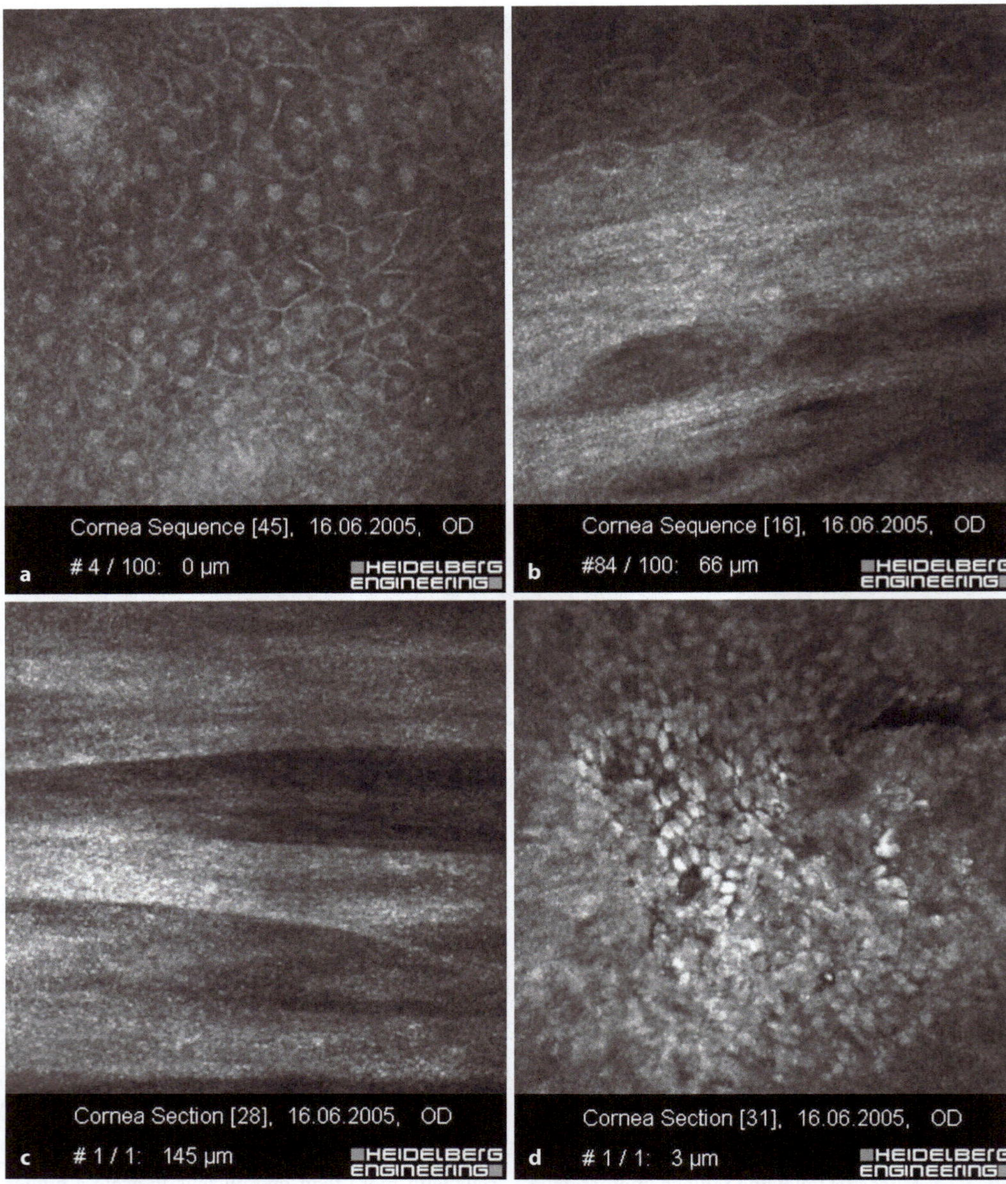

Fig. 7.13 Normal epithelia of the true vocal cord (**a–c**) and the supraglottis (**d**). a Regular architecture is apparent and both nuclei and cell membranes are clearly identifiable at a depth of 5 µm. **b** Transition to the subepithelial space at about 100 µm. **c** Parallel elastic fibers are visible in the subepithelial space. **d** Ciliated epithelium of the supraglottis

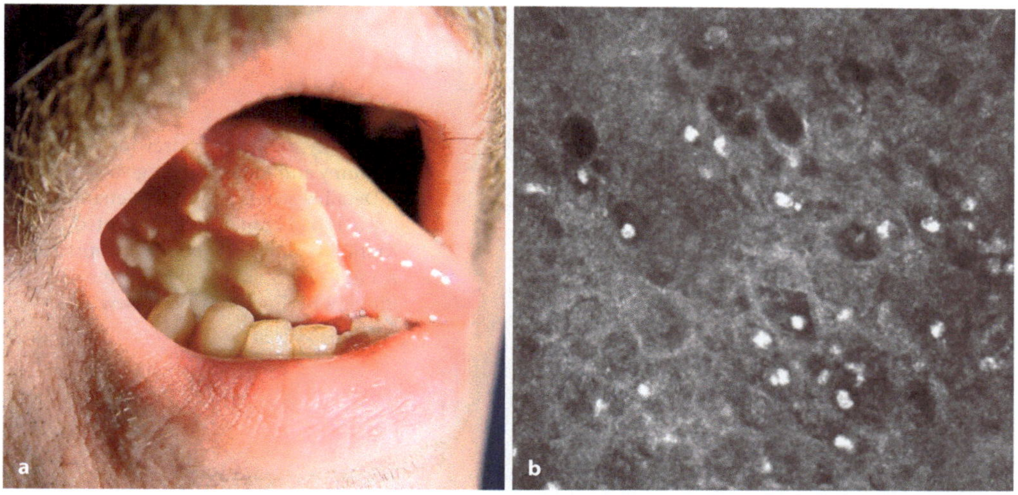

Fig. 7.14 Cancer of the tongue. **a** Photograph of tongue cancer. **b** Image of tongue cancer

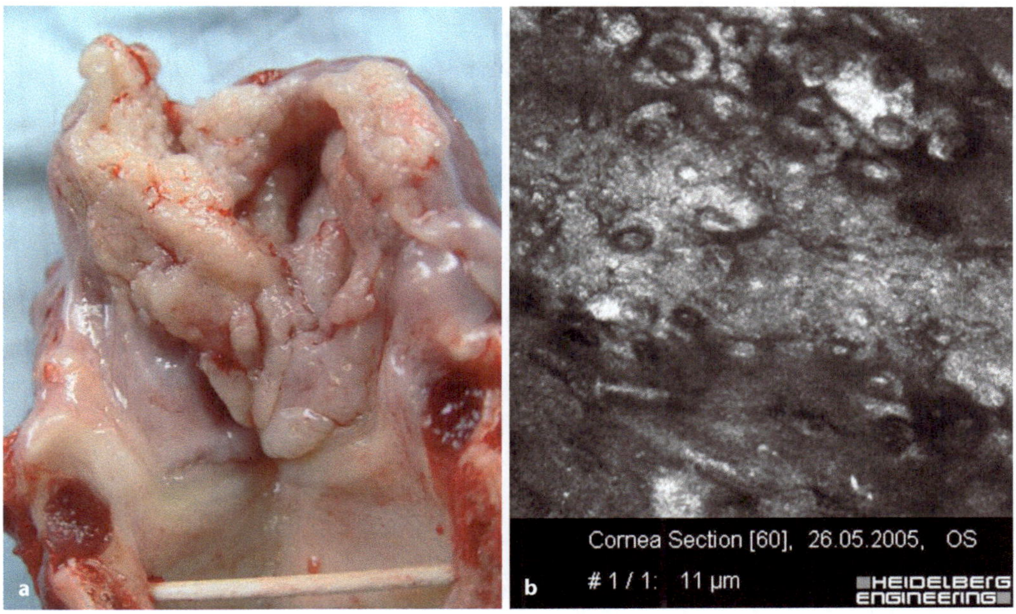

Fig. 7.15 Cancer of the larynx. **a** Photograph of larynx cancer. **b** Enlarged nuclei, enlarged cells with variable shapes (*arrow*), cell clustering, increased nucleus/cytoplasm ratio, and irregular cell architecture are characteristic of laryngeal carcinoma

7.3 Gingiva and Teeth

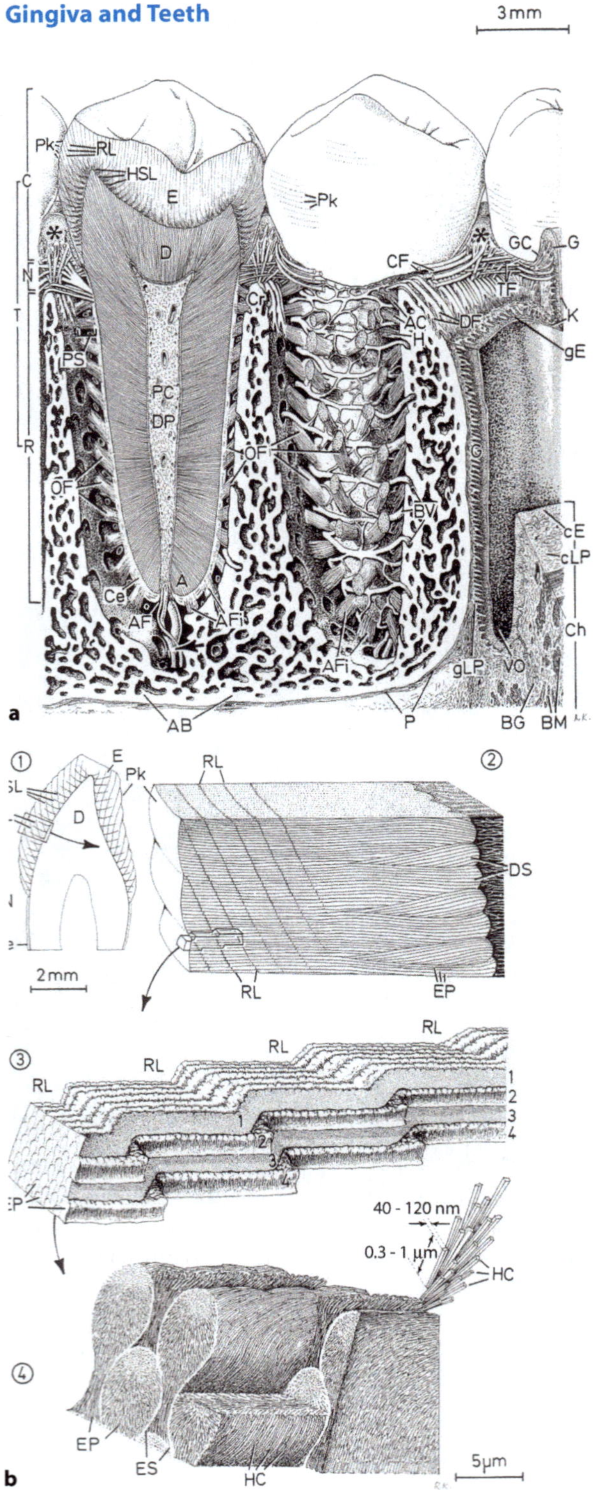

Fig. 7.16 a Schematic drawing of tooth anatomy and tooth and gingiva anatomy. b Detailed schematic view of enameled crystals of a human tooth (adapted from Krstić RV. *Human Microscopic Anatomy*: An Atlas for Students of Medicine and Biology. Berlin, Heidelberg, New York: Springer-Verlag, 1991 [39])

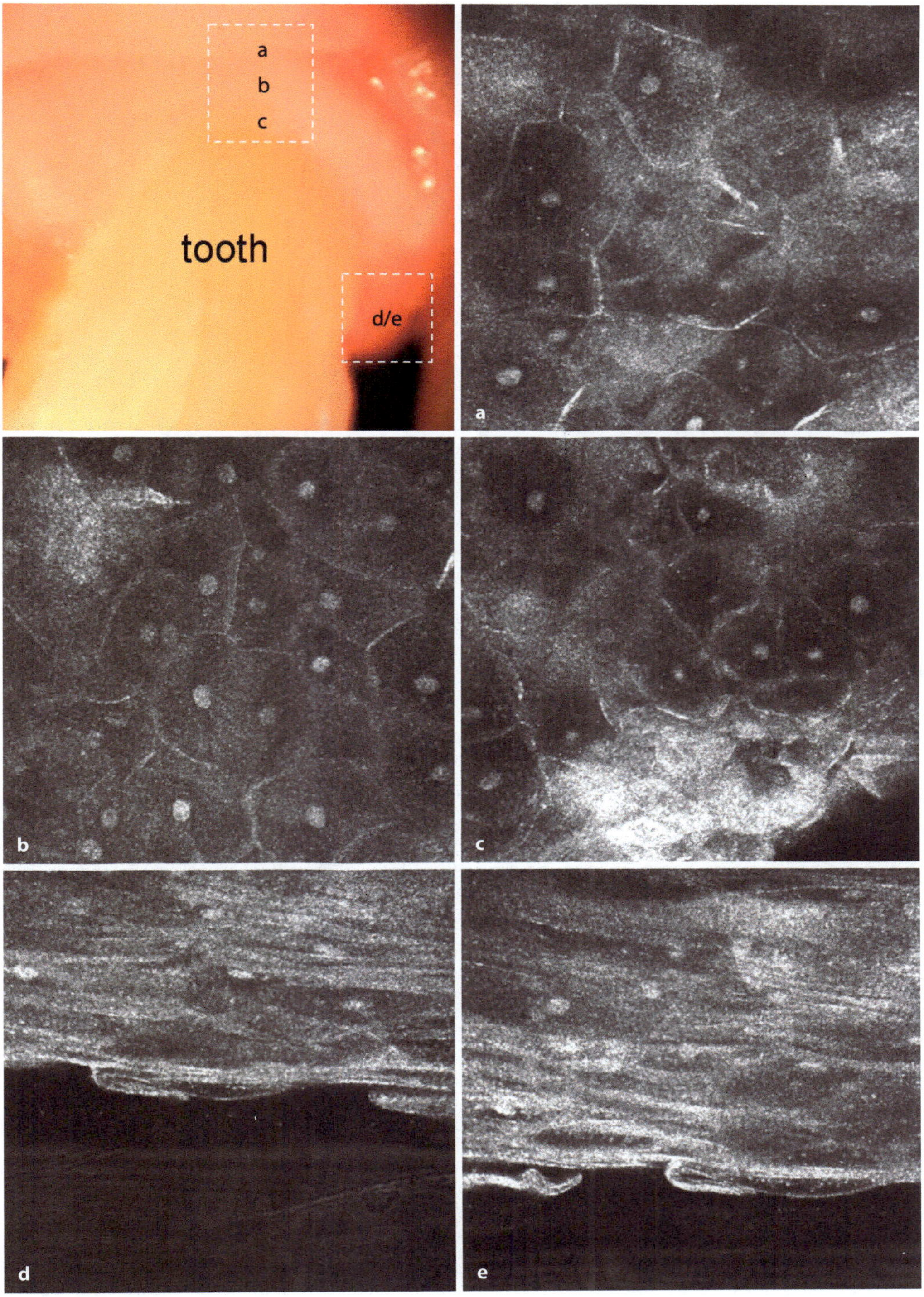

Fig. 7.17 Gingiva adjacent to the dental neck (**a–c**) and interdental papilla (**d, e**); helpful when checking for metal allergy

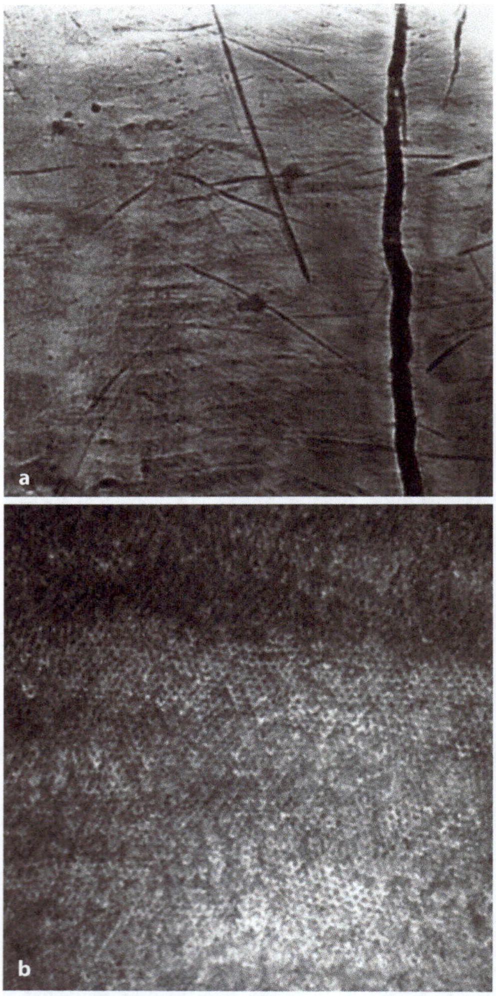

Fig. 7.18 **a** Tooth surface with scratches and large crack. **b** Tooth material with ceramic-like structure

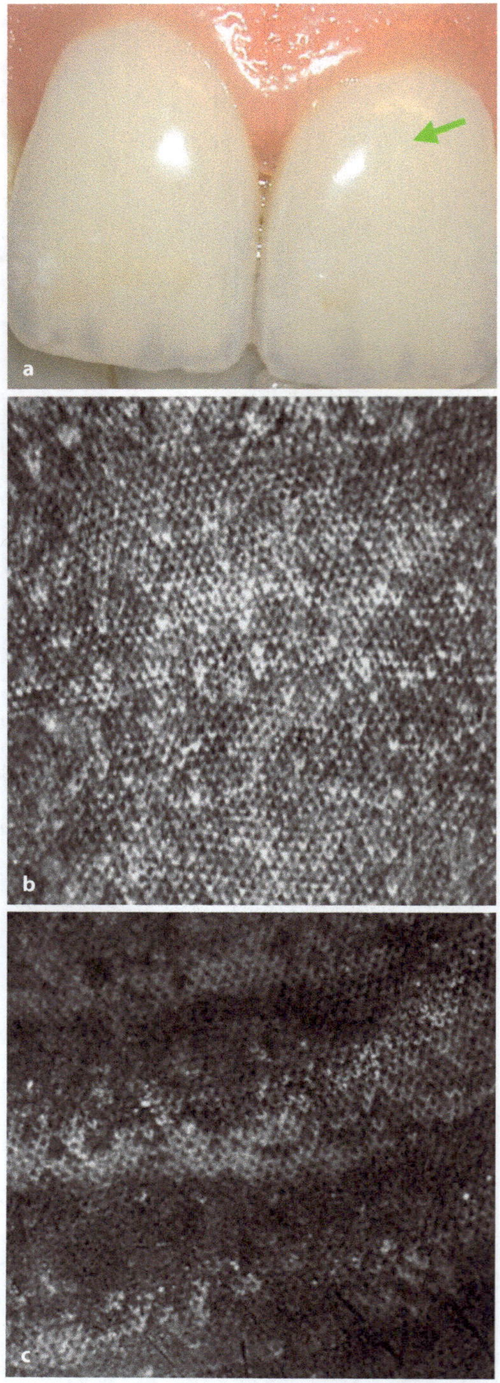

Fig. 7.19 **a** Clinical picture: healthy hard substance of the tooth. **b** Healthy dental enamel with regular arrangement of enamel prisms. **c** Hunter-Schreger bands indicating healthy dental enamel

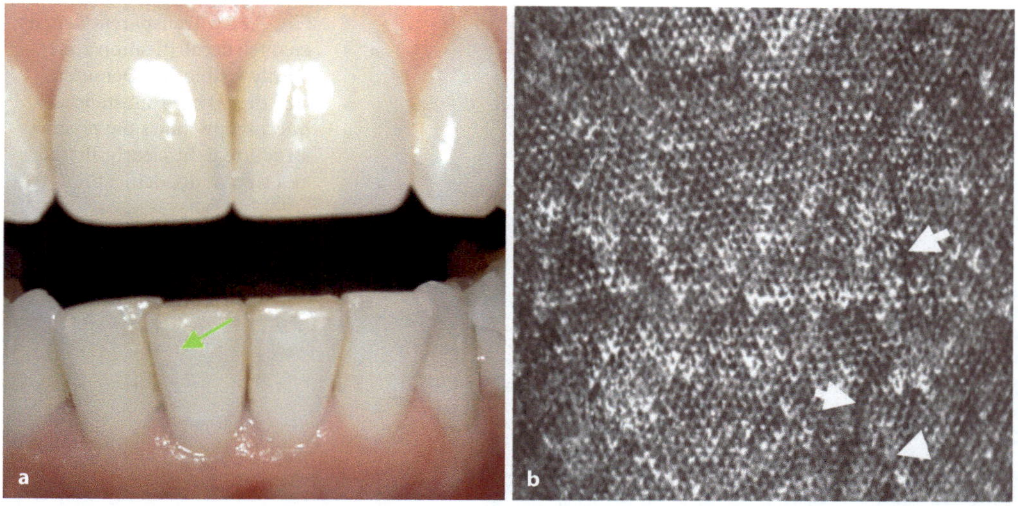

Fig. 7.20 Cracks in dental enamel. a Clinical picture: enamel cracks almost unidentifiable. b Enamel cracks on confocal microscopy

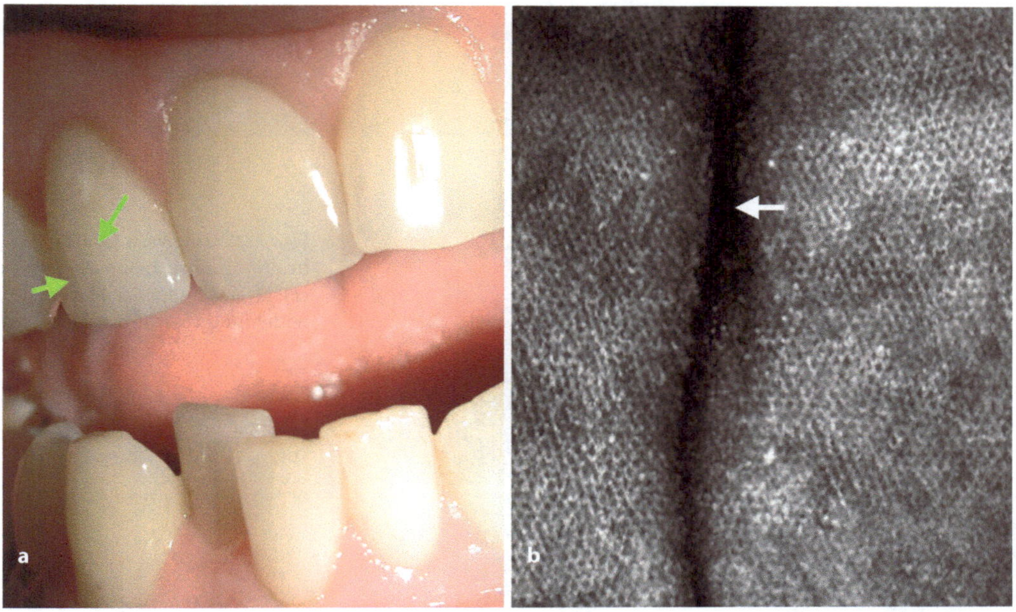

Fig. 7.21 a Clinical picture of a relatively major crack in the enamel. b Relatively large enamel crack on confocal microscopy

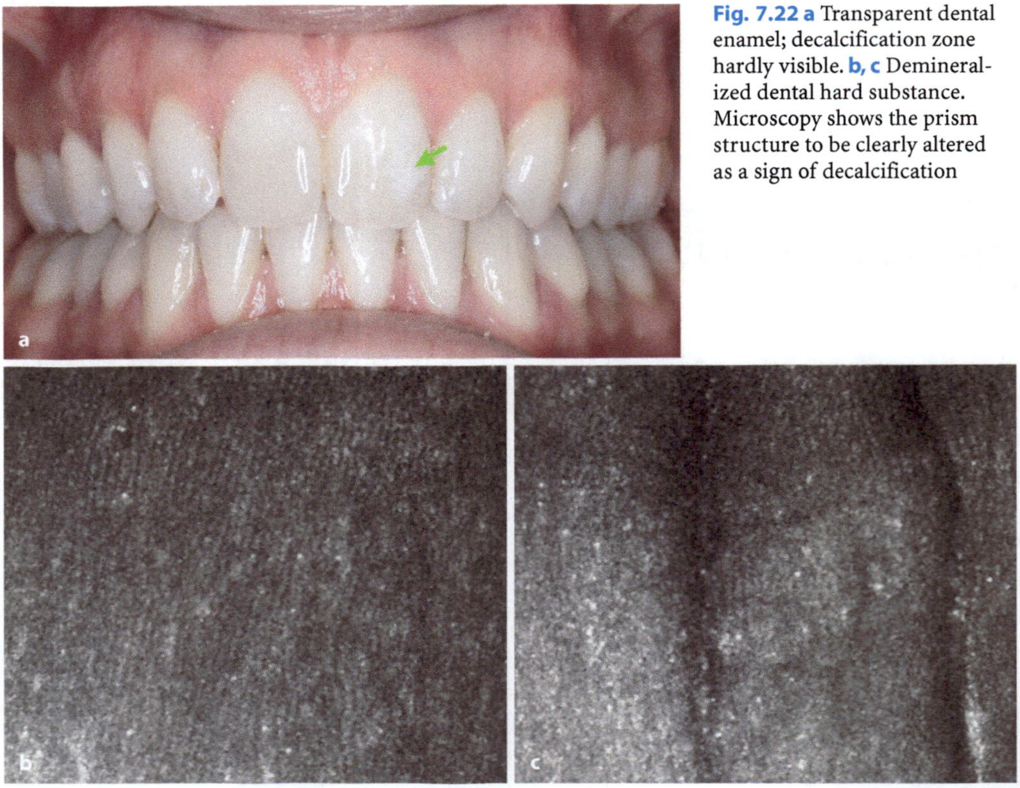

Fig. 7.22 a Transparent dental enamel; decalcification zone hardly visible. **b, c** Demineralized dental hard substance. Microscopy shows the prism structure to be clearly altered as a sign of decalcification

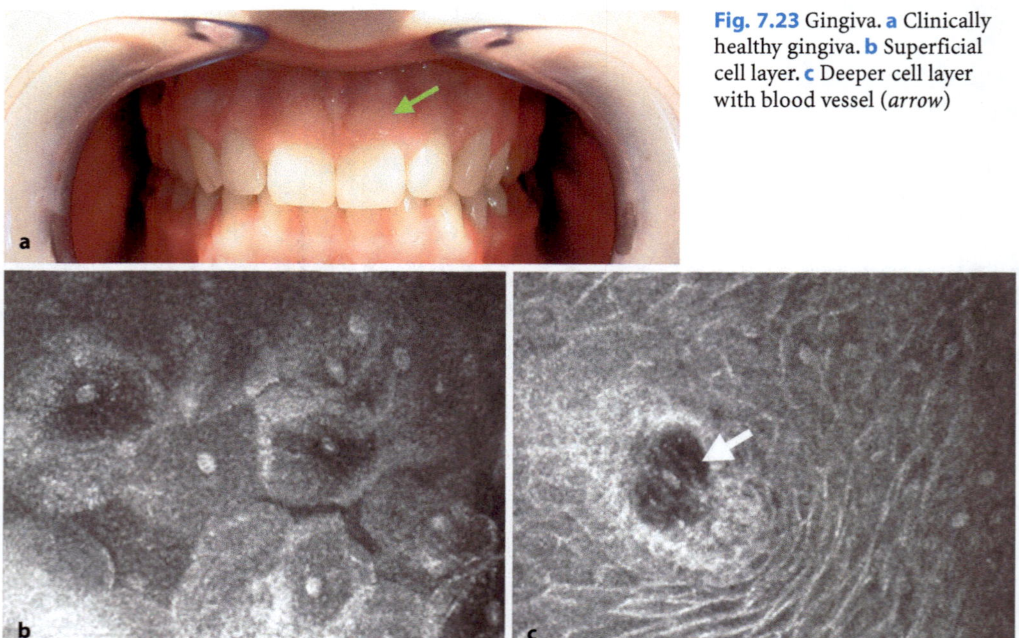

Fig. 7.23 Gingiva. **a** Clinically healthy gingiva. **b** Superficial cell layer. **c** Deeper cell layer with blood vessel (*arrow*)

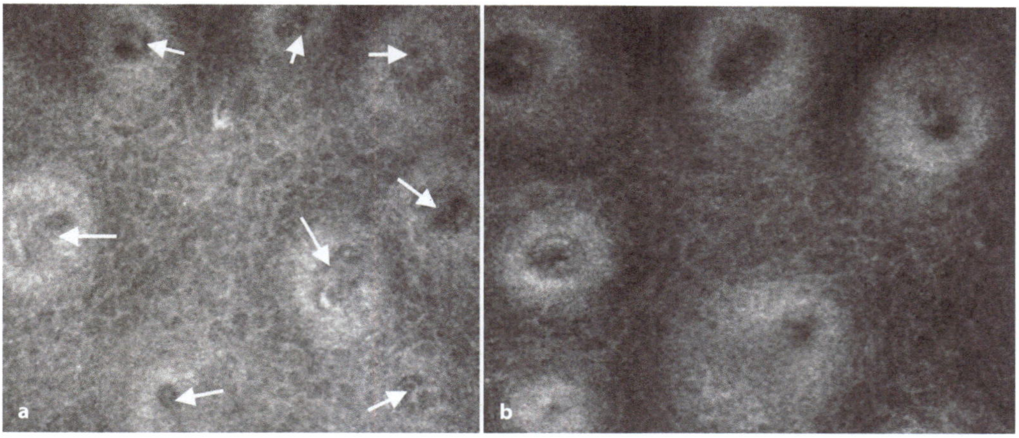

Fig. 7.24 Gingivitis. Inflamed gingiva (gingivitis) with hyperemic zones. Dense arrangement of blood vessels in tissue

Fig. 7.25 Material quality: discolored synthetic filling in tooth 21

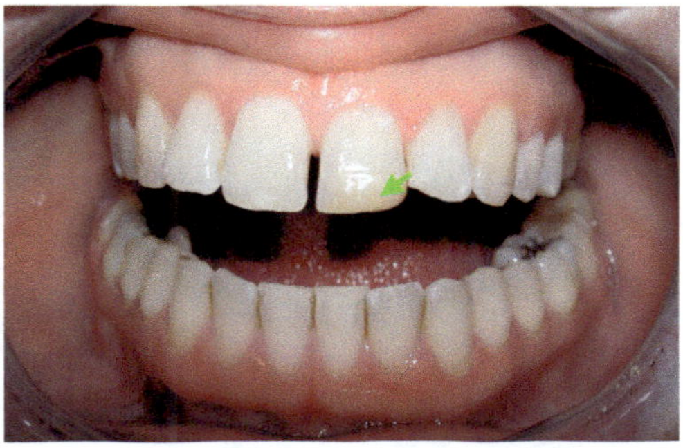

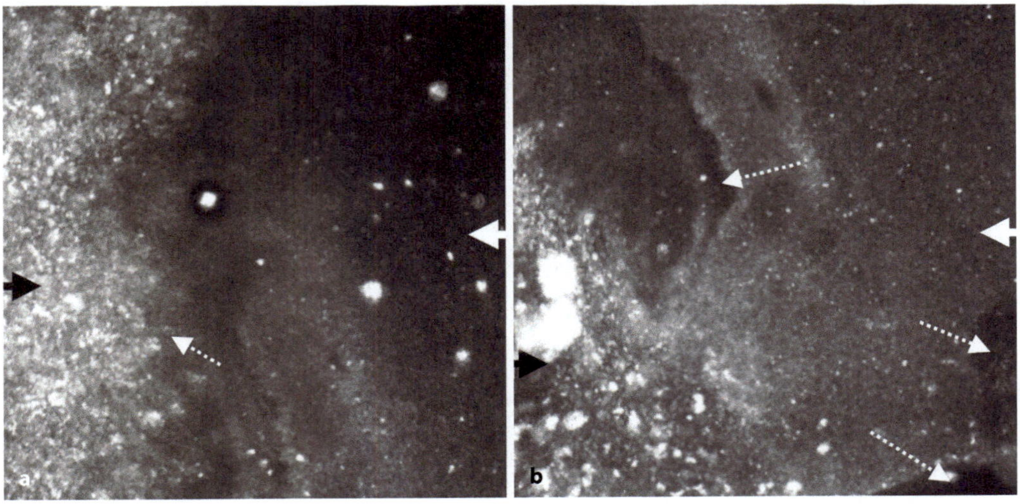

Fig. 7.26 **a** Interface zone (*dashed-line arrow*), composite (*white arrow*) and enamel (*black arrow*). **b** Air inclusions in filling material and in vicinity of interface zone (*dashed-line arrows*)

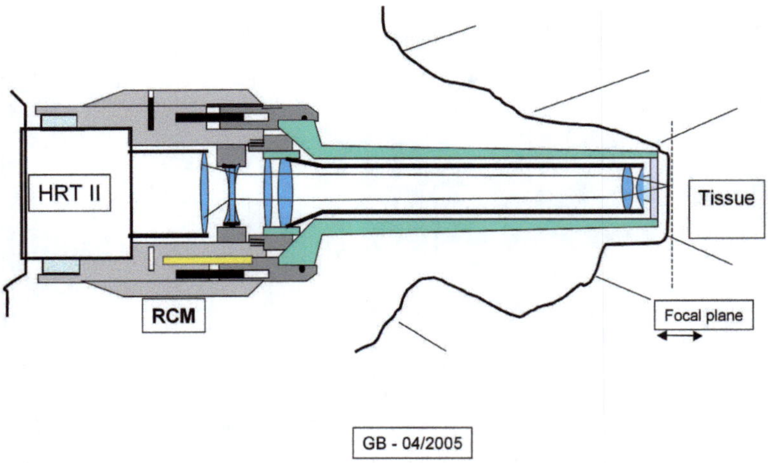

Fig. 7.27 Heidelberg Retina Tomograph II–Rostock Cornea Module: endoscopic confocal microscopy

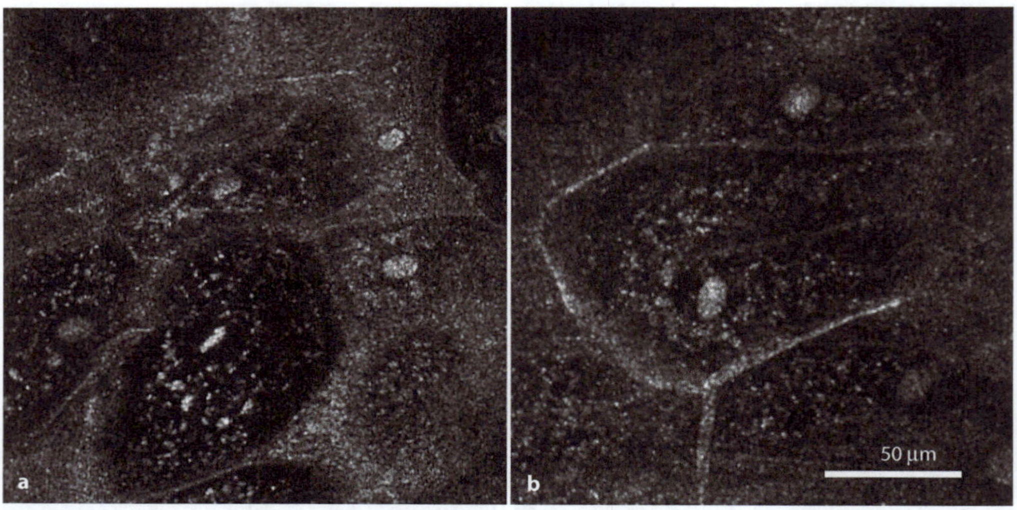

Fig. 7.28 Heidelberg Retina Tomograph II–Rostock Cornea Module: endoscopy of different cells in the pharyngeal mucosa (laser scanning contact endoscopy)

Confocal In Vivo Microscopy in Animal Experiments

Experimental studies using confocal in vivo microscopy can be performed in laboratory animals. The different corneal layers, the conjunctiva, the iris, and the lens can be evaluated in animals at the cellular level, under normal or pathological conditions. The advantage of this technique is that the animals do not need to be sacrificed.

8.1 Rabbits

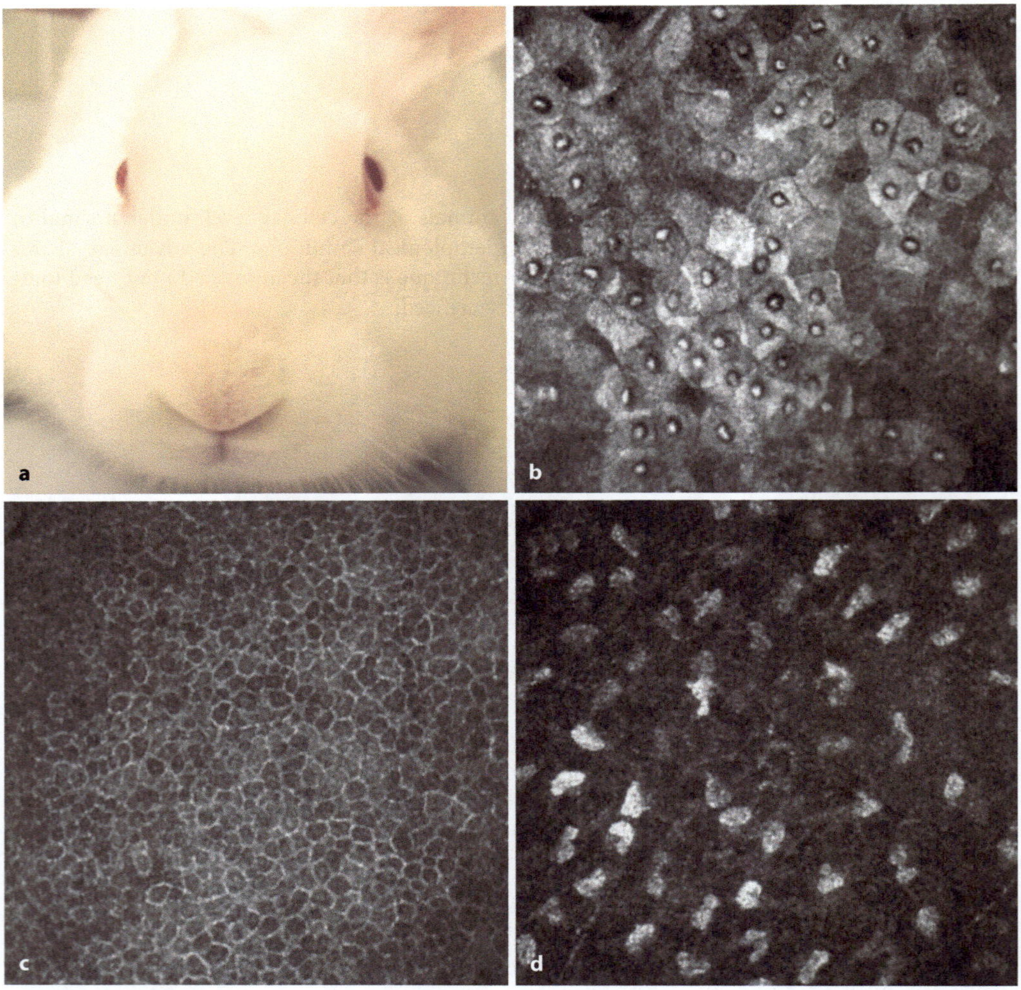

Fig. 8.1 a–h Rabbits. a Photograph of a New Zealand white rabbit. b Superficial corneal cells. c Basal corneal cells. d Anterior stroma.

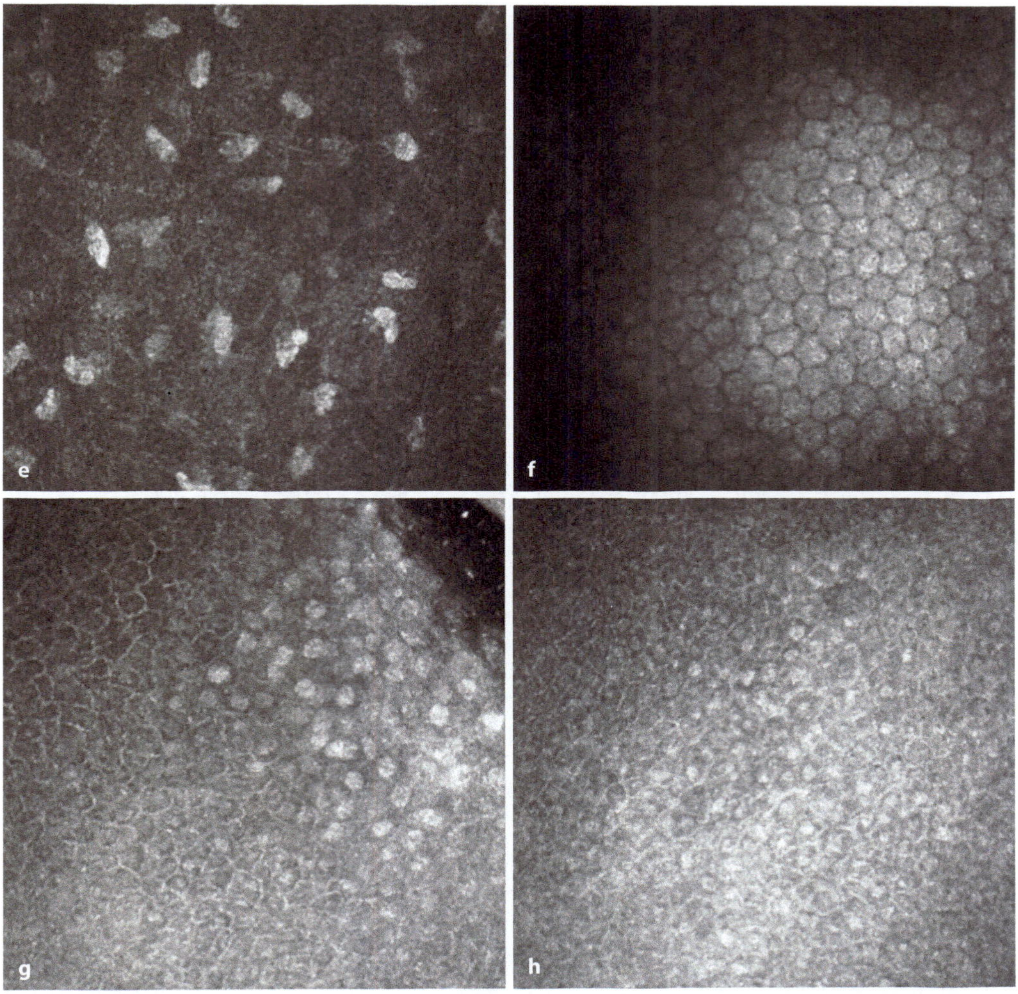

Fig. 8.1 (*continued*) **e** Posterior stroma. **f** Endothelium. **g, h** Conjunctival superficial cells and goblet cells (*round reflective cells*)

8.2 Rats

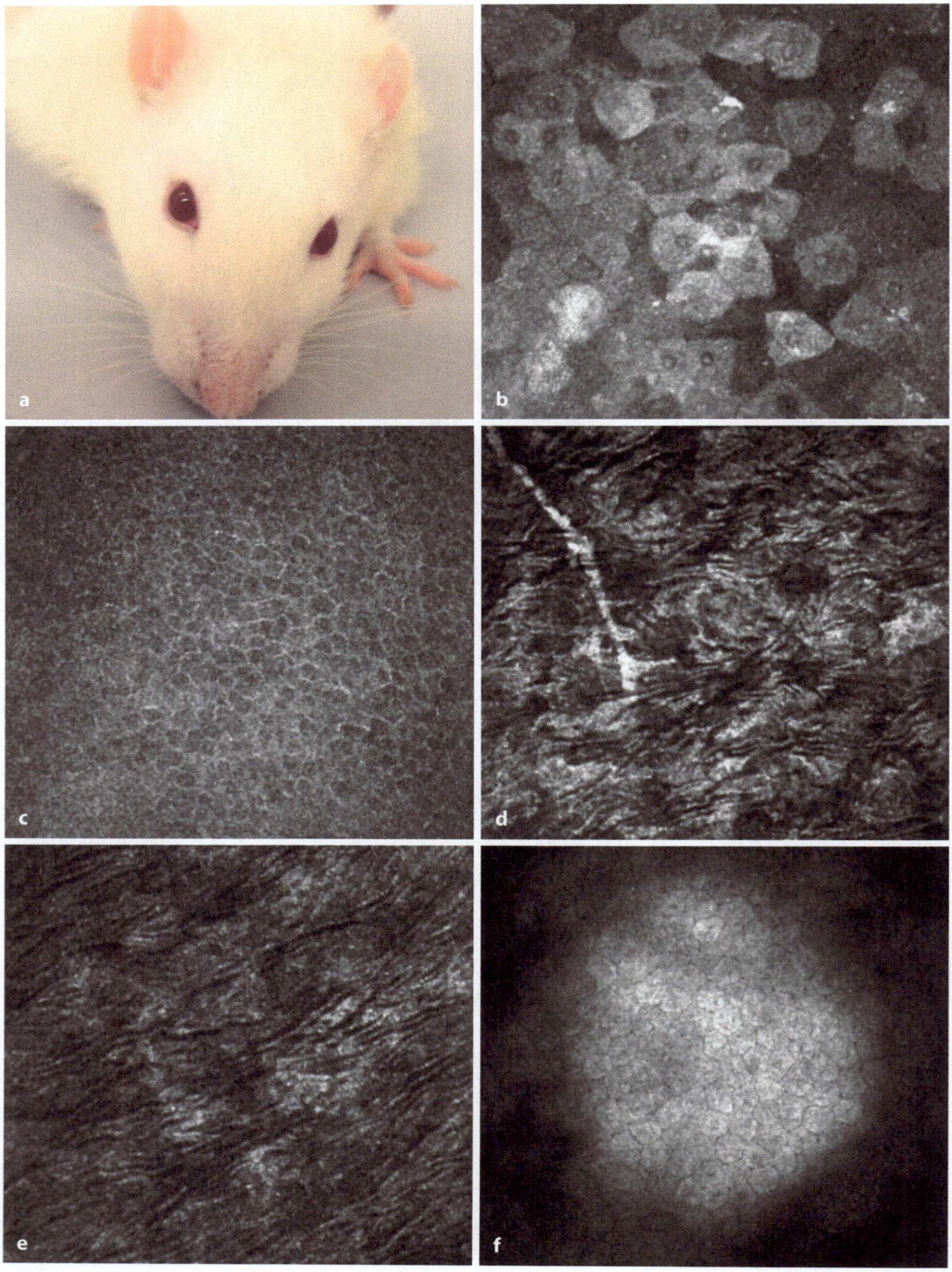

Fig. 8.2 Rats: normal corneal structures. **a** Photograph of a Lewis rat. **b** Superficial corneal cells. **c** Basal corneal cells. **d** Anterior stroma with a corneal nerve (reflective). **e** Posterior stroma **f** corneal endothelium

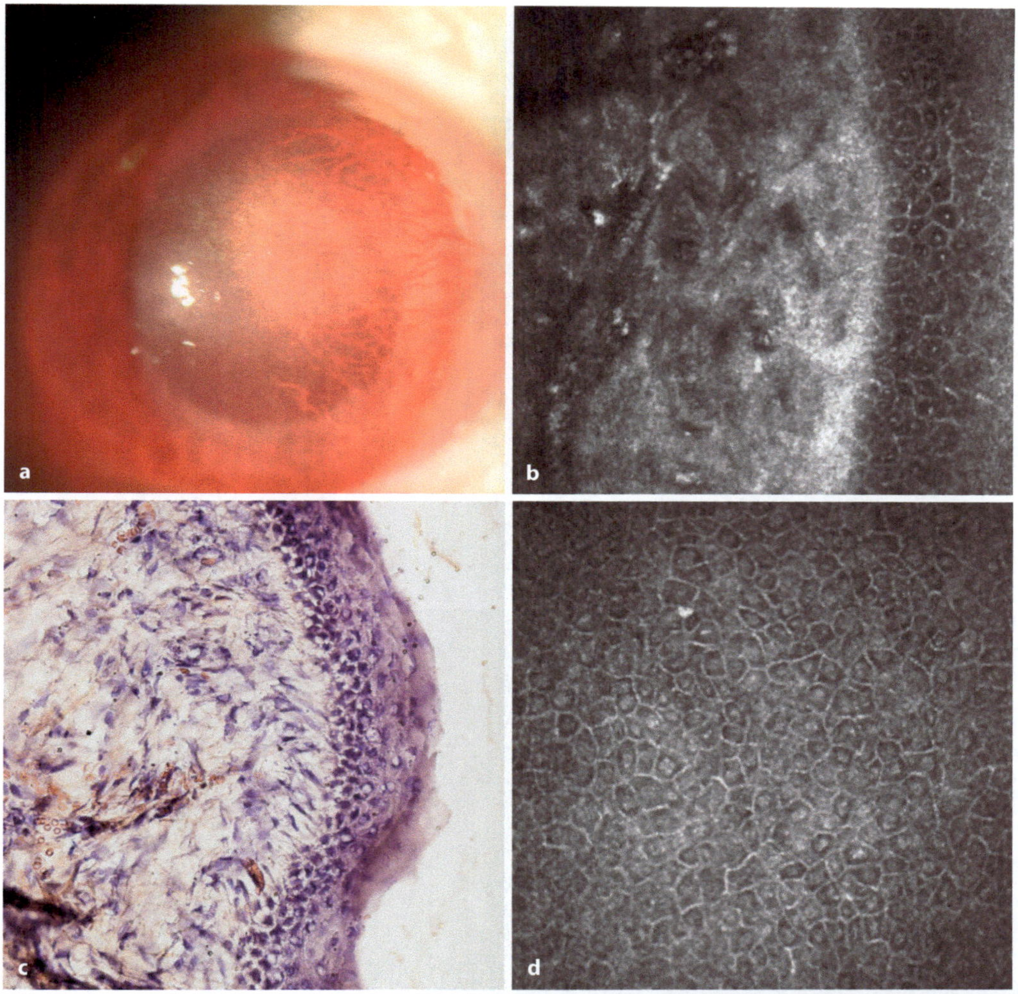

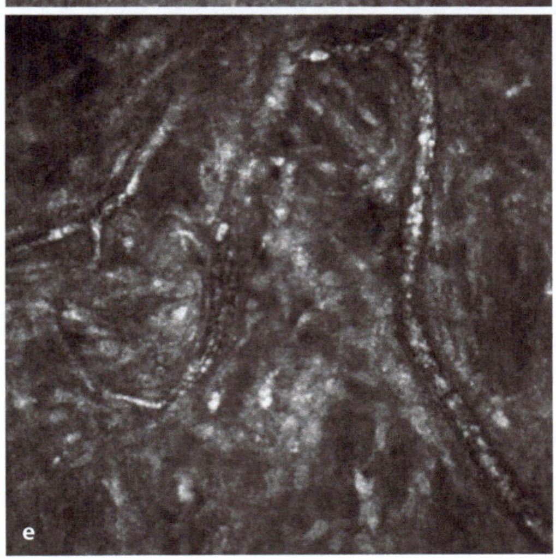

Fig. 8.3 Experimental studies: rat model of ocular surface toxicity. **a** Slit-lamp photograph of a rat cornea after 11 days of benzalkonium chloride (0.5%) instillation. Corneal opacification and neovascularization. **b** Confocal in vivo microscopy of the cornea, oblique section. Corneal metaplasia with irregular basal cells and reflective nuclei. Stromal infiltration and neovascularization. **c** Histology: excellent correlation with confocal in vivo microscopy. **d** Abnormal corneal basal cells. Reflective nuclei, polymorphism and flattened cells. **e** Abnormal stroma. Neovascularization and infiltration by inflammatory cells

8.3 Mice

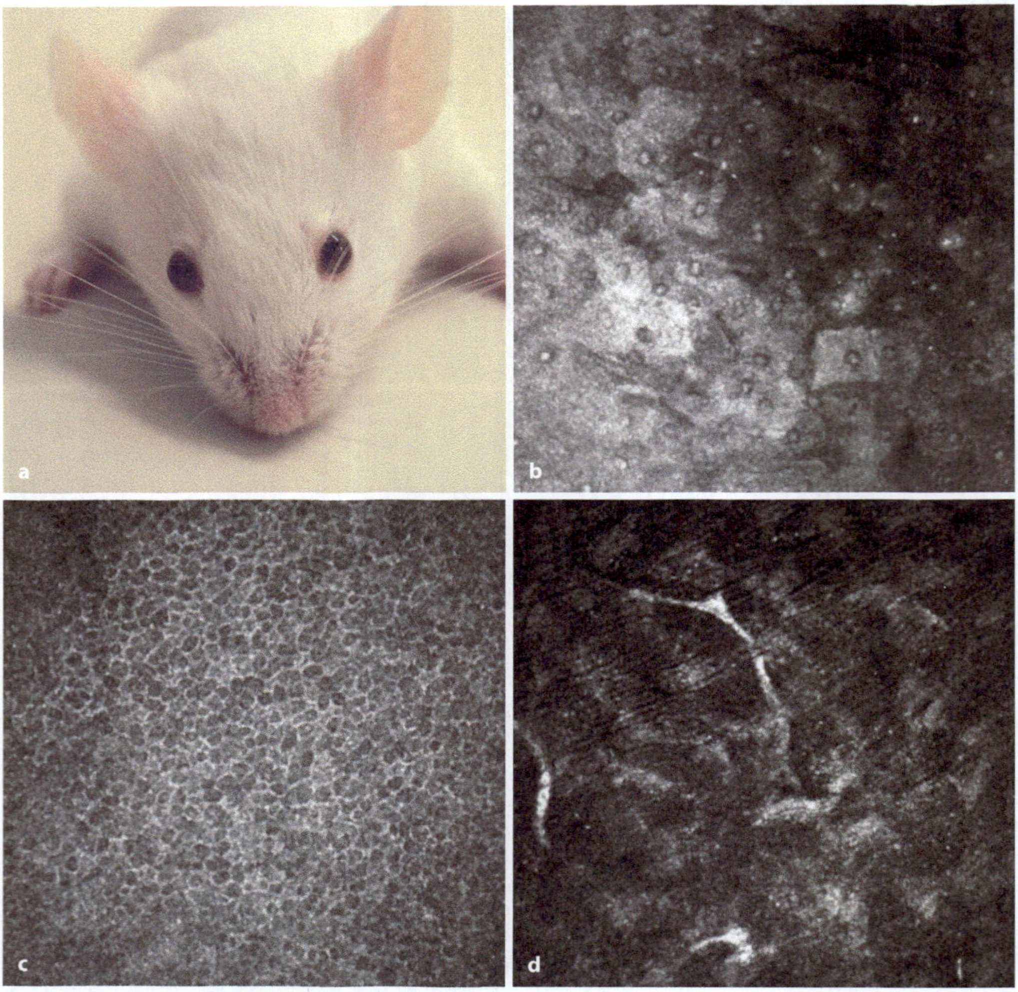

Fig. 8.4 a–g Mice. a Photograph of a Swiss mouse. b Superficial corneal cells. c Basal corneal cells. d Anterior stroma with corneal nerves (reflective structures)

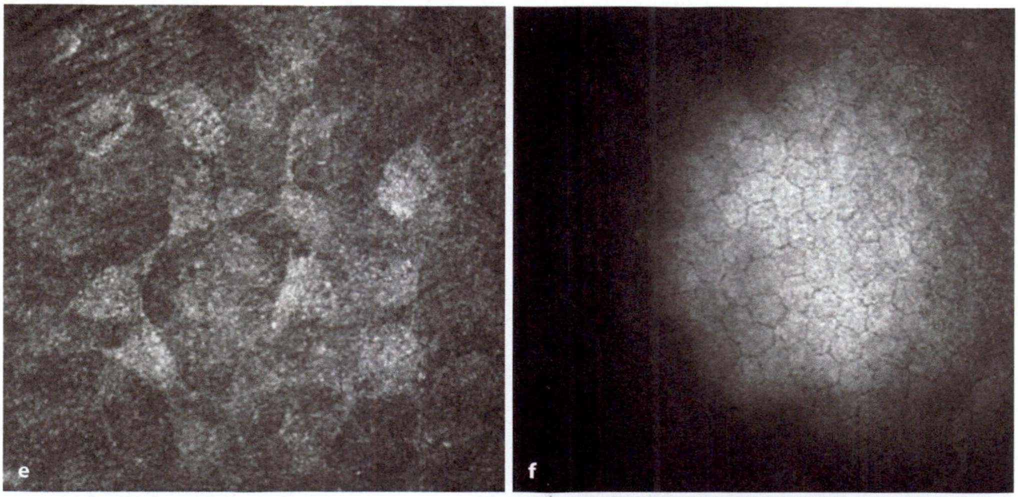

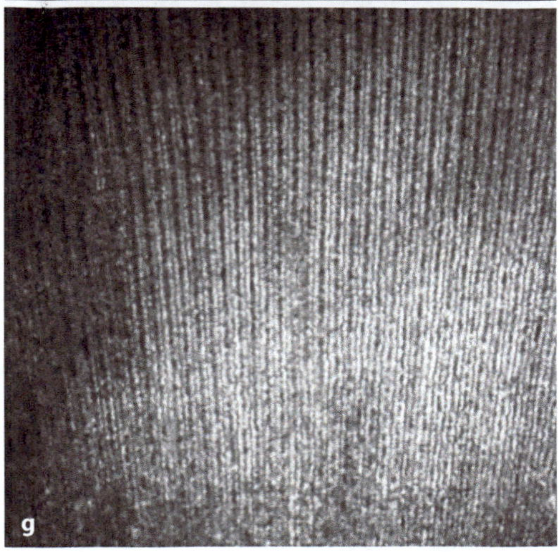

Fig. 8.4 (*continued*) **e** Posterior stroma. **f** Endothelial cells. **g** Lens fibers

8.4 Monkeys

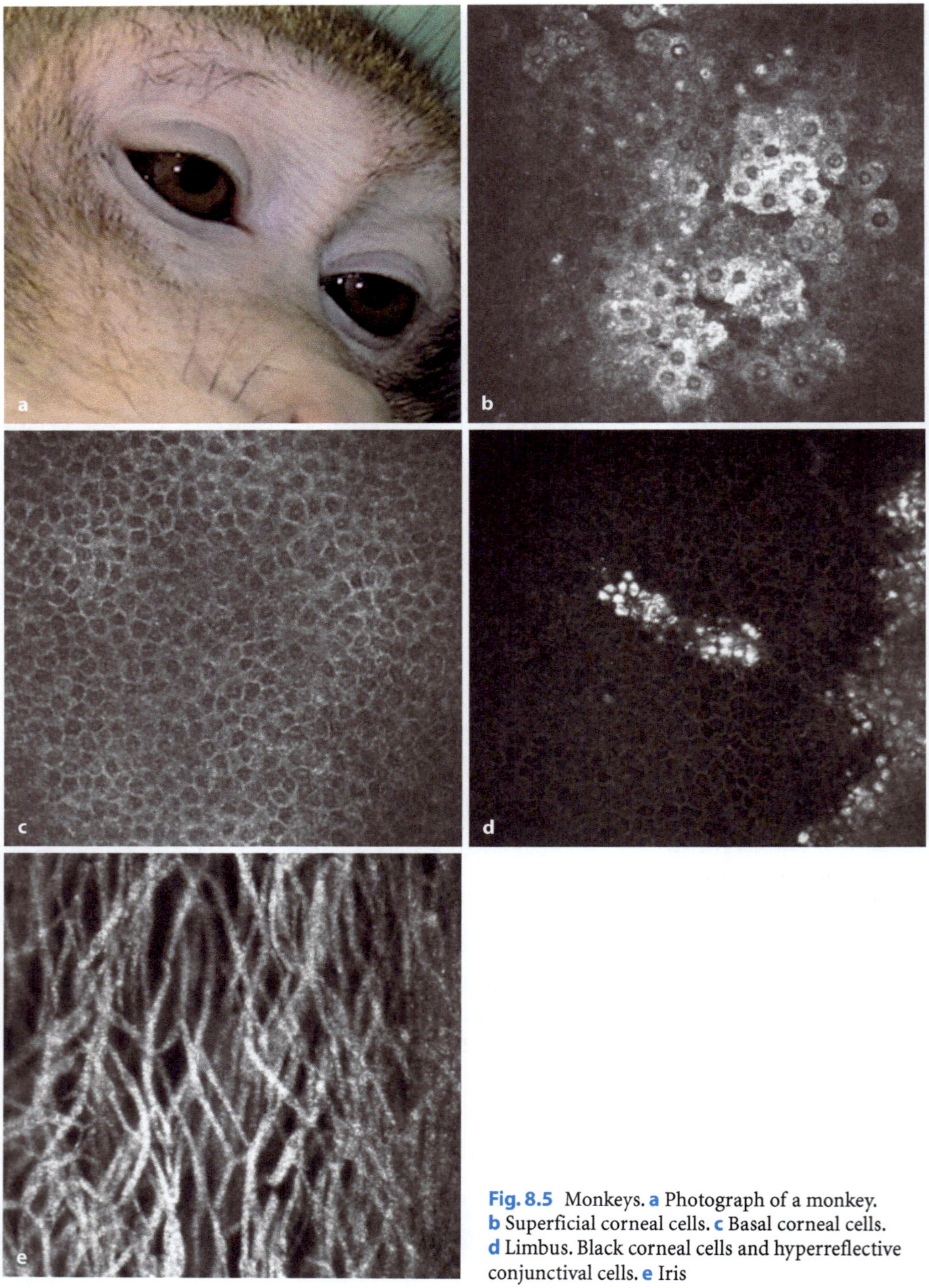

Fig. 8.5 Monkeys. a Photograph of a monkey. b Superficial corneal cells. c Basal corneal cells. d Limbus. Black corneal cells and hyperreflective conjunctival cells. e Iris

Current and Future Developments

9.1 Three-Dimensional Confocal Laser Scanning Microscopy

Many researchers have investigated the cornea with confocal in vivo microscopy [4, 27, 28, 29, 49, 84]. This sophisticated tool has been useful in augmenting our understanding of anatomy in the healthy and diseased human cornea. The limitations in magnification due to slight, unavoidable eye movements are obvious; therefore, 3D reconstruction is restricted on practical grounds. The step size is too coarse, and magnification is too small. However, 3D visualization and modeling would improve our understanding of the morphology of corneal architecture (epithelial nerve structure, for example).

This was our motivation for developing a fast, noninvasive, high-resolution method for detailed 3D investigation of the human cornea. The further development of the confocal microscope [92] took the form of a modified confocal laser scanning ophthalmoscope [80] based on a commercially available instrument, the HRT II [62]. A water-immersion microscope lens (Achroplan ×63/0.95 W/AA 2.00 mm, Carl Zeiss, Germany) with a long working distance and high numerical aperture was used and coupled to the cornea via a PMMA cap by interposing a transparent gel (Vidisic; Dr.Mann Pharma Germany) for in vivo imaging [80]. For 3D imaging, an internal scanning device moves the focal plane perpendicularly to the x–y plane, in the same way as in optic disk tomography performed with the original HRT II configuration. During image capture, the z-movement is stopped, and the image plane is exactly perpendicular to the z-axis. For the investigations presented, an acquisition time of 1 s with a scanning depth of 30 µm was used for all subjects; this is currently thought to be the maximum when patient and examiner movements are taken into account.

Images are presented in the form of a series of 2D grayscale images (384×384 pixels, 8-bit) representing optical sections through the cornea. The original raw image stacks were converted using ImageJ (NIH, USA) for 3D reconstruction using Amira 3.1 (TGS, USA). The voxel size is around 0.8×0.8×0.9 µm using the above-mentioned acquisition parameters. The Amira volume-rendering software package provides an interactive environment allowing features such as volume orientation for viewing planes and 3D perspectives, segmentation, and determination of distances and surfaces. The image stacks were carefully aligned and modified to eliminate unspecific information by adapting the gray values in the depicted spectrum. Shadows and illumination were manipulated after assigning density values to gray values to more clearly visualize the spatial arrangement without losing information.

As a first in vivo application of the new device in combination with 3D reconstruction techniques, nerve fiber distribution was characterized in healthy human corneal epithelium. The spatial arrangement of epithelium, nerves, and keratocytes was visualized by 3D in vivo confocal laser scanning microscopy (Fig. 9.1). The 3D reconstruction of the cornea in healthy volunteers yielded a picture of the nerves in the central part of the human cornea. Thick fibers arise from the subepithelial plexus, and the nerves further subdivide dichotomously and

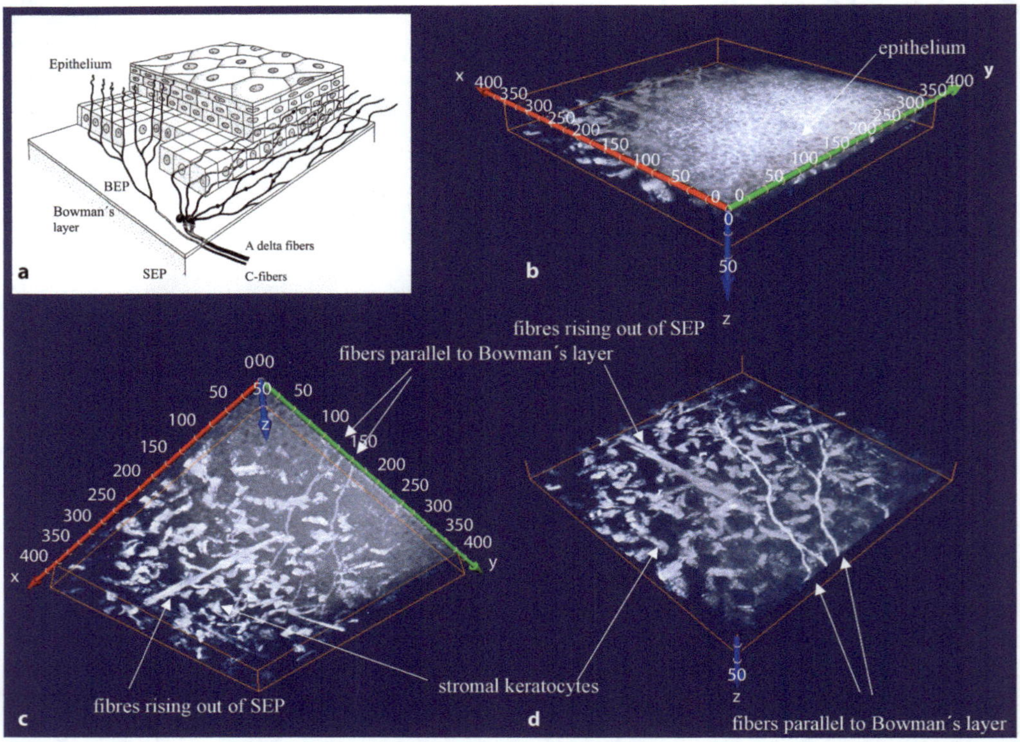

Fig. 9.1 Schematic illustration (**a**) and three-dimensional reconstruction (**b–d**) of the corneal epithelium with anterior stroma and nerves (healthy human subject). **b** Anterior view. **c** Posterior view. **d** Anterior view with virtual removal of epithelium. Thin nerves run parallel with Bowman's membrane in the basal epithelial plexus, with thicker fibers originating from the subepithelial plexus

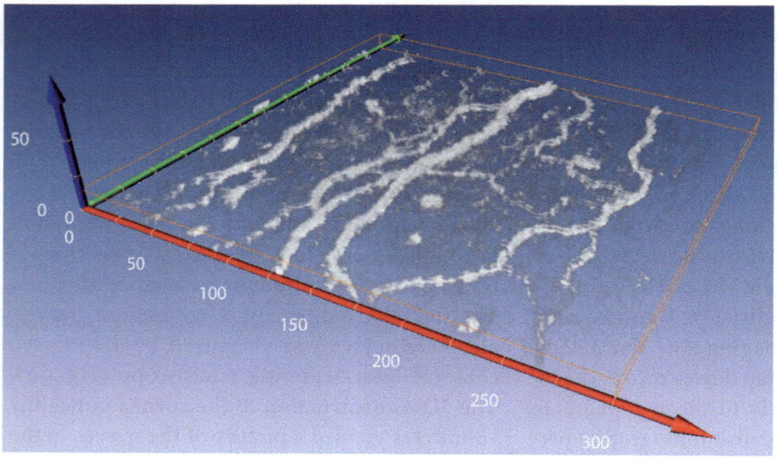

Fig. 9.2 Nerve fibers in the basal epithelial plexus, in strict alignment parallel with Bowman's membrane

trichotomously, resulting in five to six thinner fibers arranged parallel to Bowman's membrane and with partial interconnections (Fig. 9.2.) Branches penetrating the anterior epithelial cell layer cannot be visualized.

In conclusion, 3D confocal laser scanning microscopy is the first technique to permit visualization and analysis of the spatial arrangement of the epithelium, nerves, and keratocytes in the living human cornea. The method developed provides a basis for further device refinements and for studies of changes in cellular arrangement and epithelial innervation in corneal disease. For example, confocal laser scanning microscopy may help clarify gross variations of nerve fiber patterns under various clinical and experimental conditions.

9.2 Functional Imaging

In conventional microscopy, the possibility of using dyes to visualize specific anatomic structures yields major information gains. This is especially true when techniques of fluorescence microscopy or immunohistochemistry are used in combination with confocal techniques [81]. Because these methods are well suited for investigating the functional status of tissues, they are also interesting for in vivo microscopy in humans, such as for studies of wound healing or inflammation processes. However, problems arise due to the necessity for real-time investigation because of involuntary movements on the part of the subjects and due to the selection of suitable nontoxic vital stains. Nevertheless, successful initial steps have already been taken toward confocal in vivo fluorescence microscopy of the anterior eye segments [25, 38], and these may be regarded as a further enhancement of corneal assessments by slit-lamp microscopy following fluorescein or rose bengal staining [19, 20, 53, 64, 82].

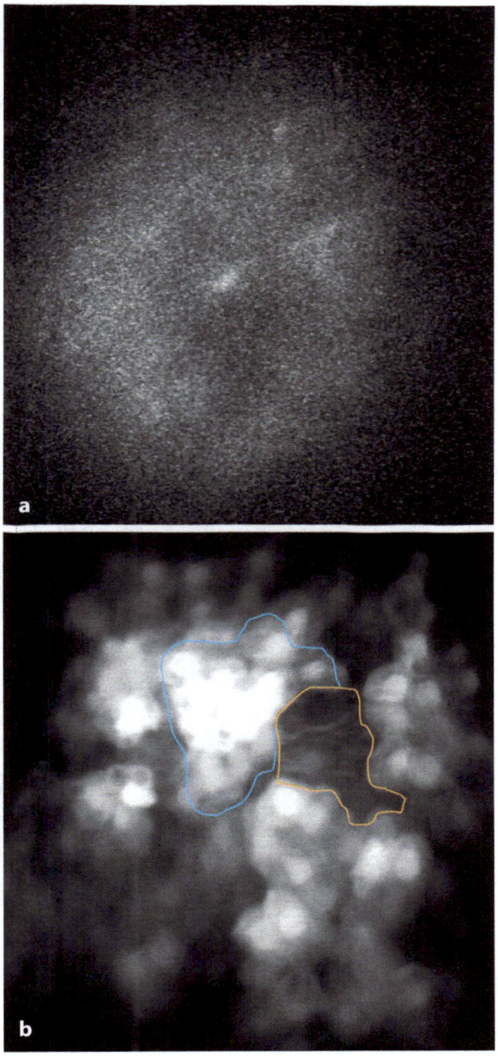

Fig. 9.3 Fluorescence micrographs of the superficial corneal epithelial cells. **a** Intact corneal epithelium without appreciable fluorescence. **b** Stippled epithelium after contact glass examination; *blue* marked area with fluorescein-stained cells, *orange* marked area with unstained intact cells

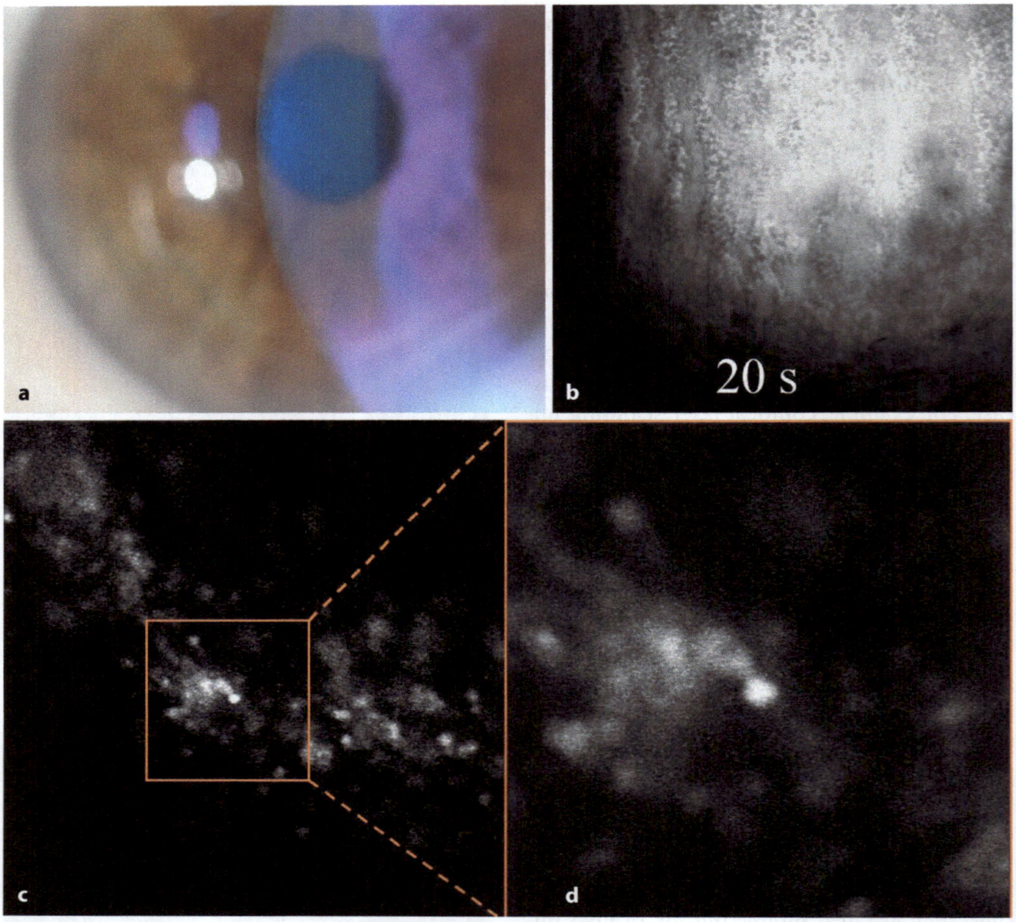

Fig. 9.4 Patient with intact corneal epithelium. **a** Intact corneal epithelium, slit-lamp microscopy photograph. **b** Reflection mode, tear film still intact 20 s after eyelid opening. **c** Fluorescence mode, superficial corneal epithelium, only minimal fluorescence of individual cells. **d** Fluorescence mode, higher magnification

To achieve this, an HRA (HRA/C, Heidelberg Engineering, Germany) has been modified with a lens attachment (RCM) so that the laser focus is shifted to the anterior eye segments. This enables fluorescence microscopy images to be obtained after staining with the nonspecific stain NaF and excitation with blue argon laser light (wavelength 488 nm) and the addition of a barrier filter (500 nm). Using a green argon laser (514 nm) in reflection mode with the same device, it is also possible to visualize break-up phenomena of the tear film [38, 88, 89] (Figs. 9.3a, b; 9.4a–d; 9.5a–d).

The result is a technique that enables further-reaching investigations of damaged corneal epithelium and of associated wound healing processes. In the future, confocal fluo- rescence microscopes specially designed for in vivo investigations in humans will perhaps permit high-quality functional imaging that is even more comprehensive and specific.

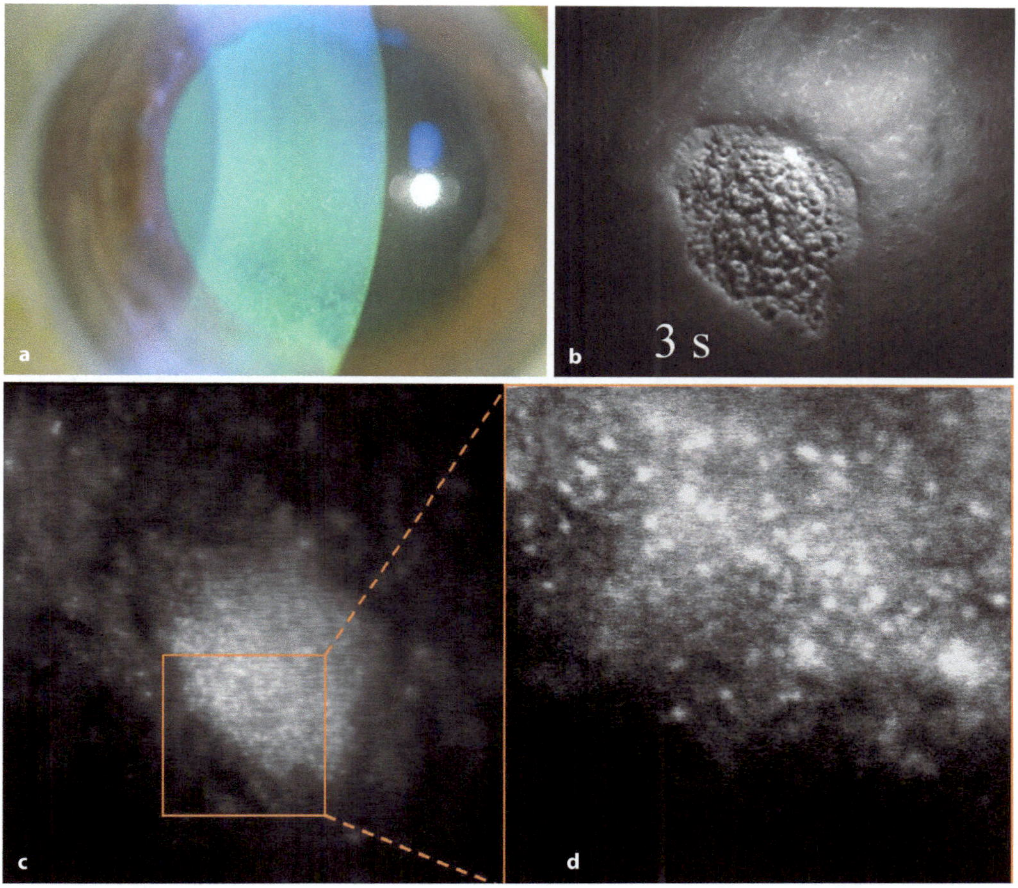

Fig. 9.5 Same patient as in Fig. 9.4 after application of a local anesthetic and following applanation tonometry. **a** Slit-lamp microscopy photograph with corneal stippling. **b** Reflection mode, tear film defect (dry spot) just 3 s after eyelid opening. **c** Fluorescence mode, same area as in b; superficial corneal epithelium, area with marked fluorescence. **d** Fluorescence mode, higher magnification

9.3 Outlook

Confocal high-resolution biomicroscopy will be used for the in vivo description of corneal pathology at the cellular level. It will enable degeneration and repair mechanisms under various conditions to be examined so that the findings can be correlated with those from conventional slit-lamp biomicroscopy. This will generate enhanced quality in clinical evaluation. The use of vital staining substances, such as NaF, ethidium homodimer, and calcein, may give insights into the metabolic activities of a variety of cells under different wound healing and degenerative conditions.

References

1. Addicks EM, Quigley HA, Green R, Robin AL. Histological characteristics of filtering blebs in glaucomatous eyes. *Arch Ophthalmol* 1983; **101**:795–798
2. Augustin AJ (ed). *Augenheilkunde* (2nd edition). Berlin, Heidelberg, New York: Springer-Verlag: 2001
3. Banchereau J, Steinman RM. Dendritic cells and the control of the immunity. *Nature* 1998; **392**: 245–252
4. Berlau J, Becker HH, Stave J, Oriwol C, Guthoff RF. Depth and age-dependent distribution of keratocytes in healthy human corneas: a study using scanning-slit confocal microscopy in vivo. *J Cataract Refract Surg* 2002; **28**:611–616
5. Bigar F, Thaer A. Untersuchung der Hornhaut intakter Spenderaugen mit konfokalem Spalt-Scanning Video-Mikroskop. *Klin Monatsbl Augenheilkd* 1994; **204**:421–423
6. Bochert R, Zhivov A, Kraak R, Stave J, Guthoff RF. Contribution to comprehension of image formation in confocal microscopy with the Rostock Cornea Module. *Br J Ophthalmol* 2005; **89**:1351–1355
7. Böhnke M, Thaer AA. Untersuchung der Kornea mit einem neuen konfokalen Mikroskop. In: Lund OE, Waubke TN (eds): *Bildgebende Verfahren in der Augenheilkunde. Methoden und Indikationen.* Stuttgart: Enke, 1994:47–53
8. Böhnke M, Masters BR. Confocal microscopy of the cornea. *Prog Retin Eye Res* 1999; **18**:553–628
9. Bourne WM, Kaufman HE. Specular microscopy of human corneal endothelium in vivo. *Am J Ophtalmol* 1976; **81**:319–323
10. Cantor LB, Mantravadi A, WuDunn D, Swamynathan K, Cortes A. Morphologic classification of filtering blebs after glaucoma filtration surgery: the Indiana Bleb Appearance Grading Scale. *J Glaucoma* 2003; **12**:266–271
11. Cavanagh HD, Jester JV, Essepian J, Shields W, Lemp MA. Confocal microscopy of the living eye. *CLAO J* 1990; **16**:65–73
12. Cavanagh DH, Petroll WM, Alizadeh H, He YG, McCulley JP, Jester JV. Clinical and diagnostic use of in vivo confocal microscopy in patients with corneal disease. *Ophthalmology* 1993; **100**:1444–1454
13. Donnenfeld ED, Solomon K, Perry HD, et al. The effect of hinge position on corneal sensation and dry eye after LASIK. *Ophthalmology* 2003; **110**: 1023–1029
14. Donnenfeld ED, Ehrenhaus M, Solomon R, et al. Effect of hinge width on corneal sensation and dry eye after laser in situ keratomileusis. *J Cataract Refract Surg* 2004; **30**:790–797
15. Doughty MJ, Naase T, Donald C, Hamilton L, Button NF. Visualisation of "Marx's line" along the marginal eyelid conjunctiva of human subjects with lissamine green dye. *Ophthalmic Physiol Opt* 2004; **24**:1–7
16. Dua HS, Azuara-Blanco A. Limbal stem cells of the corneal epithelium. *Surv Ophthalmol* 2000; **44**:415–425
17. Duke-Elder S, Wybar KC (eds). *System of Ophthalmology (vol. II).* Henry Kimpton, London, 1958: 113–127 and 543–557
18. Dunn A, Smithpeter C, Welch AJ, Richards-Kortum R. Finite-difference time-domain simulation of light scattering from single cells. *J Biomed Opt* 1997; **2**:262–266
19. Feenstra RP, Tseng SC. Comparison of fluorescein and rose bengal staining. *Ophthalmology* 1992; **99**:605–617
20. Fromm, Groenouw. Ueber die diagnostische Verwendbarkeit der Flouresceïnfärbung bei Augenerkrankungen. *Arch f Augenheilkd* 1891; **22**:247–257
21. Gray H, Standring S, Ellis H, Berkovitz BKB. *Gray's Anatomy: The Anatomical Basis of Clinical Practice.* Edinburgh, New York: Elsevier/Churchill Livingstone, 2005
22. Gullstrand A. Demonstration der Nernstspaltlampe. Verslg ophthalm. Ges. Heidelberg 1911, 374, cited in Vogt A. *Lehrbuch und Atlas der Spaltlampenmikroskopie des lebenden Auges* (2nd edition); Berlin: Verlag von Julius Springer, 1930:1

23. Guthoff RF, Wienss H, Hahnel C, Wree A. Epithelial innervation of human cornea: a three-dimensional study using confocal laser scanning fluorescence microscopy. *Cornea* 2005:**24**:608–613
24. Guthoff RF, Stave J. In vivo micromorphology of the cornea: confocal microscopy principles and clinical applications. In: Reinhard T, Larkin F (eds). *Essentials in Ophthalmology – Cornea and External Eye Disease.* Berlin, Heidelberg, New York: Springer-Verlag, 2006:173–208
25. Hahnel C, Somodi S, Weiss DG, Guthoff RF. The keratocyte network of human cornea: a three-dimensional study using confocal laser scanning fluorescence microscopy. *Cornea* 2000; **19**:185–193
26. Hamrah P, Huq SO, Liu Y, Zhang Q, Dana MR. Corneal immunity is mediated by heterogeneous population of antigen-presenting cells. *J Leukoc Biol* 2003; **74**:172–178
27. Hara M, Morishige N, Chikama T, Nishida T. Comparison of confocal biomicroscopy and noncontact specular microscopy for evaluation of the corneal endothelium. *Cornea* 2003; **22**:512–515
28. Harrison DA, Joos C, Ambrosio R. Morphology of corneal basal epithelial cells by in vivo slit-scanning confocal microscopy. *Cornea* 2003; **22**:246–248
29. Holopainen JM, Moilanen JA, Tervo TM. In vivo confocal microscopy of Fleck dystrophy and pre-Descemet's membrane corneal dystrophy. *Cornea* 2003; **22**:160–163
30. Just T, Stave J, Pau HW, Guthoff RF. In-vivo observation of papillae of the human tongue using confocal laser scanning microscopy. *J Otorhinolaryngol Relat Spec* 2005; **67**:207–212
31. Just T, Zeisner C, Stave J, Pau HW. Konfokale Laser-Scanning Mikroskopie zur Beurteilung des Zungenepithels. *Laryngorhinootologie* 2004; **83**: 108–112
32. Just T, Stave J, Boltze C et al. Laser scanning microscopy of the human larynx mucosa: a preliminary, ex vivo study. *Laryngoscope* 2005; *submitted*
33. Just T, Bombor I, Pau HW, Klautke G, Fietkau R. Paraneoplastic changes of oropharynx mucosa in breast cancer. Strahlenther Onkol 2006; **182**:112–115
34. Just T, Pau HW, Bombor I, Guthoff RF, Fietkau R, Hummel T. Confocal microscopy of the peripheral gustatory system: comparison between healthy subjects and patients suffering from taste disorders during radiochemotherapy. Laryngoscope 2005; **115**:2178–2182
35. Kaufmann T, Bodanowitz S, Hesse L, Kroll P. Corneal reinnervation after photorefractive keratectomy and laser in situ keratomileusis and in-vivo study with a confocal video microscope. *German J Ophthalmol* 1997; **5**:508–512
36. Kobayashi A, Yoshita T, Sugiyama K. In vivo findings of the bulbar/palpebral conjunctiva and presumed meibomian glands by laser scanning confocal microscopy. *Cornea* 2005; **24**:985–988
37. Koester CJ, Roberts CW, Donn A, Hoefle FB. Wide field specular microscopy. Clinical and research applications. *Ophthalmology* 1980; **87**:849–860
38. Kraak R, Stave J, Guthoff RF. In vivo Untersuchung des Verteilungsmusters von Na-Fluorescein an menschlichen Hornhäuten mittels konfokaler Laser-Scanning Fluoreszenzmikroskopie. Paper presented at the 102nd Meeting of the German Ophthalmology Society: Berlin, 2004
39. Krstic RV. *Human Microscopic Anatomy: An Atlas for Students of Medicine and Biology.* Berlin, Heidelberg, New York: Springer-Verlag, 1991
40. Labbe A, Dupas B, Hamard P, Baudouin C. An evaluation of blebs after filtering surgery with the in vivo confocal microscope. *J Fr Ophtalmol* 2004; **27**:1083–1089
41. Laing RA, Sandstrom MM, Leibowitz HM. In vivo photomicrography of the corneal endothelium. *Arch Ophthalmol* 1975; **93**:143–145
42. Latvala T, Barraquer-Coll C, Tervo K, et al. Corneal wound healing and nerve morphology after laser in situ keratomileusis (LASIK) in human eyes. *J Refract Surg* 1996; **12**:677–683
43. Li HF, Petroll WM, Moller-Pedersen T, Maurer JK, Cavanagh HD, Jester JV. Epithelial and corneal thickness measurements by in vivo confocal microscopy through focusing (CMTF). *Curr Eye Res* 1997; **16**:214–221
44. Linna TU, Vesaluoma MH, Pérez-Santonja JJ, et al. Effect of myopic LASIK on corneal sensitivity and morphology of subbasal nerves. *Invest Ophthalmol Vis Sci* 2000; **41**:393–397
45. Lund O-E, Stefani FH. Corneal histology after epidemic keratoconjunctivitis. *Arch Ophthalmol* 1978; **96**:2085–2088
46. Masters BR, Thaer AA. Real-time scanning-slit confocal microscopy of the in-vivo human cornea. *Appl Optics* 1974; **33**:695–701
47. Masters BR, Thaer AA. In vivo human corneal confocal microscopy of identical fields of subepithelial nerve plexus, basal epithelial, and wing cells at different times. *Microsc Res Tech* 1994; **29**:350–356
48. Masters BR, Böhnke M. Three-dimensional confocal microscopy of the human cornea in vivo. *Ophthalmic Res* 2001; **33**:125–135
49. Mastropasqua L, Nubile M. *Confocal Microscopy of the Cornea.* Thorofare, NJ: Slack, 2002
50. Mathers WD, Jester JV, Lemp MA. Return of human corneal sensitivity after penetrating keratoplasty. *Arch Ophthalmol* 1988; **106**:210–211

51. Mathers WD, Lane JA, Zimmerman MB. Assessment of the tear film with tandem scanning confocal microscopy. *Cornea* 1997; **16**:162–168
52. Maurice DM. Cellular membrane activity in the corneal endothelium of the intact eye. *Experientia* 1968; **24**:1094–1095
53. Maurice DM. The use of fluorescein in ophthalmological research. *Invest Ophthalmol* 1967; **6**: 464–477
54. Maurice DM. A scanning slit optical microscope. *Invest Ophthalmol* 1974; **13**:1033–1037
55. Messmer EM, Torres Suarez E, Mackert MI, Zapp DM, Kampik A. Konfokale In-vivo-Mikroskopie bei Blepharitis. *Klinische Monatsblätter für Augenheilkunde* 2005; **222**:894–900
56. Milz S, Neufang J, Higashiyama I, Putz R, Benjamin M. An immunohistochemical study of the extracellular matrix of the tarsal plate in the upper eyelid in human beings. *J Anat* 2005; **206**:37–45
57. Moller-Pedersen T, Ehlers N. A three-dimensional study of the human corneal keratocyte density. *Curr Eye Res* 1995; **14**:459–464
58. Moller-Pedersen T, Vogel M, Li HF, Petroll WM, Cavanagh HD, Jester JV. Quantification of stromal thinning, epithelial thickness, and corneal haze after photorefractive keratectomy using in vivo confocal microscopy. *Ophthalmology* 1997; **104**: 360–368
59. Moller-Pedersen T, Li HF, Petroll WM, Cavanagh HD, Jester JV. Confocal microscopic characterization of wound repair after photorefractive keratectomy. *Invest Ophthalmol Vis Sci* 1998; **39**:487–501
60. Müller LJ, Marfurt CF, Kruse F, Tervo TM. Corneal nerves: structure, contents and function. *Exp Eye Res* 2003; **76**:521–542
61. Müller LJ, Vrensen GF, Pels L, Cardozo BN, Willekens B. Architecture of human corneal nerves. *Invest Ophthalmol Vis Sci* 1997; **38**:985–994
62. Naumann OHG. *Pathologie des Auges*. Berlin, Heidelberg, New York: Springer-Verlag, 1997:22–30
63. Oliveira-Soto L, Efron N. Morphology of corneal nerves using confocal microscopy. *Cornea* 2001; **20**:374–384
64. Parazza F, Humbert C, Usson Y. Method for 3D volumetric analysis of intranuclear fluorescence distribution in confocal microscopy. *Comput Med Imaging Graph* 1993; **17**:189–200
65. Patel SV, McLaren JW, Camp JJ, Nelson LR, Bourne WM. Automated quantification of keratocyte density by using confocal microscopy in vivo. *Invest Ophthalmol Vis Sci* 1999; **40**:320–326
66. Petran M, Hadravsky M, Egger MD, Galambos R. Tandem scanning reflected-light microscope. *J Opt Soc Am* 1968; **58**:661–664
67. Peyman GA, Sanders DR, Goldberg MF (eds). *Principles and Practice of Ophthalmology* (volume 1). Philadelphia: WB Saunders, 1987
68. Picht G, Grehn F. Development of the filtering bleb after trabeculectomy. Classification, histopathology, wound healing process. *Ophthalmologe* 1998; **95**:W380–W387
69. Powers TP, Stewart WC, Stroman GA. Ultrastructural features of filtration blebs with different clinical appearances. *Ophthalmic Surg Lasers* 1996; **27**:790–794
70. Richter A, Slowik C, Somodi S, Vick HP, Guthoff R. Corneal reinnervation following penetrating keratoplasty – correlation of esthesiometry and confocal microscopy. *Ger J Ophthalmol* 1996; **5**:513–517
71. Richter A, Slowik C, Somodi S, Vick H-P, Guthoff R. In-vivo-Darstellung der Hornhautinnervation des Menschen mit Hilfe der konfokalen Mikroskopie. *Ophthalmologe* 1997; **94**:141–146
72. Ruben M, Colebrook E. Keratoplasty sensitivity. *Br J Ophthalmol* 1979; **63**:265–267
73. Ruben M. *Kontaktlinsenanpassung.* Stuttgart, New York: G.Fischer Verlag, 1978:23–45
74. Sangwan VS. Limbal stem cells in health and disease. *Biosci Rep* 2001; **21**:385–405
75. Scheuerle AF, Schmidt E. *Atlas of Laser Scanning Ophthalmoscopy.* Berlin, Heidelberg, New York: Springer-Verlag, 2004
76. Skriver K. Reinnervation of the corneal graft. *Acta Ophthalmol (Copenh)* 1978; **56**:1013–1015
77. Smithpeter C, Dunn A, Drezek R, Collier T, Richards-Kortum R. Near real-time confocal microscopy of cultured amelanotic cells: sources of signal, contrast agents and limits of contrast. *J Biomed Opt* 1998; **3**:429–436
78. Stave J, Slowik C, Somodi S, Hahnel C, Grümmer G, Guthoff R. Keratozytendichte der In-vivo-Kornea. Automatische Messung mit einem modifizierten konfokalen Mikroskop MICROPHTHAL. *Klin Monatsbl Augenheilkd* 1998; **213**:38–44
79. Stave J, Guthoff R. Darstellung von Tränenfilm und In-vivo-Kornea. Erste Untersuchungsergebnisse mit einem modifiziertem konfokalen Laser-Scanning-Ophthalmoskop. *Ophthalmologe* 1998; **95**:104–109
80. Stave J, Zinser G, Grümmer G, Guthoff R. Der modifizierte Heidelberg-Retina-Tomograph HRT. *Ophthalmologe* 2002; **99**:276–280
81. Stelzer EH, Wacker I, De Mey JR. Confocal fluorescence microscopy in modern cell biology. *Semin Cell Biol* 1991; **2**:145–152
82. Straub M. Flourescinlösung als diagnostisches Hülfsmittel für Hornhauterkrankungen. *Centralbl für Augenheilkd* 1888; **12**:75–77

83. Suzuki T, Sano Y, Kinoshita S. Conjunctival inflammation induces Langerhans cell migration into the cornea. *Curr Eye Res* 2000; **21**:530–533
84. Tervo T, Moilanen J. In vivo confocal microscopy for evaluation of wound healing following corneal refractive surgery. *Prog Retin Eye Res* 2003; **22**:339–358
85. Tervo T, Vannas A, Tervo K, Holden BA. Histochemical evidence of limited reinnervation of human corneal grafts. *Acta Ophthalmol (Copenh)* 1985; **63**:207–214
86. Thaer A, Geyer OC. Die Anwendung der konfokalen Spaltscanning-Mikroskopie auf die In-vivo-Untersuchung der Kornea-Mikrostruktur im Zusammenhang mit dem Tragen von Kontaktlinsen. *Contactologia* 1997; **19**:157–177
87. Tomii S, Kinoshita S. Observations of human corneal epithelium by tandem scanning confocal microscope. *Scanning* 1994; **16**:305–306
88. Torens S, Berger E, Stave J, Guthoff R. Darstellung der Mikroarchitektur und Dynamik der Aufreißphänomene des präkornealen Tränenfilms mit Hilfe der Laser-Rastermikroskopie. *Ophthalmologe* 2000; **97**:635–639
89. Torens S, Berger E, Stave J, Guthoff R. Light, reflecting and laser scanning microscopy in the imaging and assessment of the precorneal tear film. New perspectives on the break-up of the tear film. *Contactologia* 2000; **22**:49–67
90. Tugal Tutkun I, Akarcay K, Kozer Bilgin L, Sansoy N, Urgancioglu M. Corneal sensitivity after penetrating keratoplasty. *Eur J Ophthalmol* 1993; 3:66–70
91. Vesaluoma M, Perez-Santonja J, Petroll WM, Linna T, Alio J, Tervo T. Corneal stromal changes induced by myopic LASIK. *Invest Ophthalmol Vis Sci* 2000; **41**:369–376
92. Vogt A. *Lehrbuch und Atlas der Spaltlampenmikroskopie des lebenden Auges.* Berlin, Heidelberg, New York: Springer-Verlag, 1921 (2nd edition 1930)
93. Webb RH. Confocal optical microscopy. *Rep Prog Phys* 1996; **59**:427–471
94. Wiegand W, Thaer AA, Kroll P, Geyer OC, Garcia AL. Optical sectioning of the cornea with a new confocal in vivo slit-scanning videomicroscope. *Ophthalmology* 1995; **102**:568–575
95. Wirbelauer C, Scholz C, Engelhardt R, Laqua H, Pham DT. Biomorphometrie des Hornhautepithels mittels spaltlampenadaptierter optischer Kohärenztomographie OCT. *Ophthalmologe* 2001; **98**:848–852
96. Yanoff M, Duker JS. *Ophthalmology.* London, Philadelphia: Mosby, 1999
97. Zhivov A, Stave J, Vollmar B, Guthoff RF. In vivo confocal microscopic evaluation of Langerhans cell density and distribution in the normal human corneal epithelium. *Graefe's Arch Clin Exp Ophthalmol* 2005; **243**:1056–1061

Subject Index

MIX
Papier aus verantwortungsvollen Quellen
Paper from responsible sources
FSC® C105338

If you have any concerns about our products,
you can contact us on
ProductSafety@springernature.com

In case Publisher is established outside the EU,
the EU authorized representative is:
Springer Nature Customer Service Center GmbH
Europaplatz 3, 69115 Heidelberg, Germany

Printed by Libri Plureos GmbH
in Hamburg, Germany